Sustainable Development Goals Series

The **Sustainable Development Goals Series** is Springer Nature's inaugural cross-imprint book series that addresses and supports the United Nations' seventeen Sustainable Development Goals. The series fosters comprehensive research focused on these global targets and endeavours to address some of society's greatest grand challenges. The SDGs are inherently multidisciplinary, and they bring people working across different fields together and working towards a common goal. In this spirit, the Sustainable Development Goals series is the first at Springer Nature to publish books under both the Springer and Palgrave Macmillan imprints, bringing the strengths of our imprints together.

The Sustainable Development Goals Series is organized into eighteen subseries: one subseries based around each of the seventeen respective Sustainable Development Goals, and an eighteenth subseries, "Connecting the Goals," which serves as a home for volumes addressing multiple goals or studying the SDGs as a whole. Each subseries is guided by an expert Subseries Advisor with years or decades of experience studying and addressing core components of their respective Goal.

The SDG Series has a remit as broad as the SDGs themselves, and contributions are welcome from scientists, academics, policymakers, and researchers working in fields related to any of the seventeen goals. If you are interested in contributing a monograph or curated volume to the series, please contact the Publishers: Zachary Romano [Springer; zachary.romano @springer.com] and Rachael Ballard [Palgrave Macmillan; rachael. ballard@palgrave.com].

Wei-Ta Fang · Arba'at Hassan ·
Ben A. LePage

The Living Environmental Education

Sound Science Toward a Cleaner,
Safer, and Healthier Future

 Springer

Wei-Ta Fang
Graduate Institute of Environmental
Education
National Taiwan Normal University
Taipei, Taiwan

Ben A. LePage
Graduate Institute of Environmental
Education
National Taiwan Normal University
Taipei, Taiwan

Academy of Natural Sciences
Philadelphia, PA, USA

Arba'at Hassan
Open University Malaysia Kota
Sabah, Malaysia

Ministry of Science and Technology, MOST 109-2511-H-003-031, The work was
supported by The Ministry of Science and Technology (MOST 109-2511-H-003-031).

ISSN 2523-3084   ISSN 2523-3092 (electronic)
Sustainable Development Goals Series
ISBN 978-981-19-4233-4   ISBN 978-981-19-4234-1 (eBook)
https://doi.org/10.1007/978-981-19-4234-1

This Springer imprint is published by the registered company Springer Nature Singapore Pte Ltd.
The registered company address is: 152 Beach Road, #21-01/04 Gateway East, Singapore
189721, Singapore

Dedicated to

Prof. and Dean Frederick Steiner
(University of Pennsylvania, USA)
Prof. Dr. Douglas K.-S. Loh
(Texas A&M University, USA)
and
my family: my father, Army General Hsun-Chih Fang
(1931–2017); my mother, Madam Su-Mei Huang (1939–2021);
my wife, Chieh-Ying Ho; my sons, Cheng-Jun (June),
and Cheng-Shun (Sam)
You are my greatest encouragement and sources
of all current life.

Dedicated to

Prof. Dr. Harold Hungerford, Prof. Dr. Trudi Volk,
Prof. Dr. Charles Klasek, Prof. Dr. Arthur Aikman
(Southern Illinois University, Carbondale, USA)
and
my family: Hassan (dad), Aminah (mom),
Hasnah (wife), Hasrul and Elly, and grandchildren: Sammy,
and Sucisonia
You are the greatest inspiration of my life.

Dedicated to

Past, Present, and Future students.
You've taught me to open my mind and embrace
to new ideas that have challenged me to become
better personally and professionally.

Preface

Global population is expected to increase from 6.7 to about 9.7 billion people by 2050. Earth has gone through many icehouse–greenhouse cycles over its 4.5-billion-year history, and there is good evidence that human activities have accelerated the warming trend globally since the Industrial Revolution. Hotter temperatures, more drought and severe storms and sea-level rise are some of the significant changes that we can expect to see in the future. Nature does not discriminate and humans are just another species living on the planet. While some areas of the world will feel these impacts more than others, global change will impact us all. In addition to the global change issues that we are all certain to face, the environmental damage that has been caused by our species is impacting the environment and our way of life.

In 2000, the concept of the Anthropocene epoch was proposed by Crutzen the 1995 Nobel Laureate in Chemistry. He believed that the impact of human activities on Earth was sufficient to form a new geologic epoch and proposed that the Industrial Revolution was the beginning of the Anthropocene. While the creation of a new geologic epoch may be warranted, neither the International Commission on Stratigraphy nor the International Union of Geological Sciences has approved naming this slice of geologic time. Nonetheless, soon after this concept was put forward, many scholars posited that the beginning of the Anthropocene epoch should be earlier and based on science. Opinions are varied. James Lovelock pointed out that the Anthropocene epoch began during the Industrial Revolution, but according to William Ruddiman, it can be traced back to the beginning of human farming activities about 8000 years ago during the Neolithic epoch. At that time, agriculture and animal husbandry replaced hunting and gathering methods of survival, followed by the extinction of large mammals.

More recently, human activities since the 1700s led to an increase in the concentration of atmospheric carbon dioxide (CO_2), which is now approaching twice the pre-industrial atmospheric levels (417 vs 275 ppm). Rising sea temperature is causing a decline in marine phytoplankton biomass, which is contributing to a decrease in carbon sequestration and O_2 production. Scientists have documented 50% of the biota that once lived on Earth have become extinct since humans arrived. The continued increase in human population and excessive consumption of the planet's natural resources may lead to the Earth's sixth mass extinction event.

If environmental problems exist, then environmental protection at all levels of government and society is needed. The ecological problems humans face today come in part, from the loss of traditional knowledge, social values, human behavior, and ethics. Pro-environmental behaviors (PEBs) promote both the intrinsic value of nature and its protection. Therefore, environmental issues need to be addressed through advocacy, education, and activism. The environmental education learning elements include concepts such as ecological principles, natural resource conservation, environmental management, interaction, interdependence, and ethics, and sustainability. The goals for environmental education are lofty, but all are aimed to cultivate human environmental awareness and sensitivity, environmental knowledge, values and attitudes, and experience.

Environmental education at the postgraduate level shoulders the responsibility of cultivating national level and socially based environmental protection programs. The sustainable development talents of the students that are related to environmental quality, and resources, and sustainable development that are cultivated through university programs may contribute to the economic and social development of various countries. As such, Living Environmental Education provides a discussion of the mechanisms for sustainable environmental development in the education curriculum. It is hoped that universities in Taiwan will continue to develop the ideas presented in the Curriculum Guidelines of 12-year basic education for the training of teachers and promote the 12-year basic education program. The *Curriculum Guidelines* actively implement literacy-based pedagogy and promote a program that uses the classroom as a learning platform to build on our collective environmental knowledge base and develop education programs that enhance learning.

Today's environmental education programs are not limited to the classroom education. They are holistic, building on traditional classroom approaches, while incorporating outside experiences, knowledge, and skills, as well as the abilities of learners to develop environmentally protective and sustainable behaviors and implement changes that are protective of the environment. How to strengthen core literacies of environmental protection, which has both the nature and ideals of the discipline, requires careful consideration. Living Environmental Education can be used as a tool to train people interested in environmental education. We've tried to demonstrate teaching activities can be enhanced by experiencing the environment. There is no single right or wrong pedagogic approach because we believe each people and culture will vary, but we will all accomplish the same goal of educating the people to become responsible environmental stewards, live sustainably, and protect the environment. Readers of Living Environmental Education will note that the flow of ideas and manner that they are presented seem somewhat disconnected or unrelated and this has been done purposefully. Although the authors are relatively close in age, we've grown up in different countries with distinct cultures, social values, and norms. We are all proficient in the study of the environment and have all taught been environmental education at institutions worldwide. The point to note is that while our backgrounds and cultures are vastly different, our goals are the same. We're just taking different approaches to get to the same spot.

Living Environmental Education is based on the state of implementing teaching styles/models and a reflection on the effectiveness of these styles, from which educators can update or adjust the content of their curricula in the future. The follow-up will expand on early childhood education, non-governmental enterprises, non-governmental conservation groups, and national environmental education demonstrations to promote sustainable development of the benefit of society. The book is divided into ten chapters that cover environmental education theory to the practical analysis of environmental education. It applies to: (1) the teaching of environmental education courses in colleges; (2) Ph.D. programs associated with national development, environmental education, environmental policy, psychology, society, economics, culture, communication, tourism and recreation, hospitality management, and education for sustainable development; and (3) reference materials for examination, practice, further education, and training for environmental education staff.

Chapters 1 and 2 are an introduction to environmental education where we explore the definition of environmental education, the philosophy of environmental protection, the history of environmental education, and its development. In Chap. 3, environmental education research methods, including the connotation of environmental education research, which is divided into historical, quantitative, and qualitative researches, and metacognitive analyses. Environmental literacy and the motivation of environmental education to build environmental awareness and sensitivity, values, and attitudes, and motivation skills are discussed in Chap. 4. In Chap. 5 environmental psychology, including human environmental cognition, personality traits, social norms, environmental stress, and the healing environment are presented. In Chap. 6, we discuss issues on environmental ethics, new environmental and ecological paradigms, paradigm shifts, and pro-environmental behaviors. Environmental communication and learning, field studies, building lesson plans, developing learning models, and using environmental information to understand the media of environmental education are addressed in Chap. 7. In Chap. 8, the connotation, motivation, obstacles, fields, and implementation contents of outdoor education and leisure education are presented.

Understanding the ecology of the environments within which we live and protecting the environment is a common thread of all people and cultures. As we pointed out previously, nature doesn't discriminate or recognize differences between humans. We are just another species on the planet. Promoting sustainable development should be a goal pursued by the people of the world. We may look back on the twenty-first century and the associated sustainable development goals as a time where humans either succeeded or failed. If we failed, there may not be many humans left and "would have, could have, should have" will ring loudly in our minds. The authors sincerely hope that the people of the world work together to create a comfortable, sustainable, and peaceful environment, and move toward a sustainable future that takes environmental protection and economic development into account.

Taipei, Taiwan   Wei-Ta Fang, Ph.D.
Sabah, Malaysia   Arba'at Hassan, Ph.D.
Taipei, Taiwan/Philadelphia, USA   Ben A. LePage, Ph.D.

Acknowledgments

We all appreciate and recognize the Hetong Culture and Art Foundation for their financial support for this book so that it is made available free-of-charge to the public. We especially thank Yi-Ning Lee for her generosity as well as the virtue of giving without expecting anything in return. We acknowledge the members of the Graduate Institute of Environmental Education, National Taiwan Normal University (NTNU), for their contributions to this book. Useful suggestions from Yu-Jie Chang, Su-Hwa Lin, Yi-Hsuan (Tim) Hsu, Huei-Min Tsai, and Shih-Tsen Liu, Chinese Society for Environmental Education (CSEE), Taiwan, were incorporated into the book. This project is also financially supported by National Taiwan Normal University, one of the best program key cultivation universities of the Ministry of Education, for the book publication by the program on bilingual education for students in college, Taiwan, Republic of China (ROC). This project was supported by grants from the Ministry of Science and Technology, Taiwan, Republic of China (ROC) (MOST 105–2511-S-003–021-MY3-, 109–2511-H-003-031-, and 110-2811-H-003-530- to Wei-Ta Fang; MOST 110-2811-H-003-500- to Ben A. LePage).

Contents

Part I   Orientation

1   Introduction to Environmental Education   3
   1.1   Introduction   3
      1.1.1   The Environment is a Concept   5
      1.1.2   Lost in Translation   6
   1.2   Definition of Environmental Education   6
      1.2.1   Initial Definition of Environmental Education   7
      1.2.2   The Extended Definition of Environmental
             Education   8
      1.2.3   The Goals of Environmental Education   11
   1.3   Approaches to Environmental Education   11
      1.3.1   Outdoor Education   13
      1.3.2   Classroom Education   15
   1.4   Development of Environmental Education   17
      1.4.1   Values   20
      1.4.2   Exploration Topics   21
   1.5   Summary   22
   References   22

2   Philosophy and History of Environmental Education   25
   2.1   Philosophy of Environmental Protection   25
      2.1.1   Theory of Ideas and Empiricism   26
      2.1.2   Transcendence and Efficiency   26
      2.1.3   Preservation Theory   28
      2.1.4   Vitality Theory and Killing Theory   28
      2.1.5   Deep Ecology   29
   2.2   History of Environmental Education   32
      2.2.1   Early Era of Environmental Education   32
      2.2.2   Environmental Education in the Eighteenth
             and Nineteenth Centuries   35
      2.2.3   Environmental Education in the Early Twentieth
             Century   36
      2.2.4   Environmental Education After the Middle
             of the Twentieth Century and in the Early
             Twenty-First Century   36

2.2.5 Environmental Education in the Twenty-First
       Century   41
2.3 Summary   46
References   47

3 **Research Methods for Environmental Education**   49
3.1 What is Environmental Education Research?   49
     3.1.1 Instrumental Research Utilization   51
     3.1.2 Conceptual Research Utilization   51
     3.1.3 Symbolic Research Utilization   52
3.2 Types of Environmental Education Research   54
     3.2.1 International Periodicals   55
     3.2.2 Local Periodicals   56
     3.2.3 Research on Environmental Education   56
3.3 Research on the History of Environmental Education   64
     3.3.1 The Rise of Environmental Education   64
     3.3.2 The Construction of Environmental Education   65
     3.3.3 Ideology of the Declaration on Environmental
            Education   66
     3.3.4 The Practical Power of Environmental Education ...   66
     3.3.5 Reflection on Environmental Education   67
3.4 Quantitative Research on Environmental Education   70
     3.4.1 Preparation and Measurement of Questionnaires
            and Tests   70
     3.4.2 The Experimental Research Methods
            of Environmental Education   70
     3.4.3 Reliability and Validity Analysis   72
     3.4.4 Retrospective Research   73
     3.4.5 Relevance Research   73
     3.4.6 Data Analysis, Interpretation and Application,
            Presentation of Research Results   73
     3.4.7 The Research Method of Delphi   73
3.5 Qualitative Research on Environmental Education   74
     3.5.1 Interview   74
     3.5.2 Observation   75
     3.5.3 Grounded Theory   76
     3.5.4 Action Research   79
     3.5.5 Ethnography   81
     3.5.6 Content Analysis   82
3.6 Promotion of Environmental Education Theory   83
     3.6.1 Theoretical Expansion and Practice   83
     3.6.2 Integration of Environmental Disciplines   84
     3.6.3 Integration of Environmental Education
            Disciplines   85
     3.6.4 Learning Method of ABC Emotion Theory   86
3.7 Summary   87
References   88

Part II Contextualization: The Ultimate End of Human Ecology

4 Environmental Literacy 93
   4.1 Introduction 94
   4.2 Motivation for Environmental Education 98
      4.2.1 Pedagogic Reasons 98
      4.2.2 Social Learning Reasons 98
      4.2.3 Better Environmental Protection Behaviors 99
   4.3 Environmental Awareness and Sensitivity............... 103
      4.3.1 Environmental Awareness 103
      4.3.2 Environmental Sensitivity 105
   4.4 Environmental Values and Status 107
      4.4.1 Environmental Values 107
      4.4.2 Environmental Attitude 110
   4.5 Cognitive, Affective, and Psychomotor Skills............ 112
      4.5.1 Cognitive Domain........................... 113
      4.5.2 Affective Domain 113
      4.5.3 Psychomotor Domain 114
      4.5.4 Modify the Cognitive Domain.................. 114
      4.5.5 Correction of Skill Areas...................... 115
   4.6 Environmental Action Experience and Pro-Environmental Behavior... 116
      4.6.1 Actions and Behaviors 116
      4.6.2 Research on Actions and Behaviors 116
      4.6.3 Pro-environmental Behavior 117
      4.6.4 Responsible Environmental Behavior 118
   4.7 Environmental Aesthetics 120
      4.7.1 Dadaism 121
      4.7.2 New Expressionism.......................... 121
      4.7.3 Action Art................................. 122
      4.7.4 Landscape Art.............................. 122
      4.7.5 Installation Art 122
   4.8 Summary 123
   References... 123

5 Environmental Psychology 127
   5.1 Environmental Cognition 127
      5.1.1 Cognitive Theory of Environmental Learning 129
      5.1.2 The Exploratory Theory of Environmental Learning 130
      5.1.3 The Social Theory of Environmental Learning 131
   5.2 Personality Traits 132
      5.2.1 Personality Traits Responsibility 133
      5.2.2 The Green Personality Traits.................... 133
   5.3 Social Norms 135
      5.3.1 Specification 135

		5.3.2	Direct and Indirect Paths for Social Norms to Predict Environmentally Friendly Behavior	136
	5.4		Development Pressure	137
		5.4.1	Developing Pressure Defined by Environmental Scientists	138
		5.4.2	Marine Impact and Stress Defined by Environmental Scientists	139
		5.4.3	Climate Change Impact and Stress Defined by Environmental Scientists	140
	5.5		Healing Environment	142
		5.5.1	Natural Environment	142
		5.5.2	The Comfortable Environment	145
		5.5.3	The Postnatal Environment	145
		5.5.4	Surfing in a Safe Virtual Environment After COVID-19 Pandemic?	145
	5.6		Summary	146
	References			147
6	**Environmental Ethics: Modelling for Values and Choices**			**151**
	6.1		What is Environmental Ethics?	151
		6.1.1	Beliefs of Land Ethics	151
		6.1.2	Beliefs of Anthropocentric Value System	152
		6.1.3	Beliefs of Biocentric Ethic Value System	153
		6.1.4	Beliefs of Ecological Ethic Value System	153
		6.1.5	From Deep Ecology to Animal Rights	155
	6.2		Environmental Paradigm	157
		6.2.1	Traditional Beliefs and Values	158
		6.2.2	New Environmental Paradigm	158
	6.3		Paradigm of the Theory of Behaviors	159
		6.3.1	Theory of Planned Behavior (TPB)	160
		6.3.2	The Motivation-Opportunity-Abilities Model	162
		6.3.3	The Value-Belief-Norm Theory	163
		6.3.4	Two-Phase Decision-Making Model	165
	6.4		Paradigm Shift	167
		6.4.1	Dominant Social Paradigm	167
		6.4.2	New Ecological Paradigm	168
		6.4.3	Sustainability Paradigm	169
	6.5		Summary	169
	References			170
Part III	**Living Media Lab: Searching a Means to an End**			
7	**Environmental Learning and Communication**			**177**
	7.1		What Should We Learn?	177
		7.1.1	Learning is a Binary Opposition Process	179
		7.1.2	Teaching is the Half of Learning	181

7.2   How Should I Learn?.............................   184
    7.2.1   School Environmental Education Should
         be a Practical-Learning   184
    7.2.2   Social Environmental Education Should be an
         Adaptive Learning...........................   189
7.3   Environmental Learning Center Should be Suitable
    for the Best Quality Education   191
    7.3.1   Features of Environmental Learning Center
         Implementation   192
    7.3.2   Participatory Learning at the Environmental
         Learning Center.............................   196
7.4   Learning Plans Toward Sustainability..................   197
    7.4.1   Program of Environmental Learning Center   197
    7.4.2   Course Content of the Environmental Learning
         Center......................................   198
    7.4.3   Educational Interpretation Plan   201
    7.4.4   Teaching Activity Strategy   204
7.5   Learning.......................................   206
    7.5.1   Triadic Reciprocal Determinism Explores
         the Dependence of Teachers and Students.........   206
    7.5.2   Strengthen Voluntary Environmental Learning
         Behaviors   207
    7.5.3   Learning Through Different Types of Learning
         Methods   208
7.6   Information Transmission   210
    7.6.1   Learning Interface for Information Transfer   215
    7.6.2   The Transfer Process.........................   216
    7.6.3   Capacity Building   218
7.7   Communication Media   220
    7.7.1   Formal Media   221
    7.7.2   Informal Media   222
    7.7.3   Communication Effect........................   222
7.8   Summary   224
References...   224

8   Outdoor Education   229
8.1   Connotation of Outdoor Education   229
    8.1.1   History of Outdoor Education   230
    8.1.2   Site Planning for Outdoor Education.............   232
    8.1.3   The Content of Outdoor Learning   235
8.2   Motivation for Outdoor Education   238
    8.2.1   Hometown Learning   238
    8.2.2   Research   239
    8.2.3   Experience Learning   240
    8.2.4   Emotional Learning..........................   240
8.3   Barriers to Outdoor Education   244
    8.3.1   Barriers to School Education...................   244
    8.3.2   Barriers to Family Education...................   244

8.4   Outdoor Education Field............................ 245
        8.4.1   Field-Dependent Development................. 245
        8.4.2   Development of Field Selection............... 246
8.5   The Implementation Content of Outdoor Education 248
        8.5.1   Goals of Outdoor Education 250
        8.5.2   Implementation Methods of Outdoor Education 251
8.6   Summary 256
References... 258

**Conclusion: How Can We Maintain Our Precious Environment
and Its Resources Now and for the Future?** 261

Appendix: Environmental Attitude Scales................... 265

Index 267

About the Authors

Wei-Ta Fang was born in Kaohsiung, Taiwan, on February 14, 1966. He received a B.A. degree in Land Economics and Administration from National Taipei University (Taipei, Taiwan) in 1989. He received his first master's degree in Environmental Planning (MEP) from Arizona State University, USA, in 1994, and a second master's degree in Landscape Architecture in Design studies (MDes. S.) from the Graduate School of Design, Harvard University, USA, in 2001. He obtained a Ph.D. from the Department of Ecosystem Science and Management, Texas A&M University, USA, in 2005. He served as a specialist in the Taipei Land Management Bureau in 1991 and 1992, and a senior specialist in-charge of environmental education and environmental impact assessments (EIAs) at Taiwan's Environmental Protection Administration (EPA) Headquarters from 1994 to 2006. He was also the Co-Principal Investigator (Co-PI) for the National Environmental Literacy Survey in Taiwan during 2012 and 2020. He is currently serving as a Distinguished Professor (2019 ~) and the Director (2019 ~ 2025) of the Graduate Institute of Environmental Education, National Taiwan Normal University and is the President of the Society of Wetland Scientists (SWS) Asia Chapter and Taiwan Wetland Society (TWS). He served as one of the government officials, scientists in both natural and social sciences, university's professors, fine-art painters, editors for three journals, writers for twenty Chinese books and three English books, five non-governmental organization's leaders and served also as one of the training runners of marathons trained by Angel Wings House's nutrition and diet programs. His interest is quite broad, determined to participate in the United Nations environmental affairs by strengthening environmental security and, lay the ground based on the foundation of human peace forever to achieve for enduring social, economic, and environmental sustainability. He lives in Taipei City with his wife, Chia-Ying Ho, and two sons, Cheng-Jun (June) and Cheng-Shun (Sam). His contact e-mail is wawaf@hotmail.com.

Arba'at Hassan was born in Sungai Leman, Selangor, on May 16, 1952. He was a science teacher at Tamparuli Secondary School (1970 ~ 1979), then a science education lecturer at Kent Teachers' College, Tuaran (1980 ~ 1982). He received all degrees from Southern Illinois University at Carbondale (SIUC), USA (B.Sc. Biological Sciences (1985), M.Sc. Science Education and Curriculum and Instruction (1987), and Ph.D. in Environmental and Science Education (1992). Upon graduation, he taught environmental education in various universities in Malaysia (1994 ~ 2011). He served in scouting and gave leadership training. He received the Excellent Merit Award from the Malaysian Scout's Chief Commissioner (1979). While studying at SIUC, he involved in many environmental education activities including during Earth Summit in Rio (1990). He additionally, received a merit of Teaching Excellent Award, "Tokoh Pendidik," from the Minister of Education (2016). Since retired, he has been part-time teaching at Open University Malaysia. He is also active in editing articles for international journals, books, etc. He involves involuntary environmental works, then appointed as an associate member of the Society of Wetland Scientists Asia Chapter. He lives with his wife, Hasnah, in Tuaran. His son and family live in Kuala Lumpur. He can be reached by e-mail at and Facebook: Arba'at Hassan.

Ben A. LePage was born in Cudworth, Canada, on September 12, 1959. He earned a B.Sc. with honors degree in Biology in 1988 and a Ph.D. in Geology in 1993 at the University of Saskatchewan, Canada. He was an NSERC postdoctoral fellow in Biology at the University of Alberta, Canada. He was a Lecturer and Senior Research Scientist-Professor at the University of Pennsylvania, Philadelphia, Pennsylvania, USA, from 1995 to 2001. In 2001, he moved to the applied sector as a Senior Ecologist/Environmental Scientist in the Philadelphia area from 2001 until 2008. From 2008 until 2012, he worked for the Philadelphia Electric Company as a Senior Remediation and Environmental Project Manager. From 2012 to 2020, he was an Environmental Program Manager at Pacific Gas and Electric, San Francisco, California, USA. He has served as a Board of Director for many organizations in the Philadelphia area and served President of the Society of Wetland Scientists (SWS) from 2011–2012 and President of the SWS-Professional Certification Program from 2017 to 2018. He is currently a Distinguished Visiting Chair Professor in the Graduate Institute of Environmental Education, National Taiwan Normal University, Taipei, Taiwan. He lives in Concord, California, USA, with his wife (Carol). His e-mail address is benlepage2@gmail.com.

Abbreviations

A-B	Attitude-behavior
AC	Awareness of consequences
ACR	The Association for Consumer Research
AI	Artificial intelligence
APMP	Ambient particulate matter pollution
APRC	Asia-Pacific Regional Center
AR	Ascription of Responsibility
AR	Augmented reality
CASS	Chinese Academy of Social Sciences
CH_4	Methane
CO_2	Carbon dioxide
CRI	Long-Term Climate Risk Index
CSEE	Chinese Society for Environmental Education
Cu	Copper
DDT	Dichloro-diphenyl-trichloroethane
DESD	Decade of Education for Sustainable Development
DRR	Disaster risk reduction
DSP	Dominant Social Paradigm
EE	Environmental education
EER	Environmental Education Research
EFA	Education For All
EfS	Education for Sustainability
EJ	Environmental justice
ERIC	Education Resources Information Center
ESD	Education for Sustainable Development
Ga	Gallium
GAP	Global Action Programme on ESD
GCC	Global climate change
GCRI	Global Conflict Risk Index
GEEP	Global Environmental Education Partnership
GEMR	Global Education Monitoring Report
GIEE	Graduate Institute of Environmental Education, NTNU
GLOBE	Global Learning and Observations to Benefit the Environment Program
GREEN	Global Rivers Environmental Education Network
H_2S	Hydrogen sulfide

HEP	Human Exceptionalism Paradigm
ICT	Information and communications technology
IEEP	International Environmental Education Programme
In	Indium
INSET	In-Service Education and Training
INTECOL	International Association for Ecology
IUCN	International Union for Conservation of Nature
JEE	Journal of Environmental Education
JEER	Journal of Environmental Education Research
K-12	Kindergarten to twelfth grade (grade 12)
LoF	Level of functioning
MOAs	Motivation–Opportunity–Abilities
MOE	Ministry of Education
MQA	Malaysian Qualifications Agency
MRT	Mass rapid transit
Mt.	Mountain
NAAEE	North American Association for Environmental Education
NAM	Norm Activation Model
NEEA	National Environmental Education Act
NEP	New Ecological Paradigm
NOAA	National Oceanic and Atmospheric Administration
NSSE	National Special Security Event
NTNU	National Taiwan Normal University
OSU	Ohio State University
PBL	Problem-based learning
PCBs	Polychlorinated biphenyl
PEB	Pro-environmental behavior
Ph.D.	Doctor of Philosophy
PM	Particulate matter
PRC	People's Republic of China
PV	Photovoltaics
QoL	Quality of life
REB	Responsible environmental behavior
ROC	Republic of China
RRC-EA	Ramsar Regional Center—East Asia
sAA	Salivary alpha-amylase
SAD	Seasonal affective disorder
sC	Salivary cortisol
SCI	Science Citation Index
SDE	Sustainable Development Education
SDGs	Sustainable Development Goals
Se	Selenium
SEM	Structural equation modeling
SENR	School of Environment and Natural Resources, OSU
SERC	Smithsonian Environmental Research Center
SIUC	Southern Illinois University at Carbondale

SNR	School of Natural Resources (SNR, OSU) (now the School of Environment and Natural Resources)
S-R	Stimulus-and-response
SSAP	Social Sciences Academic Press
STS	Science, technology, and society
SWS	Society of Wetland Scientists
Taiwan EPA	Environmental Protection Administration of the Executive Yuan, Republic of China
Taiwan, ROC	Taiwan, Republic of China
TEK	Traditional Ecological Knowledge
TEM	Traditional Ecological Management
Tl	Thallium
TPB	Theory of Planned Behavior
TRD	Triadic Reciprocal Determinism
U.S. EPA	U.S. Environmental Protection Agency
UK	United Kingdom of Great Britain and Northern Ireland
UN DESD	UN Decade of Education for Sustainable Development
UNCED	UN Conference on Environment and Development
UNEP	United Nations Environment Programme
UNESCO	United Nations Educational, Scientific and Cultural Organization
USA	United States of America
USDA	U.S. Department of Agriculture
VBN	Value-belief-norm theory
VPM	Visual participatory method
WCED	World Commission on Environment and Development
WEEC	World Environmental Education Congress
WWF	World Wildlife Fund

Introduction to Environmental Education

1

All education is environmental education. By what is included or excluded, we teach students that they are part of or apart from the natural world. To teach economics, for example, without reference to the laws of thermodynamics or those of ecology is to teach a fundamentally important ecological lesson——that physics and ecology have nothing to do with the economy. That just happens to be dead wrong. The same is true throughout the curriculum.

David W. Orr, What is Education for? 1991:52.

Abstract

The concept of **education** is changing and that of the **environment** is also becoming different. Is environmental education: (1) a type of education to improve the environment; (2) education to improve the environment of education; or (3) a type of education to improve the education of people? In this chapter we focus on the ontology of the environment. In epistemology, we try to understand the nature and identity of the world around us and what environmental education is about. The purpose of environmental education is to cultivate citizens that: (1) have a working knowledge of environmental systems; (2) have concerns about environmental problems; and (3) have the capabilities to solve and actively participate in implementing solutions. Environmental problems must be solved through a root cause process, and environmental educators need to change the public's mind on environmental issues using realistic and attainable education targets to establish environmentally friendly behaviors. Through outdoor, classroom, and nature-centered education programs, our goal is to establish important curriculum goals and novel learning methods for environmental education. Our goal is to have stakeholders consider environmental issues with open minds, understand the needs of other stakeholders, take a leadership role recognizing the existing and emerging environmental issues, and internalize them into specific environmental protective action plans.

1.1 Introduction

There are many definitions of education, but for education theory, Albert Einstein, who was a leader in pioneering educational reform point, had a unique point of view. He said: *education is what remains after one has forgotten what one has learned in school* (Fig. 1.1). Before the nineteenth century, education was the process of remembering or memorization. The *San Zi Jing* (Three-character Sutras) 《三字經》 that had been passed down to the people since the Southern Song Dynasty in China-proposed:

W.-T. Fang et al., *The Living Environmental Education*, Sustainable Development Goals Series,
https://doi.org/10.1007/978-981-19-4234-1_1

Fig. 1.1 Education is what remains after one has forgotten what one has learned in school (Einstein 1879–1955) (Cheng-Jun Fang at the Busan National Science Museum, Busan, Republic of Korea, 2019) (Photo by Wei-Ta Fang)

Recite them with the mouth, and ponder over them in your hearts. Do this in the morning; do this in the evening.

Sun Zhu (孫洙) (1711–1778) of the Qing Dynasty once said:

After reading three hundred Tang poems, you can at least in tone poems even if you can't write them.

It has always been the case that students learn as much as possible until they become familiar. However, there are other theories that have always been disgusted with learning. Einstein believed that real learning is the process of internalizing information. Wang Yangming (王陽明) (1472–1529) of the Ming Dynasty stated in the Book of *Instructions for Practical Living and Other Neo-Confucian Writings* 《傳習錄》 that the most important things reading requires is self-mindedness, understanding second, and finally memorization. A friend once asked him, "How can I read a book but I can't remember," Yangming's response was:

As long as you know, how do you remember? To know that it is the second meaning, you need to know your own self-ontology. If you want to remember, you do not know; if you want to know, you don't see your own identity in your mind.

In other words, the more a human learns, the more they have yet to learn. If the purpose of learning information is because of a test requirement, then the information stored in our short-term memory serves the purpose of taking the examination, but recalling the information is often not possible because the information is not true memory. As students we've all experienced memorizing an amazingly large amount of information for an exam and flushing these data from our minds as soon as the exam was over. However, there are instances where information can be recalled for no apparent reason. These remnants of information that we internalized is what was really learned. Therefore, education and learning were intended to convey human thought through books; however, history shows that we've endured at least seventeen global pandemics since the 1300's (Piret and Boivin 2021) and two world wars in the twentieth century. Since then, all established education methods are constantly being recast. Recitation no longer has a place in the original meaning of education. Yuval N. Harari (1976–), the author of *Sapiens: A Brief History of Humankind* (2011), and *Homo Deus: A Brief History of Tomorrow* (2016), argued in *21 Lessons for the 21st Century* (2018) that the existing education system should use critical thinking, problem solving, effective communication, collaboration, and creativity to replace the current emphasis on intellectual indoctrination (Harari 2018).

If the concept of **education** is changing, then the definition of the **environment** is also changing. David W. Orr (1944–), a professor of environmental and political studies at Oberlin College, once said: *All education is environmental education* (Fig. 1.2; Orr 1991:52).

Fig. 1.2 All education is environmental education (Cheng-Shun Fang at Fulong, New Taipei City, Taiwan, 2020) (Photo by Wei-Ta Fang)

When we teach environmental education, we often ask our students and ourselves: *Is environmental education, the process to improve the environment, to improve the educational environment, or to improve the education of people?* We need to understand the *own body* of human beings and to enhance the thinking from the original human engraved thinking. In the process of interpretation, we understand epistemology to understand the nature of matter and understand what the environment is.

1.1.1   The Environment is a Concept

The environment refers to the space in which human beings can perceive their surroundings. In space, you can perceive all things that change in structure and function over time. In other words, the true nature of all things must be in a certain environment, even a vacuum is regarded as an environment (Baggini and Fosl 2003). Therefore, the environment is a concept that is bounded by space. However, in phenomenology, the environment incorporates the concept of time. The Austrian philosopher Edmund Husserl (1859–1938) believes that the human impression of the environment and the world will not gradually

disappear due to the evolution of time, and because of the memory function of the brain, the human impression of the deceased. Therefore, the *existence* of the deceased can persist in perpetuity in the world as long as the living still remembers those that came before because they are stored in the brain as a living impression. These existing memory phenomena gradually change the human imagination of the dead over time. The concept of the environment means different things to different people because every person has their own interpretation or understanding of their environment, which is based on the spatial and temporal elements that they've experienced. In an ideal world, the definition of environment would mean the same to everybody, but inevitably, each person's view of the environment is different.

For phenomenon scholars, *existence* is a base on self-awareness of all phenomena. Therefore, the person's environment is the perceptual medium of a living being to an external stimulus, including the systemic sum of space and time of the instinctive response to the external stimulus (Crowther and Cumhaill 2018). What a living organism understands about its environment includes the perception of elapsed time and distance in its space. Therefore, to understand the nature of things, we must also recognize or be cognizant of the changes in our environment through space and time (Baggini and Fosl 2003).

The definition of environment varies with context and discipline. For biological/ecological systems, the natural environment refers to the sunlight, climate, soil, hydrology, and other ecosystems in which animals and plants coexist and within which we live. The social environment refers to the constitutive state formed by the social, psychological, and cultural conditions associated with human life and culture. From the perspective of environmental protection, the environment refers to the earth on which human beings depend. Therefore, the effectiveness of achieved protection is a critical indicator for revealing environmental quality in a region (Huang et al. 2014). In addition, we also need to understand the definition of environment in terms of regulations.

1.1.2 Lost in Translation

As it turns out the words *environment* and *education* were originally considered nouns and verbs and later translated and adopted into Asian cultures about one hundred years after appearing in English culture. We began to think about the concept of environmental education after World War II. In their 1947 book *Communitas*, the Goodman brothers talked about the planning of urban space where they discussed the idea of establishing green belts around cities and the design of industrial spaces, which at the time were utopian concepts (Goodman and Goodman 1947). They believed that *a large part of the environmental education of the children would be based on technology; whereas a child brought up in a modern suburb or city might not even know what work papa does at the office*. They also criticized the idea of environmental education because they felt environmental education was very close to the *construction education* of a built environment. In fact, it was far from the concept of environmental education.

From 1965 to 1970, US industrial production grew at a rate of 18% and at the same time, it boosted the economies of its World War II allies.

However, the over-emphasis on development led to increased environmental pollution and since the 1960s, environmental problems arising from industrial development continue to emerge (e.g., love canal, Fowlkes and Miller 1987). The green agricultural revolution widely used chemical fertilizers and pesticides and among them, Dichloro-Diphenyl-Trichloroethane (DDT) hindered the reproductive ability of birds and reduced biodiversity. The book, published by Rachel Carson's book, *Silent Spring* (1962) pointed out the consequences associated with the use/abuse of pesticides, which become incorporated into food chains and webs, negatively impacting natural ecosystems and human health. Carson believed that human beings should treat the surrounding environment and animals with a life-loving vision. She said: *The public must decide whether it wishes to continue on the present road, and it can do so only when in full possession of the facts* (Carson 1962:30). In the post 1960s, the slogan of environmental protection rang through the sky and the concept of the environmental protection movement gradually espoused the definition of environmental education in conservation.

1.2 Definition of Environmental Education

The term environmental education appeared in 1947. So, when did the earliest definition of environmental education come into being? The concept of environmental education in terms of modern pedagogy and its evolutionary history feels closely tied to our understanding and development of human psychology, sociology, and how humans learn. In this context, environmental education is a relatively recent field of study and predicated on the acceptance of our hypotheses by a small community of scholars.

If we look further back in time at the development of human cultures at a time when formal education systems did not exist, our ancestors then need to be recognized. In many/most indigenous cultures, the people learned about the environment within which they lived and passed their knowledge and skills to future generations, otherwise, they all died.

We, therefore, also need to identify several elements in indigenous cultures that are related to their knowledge base and resource management systems that could be of value to western science, but the semantic issues associated with Traditional Ecological Knowledge (TEK) and Traditional Ecological Management (TEM) can be overwhelming (Song et al. 2021). We may regard that TEK and TEM should be considered to be an element of the environment associated with indigenous cultures that is defined by their relationship and interactions with the environment, including all of the other biotic, abiotic elements present in their habitats. So, let's take the definition way back and show how we used TEK and TEM to develop our knowledge systems relative to western science.

1.2.1 Initial Definition of Environmental Education

In 1962, Carson explained the importance of environmental protection and hoped to learn the ecological balance of nature through human awakening would achieve the purpose of harmonious coexistence between human and nature. In 1965 at an education seminar at the University of Keele *environmental education* was proposed as a theme, becoming the first conference in the UK to use the term *environmental education* (Palmer 1998). The meeting participants agreed that environmental education *should become an essential part of all citizens, not only because of the importance of understanding aspects of their environment, but because of its immense educational potential in assisting the emergence of a scientifically literate nation.* The meeting emphasized that teachers' participation in basic education research should be strengthened to accurately determine the teaching methods and content of environmental education that are most suitable for modern needs. Therefore, the United Kingdom held a Council for Environmental Education in 1968.

In 1969, William Stapp (1929–2001), a professor at the University of Michigan, School of Natural Resources and Environment (SNRE), first defined environmental education as a process *producing a citizenry that is knowledgeable concerning the biophysical environment and its associated problems, aware of how to help solve these problems, and motivated to work toward their solution* (Stapp et al. 1969:30–31). According to Stapp, the purpose of environmental education was to cultivate citizens who had environmental knowledge, were concerned about environmental problems, and had the ability to solve and actively participate in the resolving the issues. Environmental problems should be resolved using root cause analyses and environmental educators should change the minds of the existing education targets and establish environmentally friendly behaviors.

Stapp is considered the father of environmental education in the United States of America (USA). He helped plan the first Earth Day in 1970, drafted the National Environmental Education Act, served as the first director of United Nations Education, Scientific and Cultural Organization (UNESCO), the first director of the Environmental Education at UNESCO, and promoted the first inter-governmental meeting of 146 countries and territories in Tbilisi, the former Soviet Union, in 1978. In 1984, Stapp assisted students to investigate cases of hepatitis infections from the Huron River, identify the cause of the problem, and worked with the local government to find a solution. In view of the importance of river surveys, he founded the Global Rivers Environmental Education Network (GREEN) in 1989. He cooperated with elementary schools in Ann Arbor, Michigan, USA and conducted many field trips with local elementary students and they investigated and taught students about problems in the natural environment and how to interact with them. He cared about academic research and more about social services, and led college students to promote environmental monitoring programs and successfully rehabilitate the Rouge River. In western science, the roots of environmental education can be traced back 1960s as early as the eighteenth century when Jean-Jacques Rousseau stressed the importance of an education that focuses on the environment (Rousseau and Bloom 1979).

1.2.2 The Extended Definition of Environmental Education

Stapp and his colleagues promoted the definition of environmental education, which was based on American pragmatism. They believed that emphasizing environmental knowledge could change reality through the power of action. Therefore, practical experience in environmental education was considered important because it emphasized taking knowledge and using that knowledge and/or experience to solve problems on natural resource management (Disinger 1985; 1990). Thus, environmental action or doing was better than dogma, and environmental experience was better than rigid principles. Therefore, the concept of environmental education had evolved to become a critical and creative clarification for research questions and value clarification (Harari 2018), interpreting environmental knowledge as a process of assessing the real environment, and scientific exploration (Fig. 1.3). The spirit of humanity, the standard of conduct was then incorporated into the real environment of human beings.

To promote environmental protection, academic institutions needed to provide environmental education-related courses such as basic environmental research, science, planning, management, economics, society, culture, and engineering. At the grade school level, the students should be taught the history of environmental protection and environmental protection measures. The aforementioned courses are meant to be broad because the environment and the associated issues are complicated. The environment and its associated ecosystems are not one size fits all. Not only are ecosystems different from one another the variation within each is vast. As such, Wals et al. (2014) considered the learning content of environmental education to be multidisciplinary, based on environmental problem assessment, critical thinking, morality, creativity, and make judgments on environmental issues. The process of environmental education helps observation and problem-solving, with the opportunity for individuals to promote environmental improvement actions to ensure positive environmental behaviors (Fig. 1.4). Therefore, environmental education includes the social, abiotic, and biological aspects, of the environment including natural resource conservation, environmental management, ecological principles, environmental interactions and ethics, and sustainability (Fig. 1.5).

Environmental Education is defined in many ways, but each definition considers it to be a discipline or process that teaches us how to behave in a manner that promotes the responsible management of our environment and resources. This then helps the environment function in a more natural way, rather than healing anthropogenic wounds. We detected the Environmental Education Act in Taiwan (Republic of China), there is....*as a discipline that enhance the environmental awareness, environmental ethics, and responsibility of the nation taking as a whole, so as to safeguard the ecological balance of the environment, respect lives, promote social justice, and cultivate environmental citizens and environmental learning communities* (The Environmental Education Act) (Ministry of Justice 2017). Therefore, in terms of education content, environmental education was intended to integrate aspects of biology, chemistry, physics, ecology, earth science, atmospheric science, mathematics, and geography as an integrated discipline toward an education for sustainability (EfS) (Evans 2020). Methods of educational research include the applied social sciences such as psychology, sociology, culture, history, anthropology, economics, political science, and information science.

The First International Working Meeting on Environmental Education in the School Curriculum was organized by the UNESCO, and the International Union for Conservation of Nature (IUCN) in Nevada, USA in 1970. A participant resolution developed the statement that the elements of environmental education are not

Fig. 1.3 Academic institutions need to provide environmental education-related courses such as basic environmental research and environmental science such as a study camp, Taipei, Taiwan, 2019 (Photo by Yi-Te Chiang)

Fig. 1.4 Environmental education helps develop observation and problem-solving skills and provides opportunities for individuals to promote environmental improvement actions to ensure positive environmental behaviors (Audubon Nature Center, Rhode Island, USA, 2015)(Photo by Wei-Ta Fang)

Fig. 1.5 The target of environmental education should include a continuum from primary school to the university and include theoretical, practical, indoor, and outdoor experiences. This is a group of people enjoying nature in a Swiss environmental education program in 2009 (Matterhorn, Great north faces of the Alps, Valais, Switzerland) (Photo by Wei-Ta Fang)

Fig. 1.6 Environmental education also includes the dissemination of environmental education with outdoor media (Wei-Ta Fang, Ben A. LePage, and their students at Dagouxi Riverside Park, Neihu, Taipei, Taiwan, 2021) (Photo by Yi-Te Chiang)

completely combined by any single discipline. It is the product of partnerships built on sound science, public awareness, environmental issues, and outdoor educational methods (Fig. 1.6). UNESCO specifically stated that environmental education programs taught students a respect for the nature and natural environments and raised citizens' environmental awareness (UNESCO 1970). Therefore, the organization emphasized the importance of environmental education in protecting the society's quality of life in the future by protecting the environment, eradicating poverty, minimizing inequality, and ensuring sustainable development. Cerovsky (1971, p. 4) defined environmental education as.

> ...a process of recognizing values and clarifying concepts in order to develop skills and attitudes necessary to understand and to appreciate the interrelatedness among man, his culture, and his biophysical surroundings. Environmental education is also entailed practice in decision-making and self-formulation of code behavior about issues concerning environmental quality.

The target of environmental education includes education in the school system, and education from primary, middle, vocational, and technical schools, universities, and research institutes. However, environmental education also includes the dissemination of environmental education, including print, books, websites, and other media. In addition, aquaria, zoos, parks, and nature centers in social environmental education should all provide ways to teach citizens about the environment (Fig. 1.6) (see Box 1.1).

Box 1.1: The Legal Definition of Environmental Education, Republic of China (ROC)

The Article 3 of Republic of China's Environmental Education Act stated as Environmental education: Referring to the adaptation of educational means by which to culminate the citizens to understand their ethnical relationship to the environment, enhance the citizens' environmental protection awareness, skills, attitudes and values, and steer the citizens to emphasize the environment and adopt actions to achieve a civility education process that harbors sustainable development.

1.2.3 The Goals of Environmental Education

The attendees of the Tbilisi Conference in 1977 endorsed goals for environmental education into five categories (UNESCO 1977). They are:

- **Awareness:** to help social groups and individuals acquire an awareness of and sensitivity to the total environment and its allied problems;
- **Knowledge:** to help social groups and individuals gain a variety of experiences in and acquire a basic understanding of the environment and its associated problems;
- **Attitudes:** to help social groups and individuals acquire a set of values and feelings of concern for the environment and the motivations for actively participating in environmental improvement and protection;
- **Skills:** to help social groups and individuals acquire the skills for identifying and solving environmental problems; and
- **Participation:** to provide social groups and individuals with the opportunities to be actively involved at all levels in working toward resolving environmental problems (UNESCO 1977, p. 71).

Hungerford et al. (1980) organized and suggested that these goals should be operationalized within the school curriculum and categorized into four (4) levels (Table 1.1 and Figs. 1.7, 1.8 and 1.9).

1.3 Approaches to Environmental Education

In this section we discuss various methods/pedagogies used in the field of environmental education. Environmental education, like science education, is interdisciplinary and offers a variety of learning strategies, which are determined by learning resources, learning time, learning space, learning curriculum, and student attributes. These differences all affect education approaches in some way. We briefly describe outdoor education, classroom education, and nature-centered education. We include the following seven methods, including: school environmental education, school nature education, place-based education, and projects curricula; and nature center education in social and environmental education, science and environmental education in zoos and museums (Falk 2009; Falk and Dierking 2014, 2018; Ardoin et al. 2016) or environmental

Table 1.1 Environmental Education Goal Levels

Goal Level 1	Knowledge and Ecological Foundation To provide the receiver with sufficient knowledge in ecology, which will permit him/her to eventually make ecologically sound decisions with respect to environmental problems
Goal Level 2	Conceptual Awareness of Issues and Values To guide the development of conceptual awareness how individual and collective actions might influence the relationship between quality of life and the quality of the environment and how these actions resulted in environmental issues, which must be resolved through investigation, evaluation, values clarification, decision-making, and finally, citizenship action
Goal Level 3	Issue Investigation and Evaluation The development of the knowledge and skills necessary to permit receivers to investigate environmental issues and evaluate alternative solutions for remediating these issues. Similarly, values are clarified with respect to these issues and alternative solutions
Goal Level 4	Environmental Action Skills and Participation To guide the development of the skills necessary for receivers to take positive environmental action for the purpose of achieving and/or maintaining a dynamic equilibrium between quality of life and that of the environment

Source Hungerford et al. (1980)

Fig. 1.7 Goals of environmental education (adapted and modified after Hungerford et al. (1980) and revised by Wei-Ta Fang)

Fig. 1.8 Harold Hungerford (left), Trudi Volk (middle), Arba'at Hassan (right) (Photo by Arba'at Hassan)

Fig. 1.9 Harold Hungerford (left), mentor and advisor of Arba'at Hassan (right) (Photo by Arba'at Hassan)

education using surveys, assessments and actions on environmental issues (Hsu et al. 2018), and science-technology-society (STS) (Winther et al. 2010). Each approach addresses important curriculum goals and novel learning methods for environmental education. Therefore, environmental educators should choose and apply the most effective methods for their students and environment. We also explore sustainable development education from the perspective of environmental education. We also understand that a well-rounded curriculum aims to strengthen environmental awareness and environmental sensitivity, environmental knowledge, environmental ethics and values, environmental action skills, and environmental action experience. We explore values, topics, and learning in the context of Bamberg and Moeser (2007), Winther et al. (2010) and Dillion and Wals (2006).

1.3.1 Outdoor Education

Outdoor education is based on a place-based education and project curricula in the United States that include: the Project Learning Tree, Project WILD, and Project WET. In addition, surveys, assessments, and actions on environmental issues, as well as environmental education in STS that can be used for exploration, included the following methods (Braus and Wood 1993; Engleson and Yockers 1994).

1.3.1.1 Uses of the Senses

Let the learners use their senses to experience nature directly using their eyes, ears, nose, tongue, and body to feel the environment over the four seasons (Fig. 1.10).

1.3.1.2 Physical Exercises and Explanations

By using real examples, objects that can be obtained, and through practical methods, the natural or scientific phenomena contained in the environment are directly explained by practical performance, allowing learners to observe directly or actual experience (Fig. 1.11).

1.3.1.3 Surveys and Experiments

Let learners think about environmental issues and environmental phenomena through the steps of generating a hypothesis, survey, data collection, experiments, data collection, analysis,

Fig. 1.11 Physical exercises and explanations (Photos by Arba'at Hassan)

writing of small papers, briefings, etc., and actually discuss what happens behind various environmental phenomena problem.

1.3.1.4 Attractions Travel

Let learners go to various attractions and actually visit forests, mountains, seashores, wetlands and other areas to observe and obtain first-hand tourism and observation experiences (Fig. 1.12). Each observation and survey is a purposeful activity, and learners can learn about certain topics in advance through books, the Internet, and scenic spot information.

1.3.1.5 Research Questionnaires and Interviews

An issue questionnaire is performed through research methods for small papers. Through this approach, learners can obtain relevant environmental information. The perceptions and ideas of different interviewers, in addition to quantified research data, are to conduct interviews to understand qualitative information and to make more environmental issues for an in-depth discussion (Fig. 1.13).

1.3.1.6 Outdoor Observation of Nearby Places

Using the method of place-based education, selecting nearby places, conducting environmental surveys or observation activities, actually guiding learners to study in outdoor environments, and helping learners to understand the

Fig. 1.10 Use of the senses (Photos by Arba'at Hassan)

Fig. 1.12 Outdoor education is based on a place-based education and project. The U.S. Environmental Protection Agency (U.S. EPA), North American for Environmental Education (NAAEE), and Environmental Protection Administration of the Executive Yuan, Republic of China (Taiwan EPA) are the key to promote environmental education as partnerships. They jointly launched the Global Environmental Education Partnership (GEEP) in 2014, with the vision of creating a sustainable future where people and the environment prosper together through the power of education. The GEEP established the Asia–Pacific Regional Center (APRC) in Taiwan as a network center for environmental education in Southeast Asia in 2019 (Toucheng Leisure Farm, Ilan, Taiwan, 2021; please see https://geepaprc.org/en) (Photo by Wei-Ta Fang)

Fig. 1.13 Research questionnaire and interview (Photo by Arba'at Hassan)

natural exploration, experience, and awareness (Fig. 1.14).

1.3.1.7   Data Collection and Interviews

Let learners collect the data on specific environmental issues so that they can have a deeper

Fig. 1.14 Outdoor observation at nearby mangrove places (Photos by Wei-Ta Fang)

understanding of related environmental issues or areas of study, through the library, Internet, photographs, and interview specific people to help

Fig. 1.15 Assembly photos. Left side: People need to help clarify questions when facing environmental problems (Yongchunpi Wetland Park, Taipei, Taiwan, 2021) (Photo by Yi-Te Chiang); Right side: Data collection and interview on the charcoal making from mangrove trees, 2008 (Photos by Arba'at Hassan)

clarify questions when facing environmental problems if more information is needed (Figs. 1.15).

1.3.2 Classroom Education

Classroom education in environmental education includes campus environmental education, which can develop place-based education, project curricula, and STS (Winther et al. 2010). During the learning process, teachers are encouraged to participate in professional learning sessions, and fully understand the learner's learning role, that include:

1.3.2.1 Reading and Writing

In the classroom, students read environmental issues and events and write their thoughts and feelings in notebooks. Younger students can draw their thoughts and ideas (Fig. 1.16).

1.3.2.2 Case Study

Let learners directly collect and integrate data on environmental issues or discuss and assess the environmental impact of related issues on our day-to-day lives and to think about how to deal with environmental damage (Fig. 1.17).

1.3.2.3 Value Clarification

Let learners use each other's relationships between value and morality for discussion and communication. During the discussion, through mutual discussions, establish conclusions that everyone can accept to assist learners to establish

Fig. 1.16 Reading and writing in the classroom (Photo by Arba'at Hassan)

Fig. 1.17 A case study of urban park (Photo by Wei-Ta Fang)

correct environmental attitudes and values. That is, set up some ground rules so the environment remains safe.

1.3.2.4 Treemap and Brainstorming

Through brainstorming or treemap thinking, help learners connect different relationships, situations, ideas, and processes to understand the relationship of events (Figs. 1.18 and 1.19).

1.3.2.5 Debate

Through debate activities, learners can learn from different topics facing the environment, and learn to use data collection, communication, and critical thinking skills (Fig. 1.20).

1.3.2.6 Group Learning

Through the process of group learning, in addition to being able to face environmental issues more effectively and conduct more in-depth discussions, learners can learn to establish team tacit

understanding, self-social ethics norms, and know the thoughts deep inside themselves. This illustrates the importance of diversity and inclusion in the program (Fig. 1.21).

1.3.2.7 Environmental Arrangement

Through the environmental arrangement activities of the beginning of school, festivals, or parent-teacher talks, let learners participate in the creation and arrangement of teaching space. In addition, to help learners have a complete learning space, they can also learn to judge the overall environmental learning.

1.3.2.8 Comprehensive Discussions

Scout courses covering aspects of integrated geography, mathematics, nature, health and

Fig. 1.18 Value clarification (Photo by Arba'at Hassan)

Fig. 1.20 Debate on an environmental issue (Photo by Arba'at Hassan)

Fig. 1.19 Treemap and Brainstorming session (Photo by Arba'at Hassan)

Fig. 1.21 The group learning activity (Photo by Arba'at Hassan)

Fig. 1.22 All students take part in comprehensive discussions (Photo by Arba'at Hassan)

hygiene, or Chinese language learning areas, and in-depth research and discussion on environmental issues and issues (Fig. 1.22).

1.3.2.9   Activity Workshop

Let learners guide the demonstration and teaching of personnel, learn to operate, or produce a kind of labor course that requires hands-on work, and use hands-on operations. The process of the drill included working experience in farming, forestry, fishing, insect hotel building, animal husbandry, and the creation of handicrafts (Figs. 1.23 and 1.24).

1.3.2.10   Game Learning

Game learning is different at different levels. This is important because it brings to mind what today's youth think is important and fun. For

Fig. 1.23 An activity workshop (Photo by Arba'at Hassan)

example, we think memorization is boring and old school, what will the younger generation of student's think about these new approaches in 50 years? 100 years? How do we keep our learning methods current in the face of rapidly changing technologies, norms, and values? In game learning, open-ended play is adopted. The rich teaching materials of games are the basis of learning. In modeled-play, learn using simulated creatures and playing with pets (Fig. 1.25). In purpose-framed play, games are used for experience and teacher-student interaction is used (Cutter-Mackenzie et al. 2014).

1.3.2.11   Environmental Action

Use STS learning methods to allow learners to participate in practical environmental actions such as ecological management, persuasion, consumerism, political action, and legal action, and work together to improve environmental issues (Fig. 1.26).

1.4   Development of Environmental Education

The implementation of environmental education is to adopt an infusion method and conduct integrated curriculum across learning areas to connect the relationship between the surrounding their environment. Environmental education professionals generally believe that the environmental education be integrated into the school curriculum of each school year, from kindergarten to grade 12 (K-12). However, discipline integration of environmental education has not occurred in countries around the world. How to integrate environmental education into the subject in the school curriculum requires the use of teaching materials and methods (Fig. 1.27). This may be related to the type of teaching in each subject (Simmons 1989). If the core of environmental education is to incorporate the behavioral decisions of governments, enterprises, families, and individuals into the education process, then the development of environmental education from kindergarten to grade 12 (K-12) needs to be considered and economic development, a parallel

Fig. 1.24 We are just another bug on this planet (Ben LePage at Taiwan Insect Hall, Taipei, Taiwan, 2022) (Photo by Swing Chan)

Fig. 1.25 The outdoor game learning (Velsen Otte and the cat "Noodle") (Photo by Wei-Ta Fang)

Fig. 1.26 The Environmental action on wetland (Keita Furukawa, front person, and Jung-Chen Huang at Taijiang National Park, Tainan, Taiwan) (Photo by Wei-Ta Fang)

trend of environmental development that takes into account social development.

The teaching model of traditional environmental education is centered on environmental issues. However, this kind of teaching method only focuses on knowledge transfer. It does not consider social emotional learning. At the same time, it does not consider the formation of environmental attitudes, and it is difficult to cultivate responsibility—environmental behavior students. Furthermore, environmental education places too much emphasis on analysis of issues, so that students learn learned helpless. It has a sense of despair and helplessness about the future development of the global environment. It is impossible to learn through a position of control—motivation and perseverance to change the world. In addition, emotional changes in environmental education are not easy to change through indoor courses, students are easily frustrated in the classroom, and it is difficult to learn the true meaning of pro-environmental behavior. If we say that the past education focused on one-way narrative transmission, we should then look at environmental issues with a healthy mindset. By caring about environmental protection issues, based on teachers' pedagogical content knowledge and domain knowledge (Shulman 1986a, b;

Fig. 1.27 We have developed environmental education programs from kindergarten to grade 12 (K-12) from the supports of Ramsar Regional Center–East Asia and National Geographic Society during 2018 (Taipei, Taiwan, 2018) (Photo by Yi-Te Chiang)

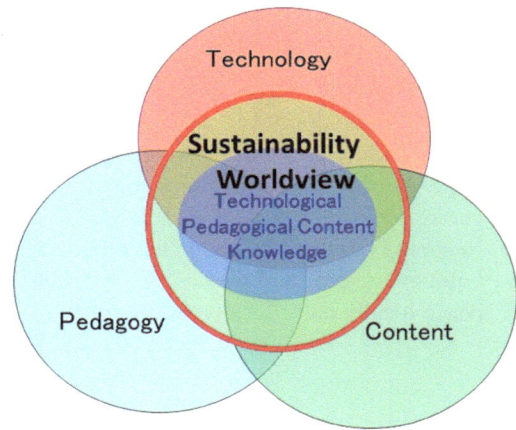

Fig. 1.28 The content of teaching content is a kind of comprehensive knowledge. It is the knowledge that teachers can use in teaching after integrating various kinds of knowledge (Illustrated by Wei-Ta Fang)

1987a, b; Fig. 1.28), supporting the idea of a sustainable worldview, strengthening the content of various disciplines in a common learning approach, and internalizing it into specific environmental protection actions.

The so-called pedagogical content knowledge model, the content includes the teacher's understanding of specific subject content, the teacher's grasp and use of specific subject content representation, and the teachers 'learning and learners' understanding. The content of teaching content

knowledge includes the content of subject knowledge and general teaching knowledge, and goes beyond the teaching material knowledge itself. The teaching content knowledge was proposed by an American educational psychologist, named Lee Shulman (1938–). He believes that the subject teaching knowledge goes beyond the scope of subject expertise and is subject matter expertise at the teaching level. Shulman pointed out that teachers' knowledge can be divided into three categories, namely, pedagogical knowledge, subject matter knowledge, and pedagogical content knowledge (Shulman 1986a, b; 1987a, b). Teaching knowledge emphasizes the principles, methods, and strategies of teaching. Disciplinary content knowledge emphasized teachers' knowledge on the facts, concepts, principles of the subject areas, and how they are organized. In addition, teaching content knowledge emphasizes that when teaching, teachers know how to use a systematic statement of their subject content knowledge, make it easy for students to understand the subject content through the most effective teaching method, and teachers can understand students' previous concepts of the subject content, Reasons for learning difficulties and strategies for remedial teaching.

Shulman said:
Teaching content knowledge means that teachers must be able to express what they are teaching. In

the category of teaching content knowledge, teachers include the most taught topics and the most effective forms of expression in the subject.

They are the most powerful analogies, examples, illustrations, demonstrations, and clarifications. That is, teachers regroup in special subjects of the subject and behave in an appropriate way to promote students to understand the content of the teaching. Knowledge of teaching content also includes teachers understanding what factors make it difficult or easy for students to learn about specific concepts when learning, and to understand the concepts and prerequisite concepts held by students of different ages and backgrounds when studying these topics. (Shulman 1986b:9).

Communication environmental and educational concepts, goals, methods, and strategies are based on the concept of immersive environmental education. Explore the in-depth fields of environmental education according to the different cultural and social backgrounds of teachers (Fig. 1.29). Therefore, based on the critical analysis of the problem, the process of learning is more important than the outcome. Moreover, the limitations of environmental, social, and economic issues, are understood, thus the teaching content can be linked to the real world.

Environmental education is not only about providing tools and technology but also necessary to cultivate students' environmental literacy. Therefore, the teaching of environmental education, in addition to teaching knowledge, also needs to inspire students' social responsibility. Therefore, environmental education needs to put forward values and strengthen the thinking of sustainable development in the curriculum. The main core lies in the fundamental values of "sustainable development education." UNESCO defined the core according to the following topics:

1.4.1 Values

- Respect the dignity and human rights of all human beings worldwide and commit to social and economic justice for all;
- Respect the human rights of future generations and promise intergenerational responsibilities (Kaplan et al. 2005; Liu and Kaplan 2006);
- Respect and care about the diversity of life in large communities, including the protection and restoration of the earth's ecosystem; and
- Respecting cultural diversity and promising to build tolerance, non-violence, and a culture of peace locally and globally.

Fig. 1.29 Exploring the in-depth fields of environmental education is crucial according to the different cultural and social backgrounds of teachers (Photo by Wei-Ta Fang)

1.4.2 Exploration Topics

1.4.2.1 Environmental Orientation

Environmentally oriented education needs to include attention on natural resources (like water, energy, agriculture, forestry, mining, air, waste disposal, toxic chemical treatment, and biodiversity), climate change, rural development, and sustainability. The purpose of mitigation and adaptation in the cities, disaster prevention, and mitigation are to strengthen the understanding of the fragility of resources and the natural environment, strengthen the understanding of the negative impact of human activities and decision-making on the environment, and include environmental factors. These factors must be considered in formulating socio-economic policies.

1.4.2.2 Economic Orientation

The Economic Oriented Education needs to focus on the issues of poverty eradication, strengthening the social responsibility of enterprises and universities, and strengthening the efficiency of the market economy. The purposes are to understand limitation, potential on an economic growth, and how they could affect the society, environment, and culture. The impact of environmental protection, culture, and social justice on the correct assessment of individual and social consumption behavior is consistent with the goal of sustainable development.

1.4.2.3 Social Orientation

Socially Oriented Education needs to include concerns about human rights, peace and human security, freedom, gender equality, cultural diversity, and cross-cultural understanding, as well as emphasis on social and personal health, and strengthening government management and people's governance. Its purpose is to understand the role of social systems and environmental change in development and to strengthen models and institutions of democratic participation. The democratic participation system provides opportunities to express opinions, adjust conflicts, decentralize government, build consensus, and resolve differences. In addition, cultural assessments in society need to be strengthened to

protect the *values, practices, languages, and knowledge systems* (Arenas et al. 2009). At the same time, the cultural foundations of social, environmental, economic, and the sustainable development, are seen as inter-connected. In other words, sustainable development emphasizes interrelationship through culture. In the process of sustainable development education, it is particularly necessary to pay attention to the diversity of culture and ethnic groups, and each ethnic group tolerates, respects, and understands each other in order to shape the values of equality and dignity.

We can know that the exploration of sustainable development education to embedding sustainability from environmental education can be an overlapping circle model, which is an intersecting system (Purvis et al. 2019). This model recognizes the intersection of economic, environmental, and social factors. Based on our research, we resized the circles to show that one factor has advantages over the other two. In the eyes of economists, economy is better than society and society is better than environment. This model means that economy can exist independently of society and environment. Therefore, we use the next more accurate system model for illustration (Fig. 1.30).

Because human beings cannot survive outside of their environment, they do not have an environment. It is just like a fish without water, which makes it difficult for them to survive. If we ask all the fishermen in the sea if overfishing the

Fig. 1.30 The environment-oriented, economic-oriented, and social-oriented rendezvous system (Illustrated by Wei-Ta Fang)

fisheries is a social disaster or an economic disaster, they will then say that it is all the above. Therefore, the nested dependency model reflects the reality of this common dependency. In other words, human society is a wholly owned subsidiary of the environment. An economic society, without food, clean water, fresh air, fertile soil, and other natural resources, we are "cooked."

1.5 Summary

Environmental Education in the twenty-first century and Education for Sustainable Development have also regarded as the key to reconstructing ecologically responsible citizens to embrace a pedagogy grounded in ecosocialism (Arenas 2021). With the adoption of the 2030 Global Education Agenda, United Nations Educational Scientific and Cultural Organization (UNESCO) is now using the United Nation's recently developed Sustainable Development Goals (SDGs) to strengthen the Global Action Follow-up Program on Education for Sustainable Development (i.e., GAP 2030). In general, the purpose of environmental education is to cultivate citizens who understand the biophysical environment and related issues, how to help solve problems, and actively understand the ways to solve problems (Stapp et al. 1969). Currently, we provided a wider range of services, strengthened appreciation of the multicultural and environmental systems around humanity, and ensure the sustainable development of human society. Shin Wang (1945–), the emeritus professor of the Department of Geographical Environmental Resources, National Taiwan University, once said: "Hometown is the beginning of learning. You need to be based on Taiwan to look at the world." The transformation of social environment and silent environmental changes to the environmental protection of the aboriginal people has produced their own views of environmental redemption (Fang et al. 2016).

At the beginning of the writing of this book, we always told ourselves in the heart: "The environment and ecology are extremely vulnerable, and only those of us who are not fame and fortune environmentalists will help the speechless environment."

In light of today's social consumerism, inequality has occurred in three areas: environmental, social, and economic. We strengthen our creativity sharing our experiences within the education system with others to develop a shared social imagination. We communicate the concepts of the environment and education based on the concept of immersive environmental education. Therefore, the environmental education concepts, implementation processes, and education policies listed in this chapter have achieved the feasibility of environmental education in various fields through teaching, research, and practice. Environmental education is not just about providing tools and technologies, it is important to cultivate a learners' environmental literacy. Therefore, the teaching of environmental education, in addition to teaching knowledge, also needs to inspire students' social responsibility.

References

Ardoin NM, Schuh J, Khalil K (2016) Environmental behavior of visitors to an informal science museum. Visit Stud 19(1):77–95

Arenas A (2021) Pandemics, capitalism, and an ecosocialist pedagogy. J Environ Educ 52(6):371–383

Arenas A, Reyes I, Wyman L (2009) When indigenous and modern education Collide. In: Zajda J, Freeman K (eds) Race, ethnicity and gender in education. Globalisation, Comparative Education and Policy Research, vol 6. Springer, Dordrecht

Baggini J, Fosl PS (2003) The philosopher's toolkit: a compendium of philosophical concepts and methods. Wiley-Blackwell, Oxford

Bamberg S, Möser G (2007) Twenty years after Hines, Hungerford, and Tomera: a new meta-analysis of psycho-social determinants of pro-environmental behaviour. J Environ Psychol 27(1):14–25

Braus A, Wood D (1993) Environmental education in the schools-creating a program that works. NAAEE Peace Corps Washington, DC

Carson R (1962) Silent spring. Fawcett, Greenwich

Cerovsky J (1971) Environmental education: yes—but how? In: Handbook of environmental education with international studies. Wiley, London, p 4

Crowther T, Cumhaill CM (2018) Perceptual ephemera. Oxford University Press, Oxford

Cutter-Mackenzie A, Edwards S, Moore D, Boyd W (2014) Young children's play and environmental education in early childhood education. Springer, New York

Dillion J, Wals AEJ (2006) On the danger of blurring methods, methodologies, and ideologies in environmental education research. J Environ Educ Res 12(3–4):549–558

Disinger JF (1985) What research says: environmental education's definitional problem. Sch Sci Math 85(1):59–68

Disinger JF (1990) Environmental education for sustainable development? J Environ Educ 21(4):3–6

Engleson DC, Yockers DH (1994) A guide to curriculum planning in environmental education. Wisconsin Dept. of Public Instruction, Madison

Evans N (2020) What ought to be done to promote education for sustainability in teacher education. J Philos Educ 54(4):817–824

Falk JH, Dierking LD (2014) The museum experience revisited. Left Coast Press, Walnut Creek

Falk JH, Dierking LD (2018) Learning from museums, 2nd edn. Rowman & Littlefield, Lanham

Falk JH (2009) Identity and the museum visitor experience. Left Coast Press Walnut Creek

Fang W-T, Wu H-W, Lee C-S (2016) Atayal's identification of sustainability: traditional ecological knowledge and indigenous science of a hunting culture. Sustain Sci 11(1):33–43

Fowlkes MR, Miller PY (1987) Chemicals and community at Love Canal. In Johnson BB, Covello VT (eds) The social and cultural construction of risk. Technology, risk, and society, vol 3. Springer, Dordrecht. https://doi.org/10.1007/978-94-009-3395-8_3

Goodman P, Goodman P (1947) Communitas: means of livelihood and ways of life. Vintage Books, New York

Harari YN (2018) 21 Lessons for the 21st century. Spiegel & Grau, New York

Huang C-W, Chiu Y-H, Fang W-T, Shen N (2014) Assessing the performance of Taiwan's environmental protection system with a non-radial network DEA approach. Energy Policy 74:547–556

Hungerford HR, Peyton RB, Wilke RJ (1980) Goals for curriculum development in environmental education. J Environ Educ 11(3):43–44

Kaplan MS, Liu ST, Steinig S (2005) Intergenerational approaches for environmental education and action. Sustain Comm Rev 8(1):54–74

Liu S-T, Kaplan MS (2006) An intergenerational approach for enriching children's environmental attitude and knowledge. Appl Environ Educ Comm 5(1):9–20

Ministry of Justice (2017) The environmental education act, Amended Date: 29 Nov 2017, Category:

Environmental Protection Administration, Taiwan, ROC. https://law.moj.gov.tw/ENG/LawClass/LawAll.aspx?pcode=O0120001

Orr D (1991) What is education for? Six myths about the foundations of modern education, and six new principles to replace them. Context 27(53):52–58

Palmer J (1998) Environmental education in the 21st century: theory, practice, progress, and promise. Routledge, London

Piret J, Boivin G (2021) Pandemics throughout history. Front Microbiol 3594. https://doi.org/10.3389/fmicb.2020.631736

Purvis B, Mao Y, Robinson D (2019) Three pillars of sustainability: in search of conceptual origins. Sustain Sci 14(3):681–695

Rousseau J-J, Bloom A (1979) Emile: or on education. Basic Books, New York

Shulman LS (1986a) Paradigms and research programs in the study of teaching: a contemporary perspective. In Wittrock MC (ed) Handbook of research on teaching, 3rd edn. Macmillan, London, pp 3–36

Shulman LS (1986b) Those who understand: knowledge growth in teaching. Educ Res 15(2):4–14

Shulman LS (1987a) Knowledge and teaching: foundations of the new reform Harvard Educ Rev 57:1–22

Shulman LS (1987b) Learning to teach. AAHEA Bull 5–9

Simmons D (1989) More infusion confusion: a look at environmental education curriculum materials. J Environ Educ 20(4):15–18

Song KS, LePage B, Fang W-T (2021) Managing water and wetlands based on the Tayal's interpretation of *Utux* and *Gaga*. Wetlands 41(92) https://doi.org/10.1007/s13157-021-01473-y

Stapp WB (1969) The concept of environmental education. Environ Educ 1(1):30–31

UNESCO (1970) International working meeting on environmental education in the school curriculum, Final Report, at Foresta Institute, Carson City, Nevada. IUCN and UNESCO, Paris

UNESCO (1977) Intergovernmental conference on environmental education, Tbilisi Declaration (USSR) 14–26 Oct 1977 (final report). UNESCO, Paris

Wals AEJ, Brody M, Dillon J, Stevenson RB (2014) Convergence between science and environmental education. Science 344(6184):583–584

Winther AA, Sadler KC, Saunders GW (2010) Approaches to environmental education. In: Bodzin A, Klein S, Weaver S (eds) The inclusion of environmental education in science teacher education. Springer, Dordrecht

2

Philosophy and History of Environmental Education

The concept of history plays a fundamental role in human thought. It invokes notions of human agency, change, the role of material circumstances in human affairs, and the putative meaning of historical events. It raises the possibility of "learning from history." And it suggests the possibility of better understanding ourselves in the present, by understanding the forces, choices, and circumstances that brought us to our current situation.

Daniel Little, Philosophy of History, 2020.

Abstract

In the first chapter, we discussed the definition of environmental education. In Chap. 2, we discuss environmental education based on the pragmatic view of environmental protection. This view is that the interpretation of the living environment is complicated, but because the process of human reasoning it is finite in nature. Our actions should be rooted in our past history, philosophy, and experience in environmental protection that may allow improvement of our physical environment. The essential part of this process is the ability to recognize the components of our ecosystems that can be managed and when adaptation is the only option or survival.

2.1 Philosophy of Environmental Protection

The philosophy of environmental protection has a long history and possesses characteristics of Eastern and Western philosophy. In the early days of agrarian-based cultures, we used and managed the planet's natural resources by storing and cultivating vegetables (crops) and raising livestock.

China's environmental education can be traced back 2,500 years to Sima Qian (司馬遷) of the Western Han Dynasty (145–86 BC), the *Records of the Grand Historian* (*Shiji*) 《史記》 is described. In the Five Emperors' Records, it was recorded that during the reign of Emperor Shun (舜), the position/role of a *Yu* (虞) officer was created to take charge of the protection of mountains, forests, rivers, plants, birds, and animals. In the Book of Yi Zhou (Lost Book of Zhou) 《逸周書》, it was recorded that Yu the Great (禹)(c. 2123–2025 BC) ordered the following regulations or methods to manage the natural resources:

> During spring, do not hack trees with an axe in the mountainous areas, so that you can see vegetation growing up. During summer, do not catch fish with fishing nets in rivers and swamps, and you can see fish and turtles growing up.

He established protocols that prohibited logging in the spring and fishing during the summer. By the Zhou Dynasty, local officials were given the responsibility of protecting the mountains, rivers, forests, and swamps, which further strengthened the protocols in place to protect these natural resources. The use of the environment and its resources by the Chinese emperors has generally been pragmatic. Although the laws and regulations that were established were not originally intended to protect the environment, these people had inadvertently established the concept and practice of environmental sustainability and inspired future generations to take note and adopt the environmental practices of other or previous cultures. For example, Mencius (孟子)(372–289 BC) said:

> Without delaying the time for the people to cultivate, the grains will be much to eat. If you can go to the pond without utilizing a fine fishing net, you can eat many fish. If you can go to the forest to cut at the right time, then you cannot run out of wood. There are many grains and fish to eat, as well as much wood to use to be satisfied by human lives.

From the perspective of eastern philosophy, the consciousness of environmental protection is based on the richness of realist products and does not resort to the conservation factors of environmental ethics.

2.1.1 Theory of Ideas and Empiricism

In the fifth century BC, ancient Greek philosophers believed that nature was a growing and changing organism. Plato (429–347 BC) put forward a *holistic* view of nature (Table 2.1). In *Ennead*, Plotinus described Plato's theory of ideas as: *the universe as a whole*. Plato believed that in natural ecosystems, there is a mutual relationship between the biotic and abiotic components of ecosystems. For example, every creature designed by the creator in nature has a special niche in nature. If one species disappears, it can then cause discord in the system. According to Plato, what can be seen by human senses is not real, but rather a form and projection of perfect rationality. Aristotle (384–322 BC) opposed the theory of ideas. He used experience to define the world, tried to observe nature, and collected huge amounts of biological

data. Aristotle believed that there are key and minor differences in the elements that comprise an ecosystem. A contemporary view of these concepts is that they are the keystone and foundation species and ecosystem engineers; concepts where an organism defines the entire ecosystem and without that species the ecosystem would probably not exist (Soga and Gaston 2016; Gaston and Soga 2020). Once a key element is lost or changed, it causes the overall ecosystem to change, although the disappearance of a minor part will not affect the integrity of the ecosystem. For example, Aristotle believed that rats cause ecological harm. Therefore, it was necessary to rely on the power of nature, such as the creator to create natural enemies for these rodents to reduce their impact to ecosystems and people.

In the Middle Ages, Europe was similar to ancient China because of the influence of religion. Through forest regulations and/or hunting laws, hunting was prohibited at specific times. Some areas were designated as sacred sites for geographical or religious reasons and were protected from being used for anything but ecological purposes. In medieval Japan, strict laws banned the cutting trees or harvesting of forest products. In the Americas, the traditional Indian belief is that there is a spiritual relationship between humans and prey. Such a relationship will restrict their hunting behavior and the people will not excessively hunt wild animals.

2.1.2 Transcendence and Efficiency

In modern times, due to the natural view of Christianity, humanitarian thought and the natural view of romanticism, there has been an environmental awareness was incorporated into the belief that the creator has given the environment and its resources to mankind *trust nature*.

This consciousness is not new, because humans have long used the natural environment, but protecting the environment and its resources seems difficult. As a result, people with lofty ideals are worried about the gradual destruction of the environment and the natural resource protection thoughts that result. This idea of

Table 2.1 Western environmental philosophy

Time	Scholars	Theory	Overview
Fifth century BC	Plato, 429–347 BC	Rationalism	In *Ennead*, Plato portrays the universe as a whole
Fourth century BC	Aristotle, 384–322 BC	Empiricism	Aristotle believes that there are critical and secondary differences in ecosystems. Once the criticality is lost, it will cause changes in the overall ecosystem
1836	Ralph Waldo Emerson, 1803–1882	Transcendence	Emerson published *Nature*, which emphasizes the direct communication between man and God, and explores divinity in human nature
1903	Gifford Pinchot, 1865–1946	Efficiency theory	Pinchot advocates ecological best interests and is considered a realist
1949	Aldo Leopold, 1887–1948	Preservation theory	Leopold's *A Sand County Almanac* advocates the need for resource conservation and habitat preservation
1972	James Lovelock, 1919–	Vitalism	Lovelock proposed the Gaia hypothesis that the earth's biosphere, including the inanimate environment and living things, constitutes a self-regulating function
1973	Arne Næss, 1912–2009	Fundamentalism	Gaia's hypothesis influenced the deep ecology advocated by Næss. Deep ecology advocates environmental philosophy and is committed to changing current economic policies and natural values
2009	Peter Ward, 1949–	Killing theory	Ward proposed the Medea Hypothesis, saying that in the process of self-evolution, the earth has not achieved the optimization of the earth, is conducive to life, has not formed a steady-state mechanism, and even caused the phenomenon of killing living things

conservation has gradually become the mainstream human approach to environmental protection (Marsden 1997).

In the nineteenth century the concept of natural resource conservation came into being in the United States. However, after Westerners entered the New World as conquerors, wilderness preservation and resource conservation became issues in nature conservation because wilderness preservation and resource conservation were not a priority. In 1836, Ralph Waldo Emerson (1803–1882) published *Nature* to emphasize the direct communication between man and God with Transcendentalism and explored the divinity in human nature. Henry David Thoreau (1817–1862) published *Walden; or, Life in the Woods* in 1854, *Marsh's Man and Nature*, were published by George Perkins (1862–1920) in 1864. Through these books, one can see the dialogue or relationship between nature and man from the point of view of a nineteenth century naturalist. Among them, scholars who advocated for wilderness preservation included Emerson, Thoreau (Thoreau 1927, 1990), and Muir (John Muir, 1838–1914). In addition, Gifford Pinchot (1865–1946) advocated for resource conservation (Pinchot 1903).

Emerson and Thoreau, advocated for wilderness conservation and were elite intellectuals in the New England region. They embraced the New England Puritans' *mission* and are full of the ideal character for the protection of wilderness ecology. Pinchot, who advocated for the conservation of natural resources, conducted material management in a *wise use* manner. He promoted the study of conservation biology, applied ecology, public economics, conservation, and utilization, renewable resources were achieved to the highest yield on a sustainable basis.

However, Pinchot's idea of sustainability mirrored that of the fourth century Chinese Confucian philosopher, Mencius (孟子) (372–289 BC), who said:

Without delaying the time for the people to culti-vate, the grains will be much to eat. If you can go to the pond without utilizing a fine fishing net, you can eat a lot of fish. If you can go to the forest to cut at the right time, then you cannot run out of wood. There are many grains and fish to eat, as well as much wood to use to be satisfied by human lives; There is no regret for health and wellbeing, and the beginning of the art of ruling as a king.

The principle of profit distribution seeks the greatest benefit for the largest number of people within the longest period (Pinchot 1903). Pinchot was considered a realist because he advocated for ecology's best interests based on the anthro-pocentrism's *enlightened selfishness*. The point to be made here is that even during the industrial revolution and humanity's rise to power, a subset of individuals had not forgotten how dependent humanity had become on nature and protecting/preserving these resources were in the best interests of the survival of the living world, including our species.

2.1.3 Preservation Theory

In the early twentieth century, John Muir (1838–1914), Enos Mills (1870–1922), Robert Marshall (1901–1939), and Aldo Leopold (1887–1948) advocated for the need for resource protection and habitat preservation, instead of focusing on environmental quality, environmental awareness, and environmental literacy through their publi-cations, public lectures, and appearances (Leo-pold 1933, 1949; Gottlieb 1995).

Leopold died in 1948 because of a heart attack while fighting a fire on a neighbor's farm. After his death, the publication of Leopold's posthu-mous work *A Sand County Almanac* in 1949 caused a sensation in the book market and his work remains a cornerstone of the American environmental movement and modern environ-mental thinking. He questions the mainstream value of pursuing an affluent life at the expense of the environment (Leopold 1949). This think-ing laid the foundation for environmental awak-ening and movements in the 1970s. He questioned whether the mainstream value of

pursuing an affluent life was appropriate at the expense of the environment? This thinking laid the foundation for environmental awakening and reconciliation. By 1970, the American people were fighting for civil rights, and with the Viet-nam and Cold Wars, people began to worry about the effects of radiation, chemical disasters, air pollution, and radiation pollution, and the general public was exposed to environmentalism. As a result, enthusiastic citizens began to sup-port, promote, and participate in environmental education. On April 22, 1970, Earth Day laun-ched in the United States opened a new milestone for the modern environmental education movement.

In 1971, the National Environmental Educa-tion Association was established. It is now known as the North American Association for Environmental Education (NAAEE), hoped to provide teachers with sufficient teaching resour-ces to improve students' environmental literacy. As one of the key roles for an Executive Director at NAAEE's Board, Judy Braus also served as many roles as possible to guide ee360 Partner and eeBLUE programs for NAAEE. She tried to create healthier communities that empower local communities, stakeholders, and individuals to restore and protect their environment in the world.

2.1.4 Vitality Theory and Killing Theory

In 1972, ecologist James Lovelock (1919–) de-veloped the **Gaia hypothesis** (Lovelock 1972). Gaia is the goddess of the earth in Greek mythology and Lovelock believed that the Earth's biosphere, including the biotic and abiotic com-ponents constitute a new property of self-regulation (emergent property). Biologist Lynn Margulis (1938–2011) supports this hypothesis that this synergistic and self-regulating function helps maintain living conditions and ecosystems on Earth. Lovelock said:

Gaia is not an organism, but an emergent property of interaction among organisms.

The Gaia hypothesis was an inspiration for animistic religious scholars and environmentalists. Towards the twenty-first century, human needed a strong philosophical sense of environmental protection. Many scholars of ecological philosophy and environmental education have accepted this hypothesis to some extent. As a result, the Gaia hypothesis peaked in the 1990s as major environmental issues were being recognized such as acid rain and polychlorinated biphenyl (PCBs) and the public wanted solutions that the scientific community was not able to provide. To put forth the idea that the planet would heal itself in time bought the scientific community much needed time to study these problems more deeply and develop policy to minimize the damage these issues had caused.

The topics discussed by scientists included on how the co-evolution of the biosphere and organisms affected the stability of global temperatures. Ocean precipitation and rocks release salt, how to maintain the salinity of seawater. Plants absorb oxygen and release carbon dioxide (CO_2), how to maintain the oxygen content in the atmosphere, and the circulation hydrosphere formed by the ocean, freshwater, and groundwater on the earth's surface, and how it affects the livable environment of the earth. However, biologists have criticized life and the environment for development in a coupled way; they have even criticized the Gaia hypothesis as wishful thinking of human beings (Kirchner 2002).

In 2009, paleontologist Peter Ward (1949–) proposed the *Medea hypothesis* (Fig. 1.10). Medea, a witch in Greek mythology, killed her child. Ward listed that throughout geologic time, the Earth' atmosphere was devoid of oxygen and at times was composed of methane and hydrogen sulfide, both of which are toxic to life and induced extinction. The impact of this biocidal phenomenon is directly opposite to the Gaia hypothesis (Ward 2009). Therefore, in the process of self-evolution, the earth has not reached Earth optimal, nor has it been favored for life, or formed a homeostatic mechanism.

Earth system scientist Tyrrell believes that the earth can be said at the best to form the process of Gaia-Co-evolutionary and Influence Gaia.

This means that during the evolution of life and the environment, there is a connection between the biological, geophysical, and chemical environments (Tyrrell 2013).

2.1.5 Deep Ecology

Arne Næss (1912–2009), a proponent of deep ecology, advocates environmental philosophies that embrace the intrinsic value of living things and reject what Banjo calls the ecosystem's instrumental value to humans (Pinchot 1903). Deep ecology is the balanced relationship between the earth's evolution of biotic and abiotic processes. He believed that nature is based on complex and delicately balanced relationships. Therefore, human interference in nature not only affects human survival, but also poses a threat to all living things. The Gaia hypothesis is an explanation of deep ecology.

Deep ecology transcends the nature of biology and uses a social conceptual framework and denies human-centered environmentalism through the exploration of human morals, values, and philosophical perspectives. As a result, deep ecology strengthens the theoretical foundations of the environment, ecology, and green movements, advocating wilderness conservation, population control, and simplicity in life (Næss 1973, 1985a, b, 1989). From this philosophical point of view of environmental protection is in of itself a process within which individuals and society recognize the environment in which they live and the interaction between the biological, physical, and social and cultural components that comprise the environment. Individual or collective knowledge, skills and values may be able to address present and future environmental issues (Fig. 2.1).

Environmental conservation must not stop at the biological species level. It needs to consider genetics, ecology, and cultural landscapes. The most basic level the educational process of environmental protection, includes value clarification, knowledge, attitude and skills, and problem solving. The philosophy of these elements also needs to be considered to lay the foundation for

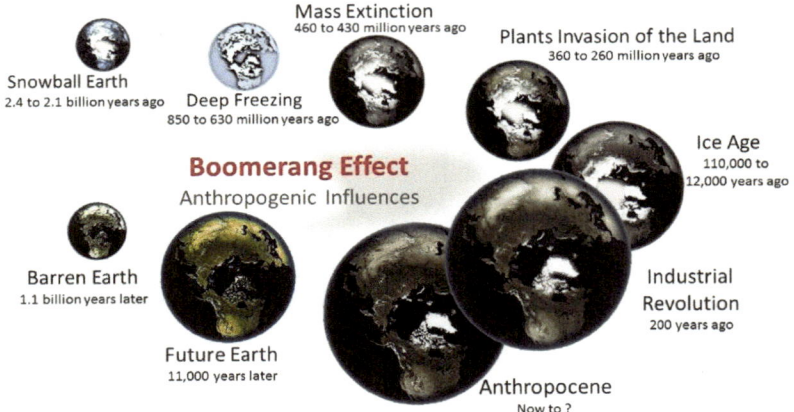

Fig. 2.1 Ward's (2009) Medea hypothesis illustrates that throughout geological time the Earth's atmosphere was once reducing and consisted of poisoning methane (CH_4) and hydrogen sulfide (H_2S) gas, which induced mass extinctions at times (Illustrated by Wei-Ta Fang)

the conservation of basic biodiversity and a model of land ethics. Therefore, the foundation of environmental protection lies in the implementation of environmental education that builds an appropriate foundation in environmental literacy. Therefore, environmental education needs to provide students with the correct environmental knowledge that is based on sound science and to develop attitudes and values towards the environment that cultivate student awareness of the surrounding environment, accountability for their actions, and take to take actions to solve environmental problems (Fig. 2.2).

The United Nations adopted five conservation principles of the World Charter for Nature in 1982 to guide and judge human behavior that affects nature. The Charter states:

> Mankind is a part of nature and life depends on the uninterrupted functioning of the natural system that provides living organisms with a supply of energy and nutrients." Civilization is rooted in nature, which has shaped human culture and influenced all artistic and scientific achievements. Living in harmony with nature gives humanity the best opportunities for the development of creativity, rest, and recreation.

Between 1985 and 1987, Næss published a series of book chapters on deep ecology, vociferously calling for changes in human lifestyles such as: *Identification as a Source of Deep Ecological Attitudes* (Næss 1985a), Ecosophy T:

Deep versus Shallow Ecology (Næss 1985b). *The Deep Ecological Movement: Some Philosophical Aspects* (Næss 1986), *Self-Realization: An Ecological Approach to Being in the World* (Næss 1987). These papers have had an impact on the environmental and ecological movements of the twenty-first century.

The concept of deep ecology has been combined with environmental movements such as: ecofeminism, social ecology, and bioregionalism. However, the environmental problems caused by human beings are increasing (Fig. 2.3). The net weight of materials and fuels used by humans had increased by 800% in the twentieth century. In addition, the amount of waste returned to the environment has increased substantially as biomass and pollution. By 2023, the global population will exceed 8 billion. By 2050, global population is estimated to reach to 9.7 billion people, and could peak at nearly 11 billion around 2100 (United Nations Department of Economic and Social Affairs 2019). How can we limit economic development, protect the environment, and human health and reconcile these elements so that they function in harmony with one another without exceeding earth's carrying capacity have become core themes in environmental development and sustainable growth (Victor 2010)? Coronary artery disease and strokes caused by excessive or poor nutrition

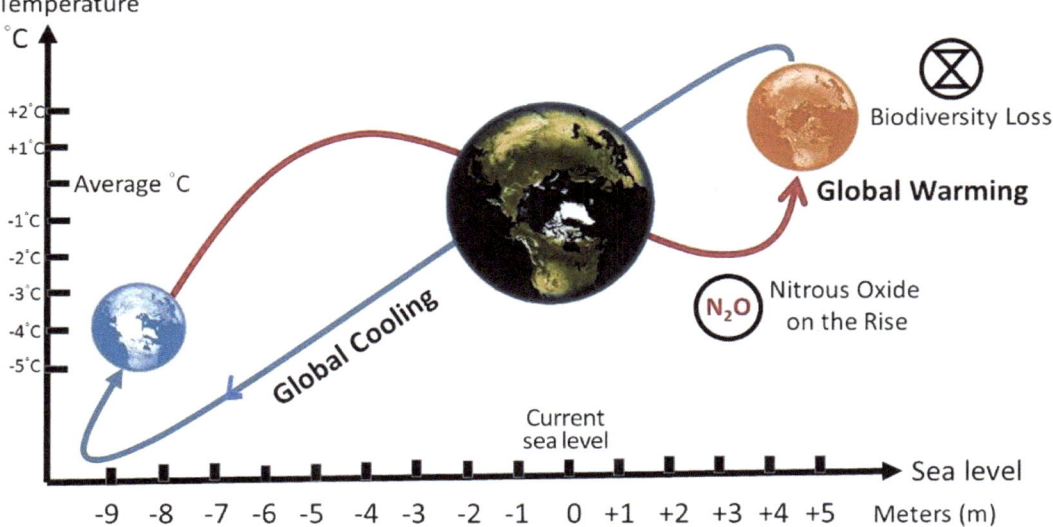

Fig. 2.2 Environmental education needs to provide students with environmental knowledge that is based on sound science, accountability, and takes environmental actions to solve environmental problems (National Museum of Modern and Contemporary Art, Seoul, Republic of Korea, 2018) (Photo by Wei-Ta Fang)

Fig. 2.3 Human development exceeds the limits of earth's carrying capacity, becoming a core theme for environmental development and sustainable growth (Illustrated by Wei-Ta Fang)

account for 32% of all global deaths in 2019 (World Health Organization 2021) and ranks first in human mortalities whereas deaths due to air pollution, i.e., exposure to ambient particulate matter pollution (APMP), such as respiratory infections, chronic obstructive pulmonary disease, and lung, tracheal, and bronchial cancers, account for 11.65% of the deaths globally in 2019 (Ritchie and Roser 2021). The side effects of economic growth and development come with environmental and human costs. For example, deaths due to Covid-19 are expected to reach 6.24 million by April 28th, 2022 (Statista, Worldometer 2022). Unfortunately, Covid-19 is probably a prelude to future pandemics that may occur and we're not prepared to manage these types of problems in systematic and efficient means.

2.2 History of Environmental Education

2.2.1 Early Era of Environmental Education

When or should environmental education be defined and from what point of view? Can the history of environmental education be defined only after the definition was created and what does the definition encompass in scope? If we define environmental education in terms of the history of education, then we need to clarify or identify when and where humankind has evolved from a systematic teaching and learning process. From the transformation course of human civilization, the eastern and western academies have been singing about the natural environment—*The Analects of Confucius*《論語》, once talked about the dialogue between Master Kong (Confucius)(孔子) (551–479 BC) and his students, including Zeng Dian (曾點). Zeng Dian said:

> In the late spring, after the spring clothes have been newly made, I would like, in the company of five or six young men and six or seven children, to cleanse ourselves in the Yi River, to revel in the cool breezes at the Altar for Rain, and then return home singing (Ni 2002: 133).

They would bathe in the river, follow the airs of the rain dance, and return home singing, after sharing in a sacrificial meal in the late spring (Ching 1997). When Confucius heard this, he heaved a long sigh at the time and said: *I am with Dian*!

Confucius' social values for learning in the environment reflect the rich learning process of outdoor activities (Fig. 2.4). The historical changes of these education courses not only reflect the history of environmental education, but also the knowledge, beliefs, skills, values, and cultural heritage of the environment that scholars experienced.

Fig. 2.4 Confucius loves *outdoor education*, and he preferred his student's behaviors that: *In the late spring, after the spring clothes have been newly made, I would like, in the company of five or six young men and six or seven children, to cleanse ourselves in the Yi River* (Photo by Wei-Ta Fang)

As such, we should expand the horizon of environmental education and promote the awakening of human culture because the environment is us and we are the environment (Figs. 2.5 and 2.6). The study of environmental education is currently based on the perspective of material, institutional, and spiritual cultures. Therefore, while we admire the liveliness and prestige of Confucius' teachings and philosophy, it is difficult to discuss when environmental education began because we really can't narrowly regulate or define what is considered environmental education. These self-limiting frameworks are all based on what scholars believed in the professional field of the self, because society is more psychological, and rejects the ideology of other academic schools to represent the mainstream of scholars in their own education field or institution. That is, the definition of environmental education is not only spatially and temporally controlled, but social norms and values play a substantial role. The definition of environmental education is/or should no longer be based on the identification, understanding, and definition of an experience that has been assessed differently based on the time, place, and philosophy. This is because human beings impart environmental values, environmental research methods, and skills to survive in the environment to the next generation, not only based on the theoretical teaching of teachers, but also emphasize the observation and learning of students themselves (2.7).

The learning results produced by these teaching models are probably not similar to the teaching models used today because each person learns and processes information differently. In addition, to being inspired by a teacher, the knowledge and skills each student possesses and the reason for learning these data are different. As such, environmental protection technologies and professional attainment are different.

The Indian Buddha Śākyamuni (566–486 BC) held up a white flower without a word in his teaching. The other disciples were at a loss because no one in the audience understands the *Flower Sermon*. Only His Holiness Mahākāśyapa (550–549 BC) and the Buddha was

Fig. 2.5 The environment is us and we are the environment (Model by Laura Chung; Photo by Max Horng)

Fig. 2.6 We should expand the horizon of environmental education and promote the awakening of human culture (Qixing Mountain, Yangminshan National Park, Taipei, Taiwan) (Photo by Max Horng)

indulged in heart and smile with the silence of Zen. Han Yu (韓愈)(768–824) of the Tang Dynasty once said in his article *Master Disciples* 《師說》:

> A disciple does not need to be inferior to his teacher, nor does a teacher necessarily need to be more virtuous and talented than his student.

He also said:

> The real fact is that one might have learned the doctrine earlier than the other, or might be a master in his own special field.

In the historical development of environmental education for our *special field*, we can see examples similar to those realized by the *Flower Sermon*. This is a kind of tacit knowledge, that is, *looking outside the scene, looking inside the heart*, and we shall find between the sensation of touch and the idea of extension is inexplicable reflected on our perception.

The philosopher Michael Polanyi (1891–1976) proposed *tacit knowledge* in 1958 (Polanyi 1958, 1966). He said: *I shall reconsider human knowledge by starting from the fact that we can know more than we can tell*. Environmental education goes beyond the inner perception of *The heart of fools is in their mouth, but the tongue of the wise is in their heart*. If the knowledge of environmental education cannot be taught through the book descriptions, then we will discuss it through the history of environmental education, the ideas and methods of environmental education, and the development process records of environmental action. Through indwelling, reorienting, and recognizing we will know the stories. Perhaps these stories have happened before environmental education has been defined. However, for us, they are all important stories. We discuss the history of environmental education since the eighteenth century.

Fig. 2.7 The definition of environmental education is not only spatially and temporally controlled, but social norms and values play a substantial role (Shesan, Taiwan) (Photo by Max Horng)

2.2.2 Environmental Education in the Eighteenth and Nineteenth Centuries

The roots of contemporary environmental education can be traced back to the eighteenth century. At the time, the novel *Émile,* written by Jean-Jacques Rousseau (1712–1778), emphasized the philosophical argument of civic education and emphasized the concern that children's natural education was important (Rousseau and Bloom 1979). He proposed three types of education in the book, arguing that educators need to teach according to the natural nature of human beings. The connotation of education includes natural education, things education, and human education.

Rousseau's educational thought was influenced by people, like Plato, Michel de Montaigne (1533–1592), John Locke (1632–1704), and others. However, he was believed to have pioneered the ideological tradition of naturalistic education and further influenced later generations of thinkers like Immanuel Kant (1724–1804), Johann Pestalozzi (1744 –1827), Friedrich Fröebel (1782–1852), John Dewey (1859–1952), and Maria Montessori (1870–1952).

In the early nineteenth century, Pestalozzi, the *father of civilian education* in Europe, set up a school for the poor in Switzerland at his own expense. He used observation to conduct nature education. Pestalozzi talked about first-level observations, perceptual activities, telling, and then advanced learning with measurement, drawing, writing, numbers, and calculations.

Friedrich Fröbel (1782–1852) founded the first *kindergarten* or the *children's garden* at Blankenburg, Thuringia, Germany in 1837. It used games and manual labor to promote gardening activities in the flowerbeds, vegetable gardens, and orchards. By 1907, Maria

Montessori (1870–1952) opened the Casa dei Bambini in her home in Rome. She designed her school in a prepared environment based on human tendencies called *Land-based education*. Montessori Education used tailor-made education methods to teach students at different stages and with different personalities. At the university level, Swiss naturalist Louis Agassiz (1807–1873) responded to Rousseau's philosophy because he encouraged students to *learn about nature, not books*. In 1847, Agassi applied to Harvard University as a professor of zoology and geology, and founded the Harvard Museum of Comparative Zoology. Agassiz believed in experimental knowledge, not knowledge in worthless books.

2.2.3 Environmental Education in the Early Twentieth Century

Western scholars promoted natural studies around 1890, and began to lead students in natural studies in the early twentieth century. The Cornell University's professor of natural studies, Anna Botsford Comstock (1854–1930) was an outstanding figure in the natural research movement. She authored the *Handbook of Nature Study* in 1911 and said: *The object of the nature study teacher should be to cultivate a child's power of accurate observation and to build up within them, understanding* (Comstock 1986). Comstock assisted community leaders, teachers, and scientists in transforming science education curricula for American children.

At that time, the people were beginning to feel that environmental damage was getting worst. Scholars were realizing the environmental damage caused by human beings and paid more attention to the topic of environmental education in promoting global conferences (Marsden 1997). Environmental education responded to the Great Depression and sandstorms in the USA economy, and formed the *Conservation Education* that emerged in the 1920s.

Conservation education is very different from pure natural research. The learning process focuses on monitoring the data generated by

rigorous scientific training, not the philosophical study of natural history. Conservation education forms an important scientific management and planning tool, which helps to solve contemporary social, economic, and environmental problems.

A Scottish geologist, Sir Archibald Geikie (1835–1924), believed that human beings could learn endless knowledge from the natural environment, so he included *love of nature* as an educational goal (Marsden 1997: 11–12). Later, Sir Scott Patrick Geddes (1854–1933), a Scottish botanist, promoted the study of civic areas, critically sought to improve the actual living environment, and carried out local town research courses.

2.2.4 Environmental Education After the Middle of the Twentieth Century and in the Early Twenty-First Century

At the end of World War II outdoor education was introduced in the 1950s. The 1960s gave birth to the modern environmental movement. To protect the environment, the US has established many international conservation organizations, such as the International Union for Conservation of Nature (IUCN). The first Secretary-General of UNESCO, Sir Julian Huxley (1887–1975), was hoping to provide an academic platform for the International Union for Conservation of Nature, initiated a conference. The first conference held at the Palace of Fontainebleau or Château de Fontainebleau in Paris, France (Fig. 2.8). As a result, France, the host country, put nature conservation and habitat protection in its policy framework in 1948.

Later, the inter-sessional meeting, that first used the term *environmental education*, led to the 1949 International Union for Conservation of Nature (IUCN). By the 1960s, the United States Congress enacted legislation required that knowledge of natural resource conservation had to be taught at the primary and secondary school levels. In 1968, the United Nations convened a biosphere conference in Paris to promote the

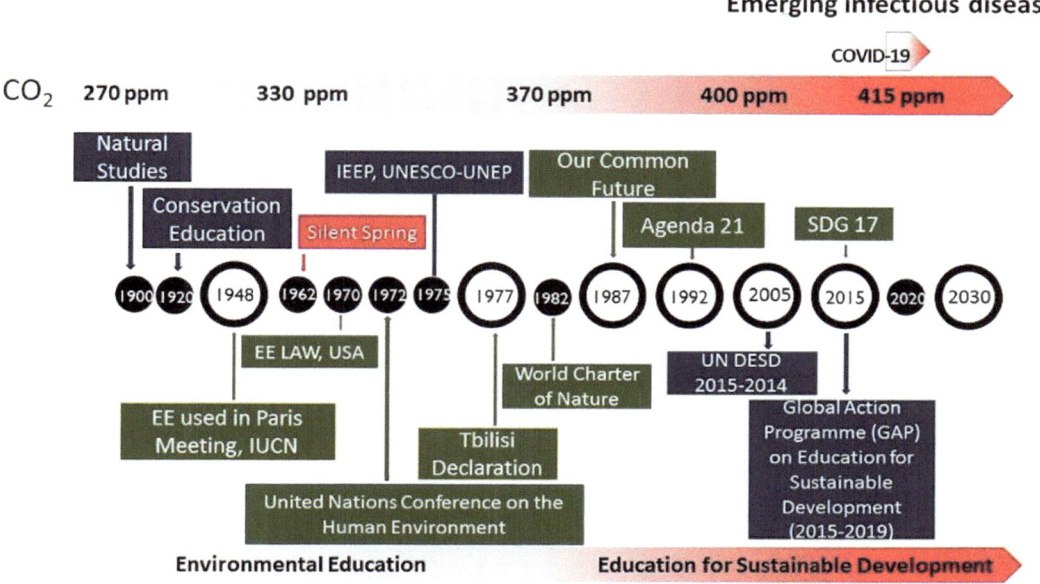

Fig. 2.8 The History of Environmental Education (Illustrated by Wei-Ta Fang)

meaning of *environmental education*. By 1969, James A. Swan, under the guidance of William Stapp, published the first article on environmental education in *Phi Delta Kappa*, within which he discussed the importance of environmental education in caring for the natural and human environment (Swan 1969). Stapp also confirmed the definition of environmental education in that year in the *Journal of Environmental Education* (Stapp 1969). He defined environmental education as *a subject that aimed at producing a citizenry that is knowledgeable concerning the biophysical environment and its associated problems, aware of how to help solve these problems, and motivated to work and participate their solution actively* (Stapp 1969: 31).

An important milestone in the history of environmental education of curriculum field was reached in the 1970s (Disinger and Monroe 1994; Reid 2019). Governments of various countries had begun to formulate environmental protection regulations in an effort to solve environmental protection issues.

John F. Disinger, undoubtedly one of the pioneers of environmental education in the United States, retired from the School of Natural Resources, Ohio State University (OSU), and spent half his life on his faculty position leading OSU's ERIC clearinghouse, established in 1966, known as a resource center for science education, mathematics, and environmental education. From 1979 to 1991, academics in science and environmental education also led the U.S. EPA's information outreach program in USA. Among them, John F. Dinsinger (1930–2005)(see Fig. 2.9), Robert E. Roth (1937–2021)(see Figs. 2.9 and 2.10), Rosanne W. Fortner (1945–), and Joe E. Heimlich (see Fig. 2.9) are professors in the two fields of cross-environmental education and science education at the School of Natural Resources (SNR, OSU)(now the School of Environment and Natural Resources, SENR, OSU) (Disinger 2001; Fortner 2001; Heimlich et al. 2022). Gary W. Mullins (see Fig. 2.10) served as director of the School of Natural Resources (SNR), OSU from 1998–2004. Mullins worked as an interpretive naturalist and National Park Ranger/Planner (see Fig. 2.10). The academic programs of the Nature Center Symposium were engaged on the topics of the *World Earth Day*'s special lectures provided by the OSU's faculties during 2001 (see Figs. 2.11 and 2.12).

Fig. 2.9 From 1979 to 1991, academics in science and environmental education also led the U.S. EPA's information outreach program in USA. Some faculties may be served as a consultant to the Nature Conservancy, the USDA Forest Service, the U.S. Fish and Wildlife Service, and the National Park Service as EE and communications projects. These project results driven by OSU's faculties (Left: John F. Dinsinger (1930–2005); middle: Joe E. Heimlich; right: Robert E. Roth (1937–2021)) had been discussed with one of their students, Tim Hsu, in 1995 (Photo by Yi-Hsuan (Tim) Hsu)

The cooperation of environmental education programs originated very early. Back to 1970, UNESCO and the IUCN organized an international environmental education schools' curriculum work conference in Carson City, Nevada, United States, which stated:

> Environmental education is the process of recognizing values and clarifying concepts in order to develop the skills and attitudes necessary to understand and appreciate the interrelationships among man, his culture, and his biophysical surroundings.

Environmental education also requires self-regulation in terms of environmental quality issues and practice. This meeting set the goals of school education and detailed the content of each stage (UNESCO 1970).

In 1972, the United Nations convened a conference on the human environment in Stockholm, Sweden. During the meeting, the United Nations Declarations on the Human Environment were resolved. Among them, Principle 19 specifically requires:

> Education in environmental matters, for the younger generation as well as adults, giving due consideration to the underprivileged, is essential in order to broaden the basis for an enlightened opinion and responsible conduct by individuals, enterprises and communities in protecting and improving the environment in its full human dimension. It is also essential that mass media of communications avoid contributing to the deterioration of the environment, but, on the contrary, disseminates information of an educational nature on the need to project and improve the environment in order to enable man to develop in every respect.

The Declaration of the United Nations Conference on the Human Environment hopes that human beings will begin to pay attention to environmental issues and opens up the possibility of a positive interaction between humans and the natural environment. The declaration emphasized

Fig. 2.10 The academic programs were engaged in at OSU and NTNU faculties, while research being conducted by faculties from OSU's alumina. From left to right: Huei-Min Tsai (GIEE, NTNU), Tzuchau Chang (GIEE, NTNU), Robert Roth (School of Natural Resources, OSU), Gary Mullins (School of Natural Resources, OSU), Ju Chou (GIEE, NTNU), and Shih-Jang Hsu (National Hualien University of Education, now the National Dong Hwa University) during 2001 (Photo by Huei-Min Tsai)

that *Both aspects of our environment, the natural and the man-made, are essential to our well-being and to the enjoyment of basic human rights the right to life itself.* Humans began to attach importance to the quality of environmental life and environmental protection issues began to get attention.

In 1974, the United Kingdom held the Schools' Council's Project Environment and revealed three themes of environmental education. They are: *Education about the Environment, Education in the Environment* or *from the Environment*, and *Education for the Environment*. This attracted worldwide attention has a wide range of applications (Palmer 1998).

In 1975, UNESCO and the United Nations Environment Programme (UNEP) jointly promoted the International Environmental Education Programme (IEEP). This project discusses how to raise environmental awareness and promote the environment educational vision (UNESCO 1975).

In 1975, UNESCO introduced the Belgrade Charter at the International Environmental Workshop held in Belgrade. The Charter emphasized:

> We need nothing more than a new global ethic—an ethic, which espouses attitudes and behavior for individuals and societies, and are consonant with humanity's biosphere; which recognizes and sensitively responds to the complex and ever—changing relationships between humanity and nature and between people. Significant changes must occur in all of the world's nations to assure the kind of rational development, which will be guided

Fig. 2.11 The academic programs of the Nature Center Symposium are engaged on the topics of the *World Earth Day's* special lectures provided by the OSU's faculties (Robert Roth and Gary Mullins, two former Deans of the School of Natural Resources, OSU) for NTNU and Dong Hwa's graduate students at National Dong Hwa University, Taiwan, April 22th, 2001 (Photo by Ju Chou)

Fig. 2.12 Two former Deans of OSU's School of Natural Resources, as the key persons in environmental education were invited to Taiwan by Ju Chou, one of the OSU's alumni (Natioswnal Dong Hwa University, Taiwan, April 22th, 2001) (Photo by Ju Chou)

by this new global ideal—changes which will be directed towards an equitable distribution of the world's resources and more fairly satisfy the needs of all peoples.

The Charter distinguishes environmental education from formal education and non-formal education (UNEP 1975). The Belgrade Charter regulates the content and goals of environmental education, promotes the world's human beings to understand and care about the environment and related issues, possesses appropriate knowledge, technologies, attitudes, motivations, and commitments, and is committed to solving today's environmental problems.

In 1976, UNESCO released the Environmental Education Newsletter *Connect*, which serves as an information exchange channel for UNESCO and the official agency of the United Nation's International Environmental Education Program (IEEP). The purpose was to disseminate environmental education messages and to establish a network of environmental education institutions and individuals. In 1977, UNESCO and the United Nations Environment Programme held the Tbilisi UNESCO-UNEP Intergovernmental Conference in Belize, Georgia (UNESCO 1977). The Tbilisi Declaration's 41 Guiding Principles on environmental education included environmental education tasks, curriculum teaching, implementation strategies, and international cooperation. Participants of the Tbilisi Declaration stated:

> The goals of environmental education are: (1) to foster clear awareness of, and concern about, economic, social, political, and ecological interdependence in urban and rural areas; (2) to provide every person with opportunities to acquire the knowledge, values, attitudes, commitment, and skills needed to protect and improve the environment; (3) to create new patterns of behavior of individuals, groups, and society as a whole towards the environment (UNEP 1977).

The fundamental task of environmental education was supposed to closely link with ethics and values. For example, its purpose was to illustrate about the need for environmental education and provide everyone with the opportunity to learn the knowledge, values, attitudes, commitments, and technologies needed to protect and improve the environment. As for the goal of environmental education, it mentions the need to *help social groups and individuals to acquire an awareness and sensitivity... a set of values and feelings of concern for the environment and the motivation for actively participating in environmental improvement and protection.* Therefore, the Tbilisi Declaration put forward five goals of environmental education including awareness, knowledge, attitude, skills, and participation.

The United Nations established the World Commission on Environment and Development (WCED) in 1983. It was concerned about environmental protection and economic development. It symbolized the relationship between human beings and the environment, care for human survival, and development in the environment. In 1987, the committee issued the *Our Common Future* declaration at the United Nations General Assembly by its chair, Gro Harlem Brundtland (1939–), who officially defined sustainable development. Sustainable development was a development model that could meet our current needs without compromising future generations to meet their needs. She called for global recognition and care for the natural environment and for vulnerable groups.

In 1990, the U.S. Congress passed the National Environmental Education Act (NEEA) to solve environmental problems by improving the problems through educating people using environmental education. And in 1992, the United Nations convened the Earth Summit in Rio de Janeiro, Brazil. It adopted the world-famous Agenda 21 and planned the concept of sustainable development as a concrete action plan, emphasizing the future generation towards caring, awareness of the limited nature of natural environment resources, and assistance to disadvantaged ethnic groups (UN 1992).

2.2.5 Environmental Education in the Twenty-First Century

In 2002 the United Nations again convened world leaders and chose Johannesburg, South

Africa to host the World Sustainable Development Summit. The organizers decided to invite partner organizations to work together to resolve the issue of protecting the environment, narrowing the gap between rich and poor, and protecting the environment of human life. In 2002, the UN General Assembly passed a resolution declaring that the UN Decade (2005–2014) had passed the UN Decade of Education for Sustainable Development (UN DESD 2005–2014).

By 2005, UNESCO officially promoted the Decade of Education for Sustainable Development and strived to mobilize international education resources to create a sustainable future. Five principles are imagining a better future, critical thinking and reflection, participation in decision-making, partner relationships, and systems thinking. The adoption of Chapter 40 of *Agenda 21* emphasizes the education was a way for *capacity-building*. Although education alone cannot achieve a sustainable future, without sustainable development of education and learning, humankind will not be able to achieve this goal. The overall goal of the UN DESD is to integrate the principles, values, and practices of sustainable development into the education and learning dimension. Education should be based on sustainable development that encourages changes in human behavior and creates a more sustainable future in terms of environmental integrity, economic viability, and a just society that meets contemporary and future generations.

In 2012, the United Nations returned to Rio de Janeiro, Brazil to host the UN Conference on Sustainable Development (Rio + 20) to commemorate the twentieth anniversary of the 1992 Earth Summit. The conference took green economy as the theme and aimed to eradicate poverty and promote global development. It is hoped to promote a green economy by establishing relevant mechanisms and organizations. By 2014, the UN called for mainstreaming sustainable development education at the World Conference on Sustainable Development Education in Nagoya, Japan. UNESCO launched the Global Action Programme on Education for Sustainable Devolepment (GAP). In May 2015, at the World Education Forum held in Incheon, Republic of Korea, it

planned to implement *2030 Education* and planned to adopt the *Incheon Declaration on Education 2030*. The Global Education Monitoring Report (GEM Report) plans to ensure inclusive and fair quality education and to provide lifelong learning opportunities for all. In September 2015, the UN General Assembly passed the 2030 Agenda for Sustainable Development to promote the 17 Sustainable Development Goals (SDGs). The international community understands that in addition to promote Goal 4: Quality education, as well as other sustainable goals need to be developed through education.

Today, with the international recognition and adoption of the Global Education 2030 Agenda, its aim is to eradicate poverty through sustainable development by 2030. At present, UNESCO uses the mechanism of SDGs, the GEM Report, and the Education for All (EFA) based on Regional Comprehensive Report to develop the GAP.

As a result, formal and non-formal educators around the world are continuing their efforts to promote education for sustainable development, especially in reimagine urban EE and in restoring the Nature in EE (Reid et al. 2021). One of the key authors in Reid et al. (2021), Justin Dillon, the UK Professor at the Science and Environmental Education, Director for Research in STEM Education at the South West Institute of Technology Observatory, University of Exeter, also taught at the University of Bristol before, has developed urban EE in science education working with schools, museums, science centers, aquariums, and botanic gardens. As the President of the National Association for Environmental Education (UK), Dillon emphasized the importance of the STEM education as well as encouraged people to learn outside beyond school environment (Dillon 2014, 2019).

Many EE conferences have been organized around the world. The World Environmental Education Congress (WEEC) congresses are the most significant existing experience of connecting all actors at the international level in the field of environmental education as well as the sustainability education. Since 2003 the international network has organized eleven World Congresses, attracted the participation of

Fig. 2.13 Education based on sustainable development encourages changes in human behavior and creates a more sustainable future in terms of environmental integrity, economic viability, and a just society that meets contemporary and future generations to save limited resources. From left to right: Chang-Po Chen (an ecosystem ecologist and ecological engineer, Academia Sinica, Taiwan, ROC); Patrick Megonigal (a soil ecologist of tidal wetlands, Smithsonian Environmental Research Center, SERC, USA); William J. Mitsch (an ecosystem ecologist and ecological engineer, OSU & Florida Gulf Coast University, USA); and Hwey-Lian Hsieh (an ecosystem ecologist and ecological engineer, Academia Sinica, Taiwan, ROC) (2019 Society of Wetland Scientists (SWS) Asia Chapter Meeting, Suncheon, Republic of Korea)(Photo by Wei-Ta Fang)

thousands of people from over 100 different countries (please see https://weec2022.org).

The content of education for sustainable development encompasses a humanistic education and development concept, and we share our responsibilities and obligations with human rights, dignity, social justice, tolerance, protection, culture, language, and ethnic diversity, and work together, such as international chapter's meetings held for wetland conservation, restoration, and education worldwide (as one of the examples: 2019 Society of Wetland Scientists (SWS) Asia Chapter Meeting, Suncheon, Republic of Korea; please see Fig. 2.13).

In 2019, Chew-Hung Chang, the Dean of Academic and Strategic Development in Nanyang Technological University of Singapore invited fifteen international scholars, including Wei-Ta Fang, Tzu-Chau Chang from the National Taiwan Normal University, and Yi-Hsuan (Tim) Hsu (Global Environmental Enhancement Inc.), Gillian Kidman (Monash University, Australia), Judy Braus (NAAEE) as well as other professors recommended by Global Environmental Education Partnership (GEEP) (https://geepaprc.org/en) to discuss the draft and issue the Singapore Declaration on Research in Education for Sustainable Development, in response to the demands for education research on sustainable development in 2030, see Table 2.2.

Table 2.2 Development history of environmental education conferences

Year	Host country/city/school	Organizers	Conference name	Meeting content
1948	Paris, France	UNESCO	The Conference of Fontainebleau	The Conference used the term environmental education for the first time. In 1949, the International Union for the Protection of Nature was established
1965	Staffordshire, United Kingdom	Keele University	The Conference in Education	The term environmental education was used for the first time in the United Kingdom
1968	Paris, France	UNESCO	International Conference on Global Biosphere Protection	Promoted environmental education at the Biosphere Conference
1970	Carson, Nevada	International Union for Conservation of Nature, IUCN	International Working Meeting on Environmental Education in the School Curriculum	The meeting accepted the definition of the environmental education
1970	Paris, France	UNESCO	UNESCO Convention	Establishment of Division of Science, Technical and Environment Education, UNESCO
1972	Stockholm, Sweden	United Nations	The United Nations Conference on the Human Environment	Convened a conference on the human environment and issued the Declaration of the UN Conference on the Human Environment, or Stockholm Declaration
1975	Paris, France	UNESCO, United Nations Environment Programme (UNEP)	International Environmental Education Programme	Creation of the UNESCO/UNEP International Environmental Education Programme (IEEP)
1975	Belgrade, Former Yugoslavia	UNESCO	International Environmental Workshop	The Belgrade Charter was proposed to standardize the content and goals of environmental education, to promote human awareness and concern about the environment and related issues in the world, to possess appropriate knowledge, technology, attitudes, motivations and commitments, and to work to resolve today's environmental problems

(continued)

Table 2.2 (continued)

Year	Host country/city/school	Organizers	Conference name	Meeting content
1977	Tbilisi, Former Soviet Republic of Georgia	UNESCO	Intergovernmental Conference on Environmental Education	The Tbilisi Declaration, which made 41 recommendations, provides countries with a complete framework for promoting environmental education. The declaration proposes five goals of environmental education including awareness, knowledge, attitude, skills, and participation
1980	Gland, Switzerland	UN Environment Programme (UNEP), International Union for Conservation of Nature (IUCN), World Wildlife Fund (WWF)	International Union for Conservation of Nature and Natural Resources (IUCN) at Gland's Meeting	Published the World Conservation Strategy: Living Resource Conservation for Sustainable Development
1982	New York, USA	United Nations	UN General Assembly	Adopted *World Charter for Nature*, to proclaim five principles of conservation, guiding and judging all human behaviors that affected nature, confirming the ethical relationship and responsibility of the international community to humans and nature
1987	New York, USA	The World Commission on Environment and Development	UN General Assembly	Adopted *Our Common Future* and put forward the concept of sustainable development
1992	Rio de Janeiro, Brazil	UN Conference on Environment and Development (UNCED)	The World Summit on Sustainable Development (Earth Summit)	Adopted *Agenda 21* and signed at the UN Framework Convention on Climate Change
2002	Johannesburg, South Africa	UN Conference on Environment and Development (UNCED)	The World Sustainable Development Summit (Earth Summit 2002)	Committed to solving the problem of protecting the environment, narrowing the gap between rich and poor, and protecting the environment of human life
2005	Paris, France	United Nations Educational, Scientific and Cultural Organization (UNESCO)	UN General Assembly Resolution (2002)	A formal education system seeks to mobilize international education resources to create a sustainable future

(continued)

Table 2.2 (continued)

Year	Host country/city/school	Organizers	Conference name	Meeting content
2009	Bonn, Germany	UNESCO	UNESCO World Conference on Education for Sustainable Development: Moving into the Second Half of the UN Decade	Discussed the Decade of Education for Sustainable Development (DESD) (2005–2014) and proposed the Bonn Declaration
2012	Rio de Janeiro, Brazil	UN Conference on Environment and Development (UNCED), Rio + 20	The World Sustainable Development Summit (Earth Summit 2002)	The goal of eradicating poverty and promoting global development is to promote a green economy by establishing relevant mechanisms and organizations
2014	Nagoya, Japan	UNESCO	World Conference on Education for Sustainable Development	Including sustainable development education in the post-2015 development agenda, launched the Global Action Plan for Sustainable Development Education which highlighting five priority areas of action
2015	Incheon, Republic of Korea	UNESCO	World Education Forum 2015	Implemented sustainable development goals and promote 2030 Education
2019	Singapore	Nanyang Technological University	Workshop on Singapore Declaration on Sustainable Development Education	Published the *Singapore's Declaration on Research in Education for Sustainable Development* in response to the demands for education research on sustainable development in 2030
2022	Prague, Czech Republic	Faculty of Social Studies, Masaryk University, Czech Republic	11th World Environmental Education Congress	The 11th World Environmental Education Congress has been held in Prague, March 14–18th, 2022. Programs have been including cross-cutting dialogues and parallel sessions

2.3 Summary

We conducted research on environmental education in the early twentieth century, conservation education in the 1920s, environmental protection education in the 1970s, and sustainable development education in the 2000s. In this chapter we proposed a theoretical and practical framework from the studies of the philosophy and history of EE, and discussed the framework of research and practice from multiple perspectives and orientations in multi-cultures. We explored the in-depth fields of environmental education according to the different cultural and social backgrounds of great teachers' and masters' in history. Therefore, according to the critical analysis method of the problem currently, the process of learning is emphasized rather than the result from lessons.

References

Ching J (1997) Mysticism and kingship in China. Cambridge University Press, Cambridge

Comstock AB (1986) Handbook of nature study (First with a Foreword by Verne Rockcastle N (ed) Comstock Associates/Cornell University Press, Ithaca

Dillon J (2014) Environmental education. In: Lederman NG, Abell SK (eds) Handbook of Research on Science Education, vol II. Routledge, Oxfordshire, pp 511–528

Dillon J (2019) Urban environmental education. Urban environmental and sustainability education. Environ Educ 122:4

Disinger J (2001) Tensions in environmental education: Yesterday, today, and tomorrow. In: Hungerford HR et al (eds) Essential Readings in environmental education, 2nd edn. Stipes Publishing Company, Champaign, pp 1–12

Disinger J, Monroe M (1994) Defining environmental education. EE toolbox—workshop resource manual. University of Michigan National Consortium for Environmental Education and Training, Ann Arbor, MI

Fortner RW (2001) Cooperative learning: a basic instructional methodology for global science literacy. In: Mayer VJ (ed) Global Science Literacy. Kluwer, Dordrecht, pp 79–92

Gaston KJ, Soga M (2020) Extinction of experience: the need to be more specific. People Nat 2:575–581

Gottlieb R (1995) Beyond NEPA and earth day: Reconstructing the past and envisioning a future for environmentalism. Environ Hist Rev 19(4):1–14

Heimlich JE, Wasserman D, Tingley K, Roberts SJ, Aloisio J (2022) An influence among influences: the perceived influence contribution scale development and use. evaluation and program planning 102091

Kirchner J (2002) The gaia hypothesis: fact, theory, and wishful thinking. Clim Change 52(4):391–408

Leopold A (1933) Game management. Charles Scribner's Sons, New York

Leopold A (1949) A sand county almanac. Oxford University Press, Oxford

Little D (2020) Philosophy of history, the stanford Encyclopedia of philosophy (Winter 2020 Edition), Zalta EN (ed) Data retrieved on April 14th, 2022 from https://plato.stanford.edu/archives/win2020/entries/history/

Lovelock JE (1972) Gaia as seen through the atmosphere. Atmos Environ 8:579–580

Marsden WE (1997) Environmental education: Historical roots, comparative perspectives, and current issues in Britain and the United States. J Curr Super 13(1):6–29

Næss A (1973) The shallow and the deep, long-range ecology movement: a summary. Inquiry 16(1–4):95–100

Næss A (1985a) Identification as a source of deep ecological attitudes. In: Tobias M (ed) Deep Ecology. Avant Books, San Diego, pp 256–270

Næss A (1986) The Deep Ecological movement: Some philosophical aspect. Philos Inq 8:10–31

Næss A (1987) Self-realization: an ecological approach to being in the world. Trumpeter 4(3):35–42

Næss A (1989) Ecology, community, and lifestyle. Cambridge University Press, Cambridge

Næss A, Ecosophy T (1985b) Deep versus Shallow Ecology. In: Pojman P (ed) Environ Ethics, pp 151–153

Ni P (2002) Confucius: the man and the way of Gongfu. Rowman & Littlefield Publishers, New York, p 133

Palmer J (1998) Environmental education in the 21st century: theory, practice, progress, and promise. Routledge, London

Pinchot G (1903) A primer of forestry. U. S. Government Printing Office Washington, DC

Polanyi M (1958) Personal knowledge: towards a post-critical philosophy. University of Chicago Press, Chicago

Polanyi M (1966) The tacit dimension. University of Chicago Press, Chicago

Reid A (ed) (2019) Curriculum and environmental education: perspectives, priorities and challenges. Routledge, Oxfordshire

Reid A, Dillon J, Ardoin N, Ferreira J-A (2021) Scientists' warnings and the need to reimagine, recreate, and restore environmental education. Environ Educ Res 27(6):783–795

Ritchie H, Roser M (2021) Air pollution. Published online at OurWorldInData.org. Retrieved on April 14th, 2022 from: https://ourworldindata.org/air-pollution

Rousseau J-J, Bloom A (1979) Emile: or on education. Basic Books, New York

Soga M, Gaston KJ (2016) Extinction of experience: The loss of human-nature interactions. Front Ecol Environ 14:94–101

Stapp WB (1969) The concept of environmental education. Environ Educ 1(1):30–31

Statista, Worldometer (2022) COVID-19/Coronavirus. Data retrieved on April 25th, 2022 from https://www.statista.com/page/covid-19-coronavirus

Swan JA (1969) The challenge of environmental education. Phi Delta Kappan 51:26–28

Thoreau HD (1927) Walden, or, life in the woods. Dutton, New York

Thoreau HD (1990) A week on the concord and merrimack rivers. University of California Libraries, Berkeley

Tyrrell T (2013) On gaia: a critical investigation of the relationship between life and earth. Princeton, Princeton University Press, p 208

UN (1992) Agenda 21 United Nations, New York

UNEP (1975) The belgrade charter. Final report, international workshop on environmental education (ED-76/WS/95). UNESCO and UNEP, Paris

UNEP (1977) Intergovernmental conference on environmental education (ED/MD/49) UNESCO and UNEP, Paris

UNESCO (1975) The belgrade charter: a global framework for environmental education. Connect 1(1):1–9

UNESCO (1970) International working meeting on environmental education in the school curriculum, Final Report, at Foresta Institute, Carson City, Nevada. IUCN and UNESCO, Paris

UNESCO (1977) Intergovernmental conference on environmental education, Tbilisi Declaration (USSR) 14–26 October, 1977 (final report). UNESCO, Paris

United Nations Department of Economic and Social Affairs (2019) Growing at a slower pace, world population is expected to reach 9.7 billion in 2050 and could peak at nearly 11 billion around 2100. Retrieved from https://www.un.org/development/desa/en/news/population/world-population-prospects-2019.html

Victor P (2010) Questioning economic growth. Nature 18:468(7322):370-371

Ward P (2009) The medea hypothesis: is life on earth ultimately self-destructive? Princeton University, Princeton

World Health Organization (2021) Cardiovascular diseases (CVDs). Data retrieved on April 14th, 2022 from https://www.who.int/news-room/fact-sheets/detail/cardiovascular-diseases-(cvds)

Research Methods for Environmental Education

3

At "Environmental education is the process of recognizing values and clarifying concepts to develop skills and attitudes necessary to understand and appreciate the inter-relatedness among man [sic], his culture, and his biophysical surroundings. Environmental education also entails practice in decision-making and self-formulation of a code of behaviour about issues concerning environmental quality."

UNESCO, International Working Meeting on Environmental Education in the School Curriculum, 1970

Abstract

Research methods are the sum of knowledge, plans, strategies, tools, steps, and processes. In this chapter, we seek to understand the "research" nature of Environmental Education (EE), define the scope of research through a systematic investigation process by gathering and understanding past facts and discovering new facts through practical investigations, experiments, and verification methods to increase or modify the contemporary know-how in our environment. After exploring the history of EE, entering quantitative research on EE and qualitative research on EE, we use this chapter to improve the level of thinking of EE theory, using the learning methods of Benjamin S. Bloom, Harold R. Hungerford, and the emotional learning theory of ABC. We aim to understand the value of post-environmental learning, strengthen our transcendental cognition of animate and inanimate objects by looking at these aspects objectively

and have a more general and mature view of the biotic and abiotic processes that shape the world around us.

3.1 What is Environmental Education Research?

In the previous chapter we mentioned that environmental educators must present the new and growing body of scientific knowledge and technologies to their students to meet changing social, economic, and cultural needs. The foundation of the environmental knowledge and the challenges facing environmental educators required for teachers and scientists today, is to re-examine the way we perform research, assess the questions that are relevant to modern issues, and to train EE professionals and educators (Fig. 3.1). We generate about 2.5 quintillion bytes of data daily (Humbetov 2021; Roque and Ram 2019), but most are out of touch with current and future societal, economic, and global

© The Author(s) 2023

W.-T. Fang et al., *The Living Environmental Education*, Sustainable Development Goals Series, https://doi.org/10.1007/978-981-19-4234-1_3

Fig. 3.1 As scientists and educators, we have the opportunity and responsibility to expand the resource base of EE. Professor Fang sharing the Asia Chapter lessons learned and progress at an annual Society of Wetland Scientists (SWS) board meeting in 2015, from far left to right-sided: Wei-Ta Fang (Graduate Institute of Environmental Education, National Taiwan Normal University), Beth A. Middleton (Wetland and Aquatic Research Center, U.S. Geological Survey), Loretta Battaglia (Center for Coastal Studies, Texas A&M University-Corpus Christi) (Photo by Yung-Nane Yang)

environmental issues. Our environmental conditions are changing daily; some are normal and have been seen throughout Earth's history. Human activities are simply accelerating a process that is occurring naturally. We may not necessarily know the exact time of a volcanic explosion or earthquake, but have a good idea on when they could happen based on past occurrences. This unfortunately does not meet society's expectations on our ability to more accurately predict these changes based on the trajectory of the amount of data we've generated, technological improvements, societal expectations of the modern environment. Society is gobsmacked when scientists and educators can't solve or explain environmental problems like global warming to the public in a way that they understand. Is it possible for the scientific community and environmental educators to distill complex environmental concepts, the problems and potential solutions, and correctly communicate them to the general public without creating panic? Therefore, in defining the goals of EE, we should strive to establish a professional standard for EE and the educators, and strengthen the standards needed for EE (Hudson 2001).

So, in terms of a basic learning plan, what is the "research" of EE? After we have defined the state of the art and the scope of our research, we then strive to validate, dispel, increase, and/or modify contemporary environmental knowledge through a systematic investigation process, by collating and understanding past facts, and discovering new facts based on actual investigations, experiments, and verification methods. Therefore, we need to understand the personnel, culture, societal norms and values in space and in time, organizations, and materials in the environment using scientific and social science methods to analyze environmental issues, and to use these tools to develop a broad-based EE research program to address current and future needs. The needs of educational research are important.

Public education for the environment should have a positive impact on life in the future. Therefore, according to the concept of sustainable development, if our current and future generations want to enjoy the benefits of the planet's natural heritage, EE must then be taken seriously. In the face of the increasingly cumbersome and complex issues of the twenty-first century, environmental problems are becoming increasingly difficult to understand and evaluate, but we need to become more heteroglossic to be effectively solve environmental problems. Heteroglossia however, has a problem of creating social controversies (Bakhtin 1981, 1994; Guez 2010). Science and EE is not immune to heteroglossia. Scientific reasoning and rational interpretation and analysis of environmental problems can be solved, but each person and/or organization has their own opinion and interpretation of complex environmental issues that are communicated at levels that are appropriate and targeted to different audiences, while others are not. Despite having messaging that is appropriate to one group of stakeholders, other groups may "hear or understand" the message differently.

Operationally and from an environmental point of view, human beings often adopt nonsustainable resource use and management methods to deal with economic development. The quality of our environment is often the victim of politically-vested interests of the public agenda. Therefore, as environmental educators our challenge now is how to express the complexity of modern environmental problems in an understandable way using simple and understandable methods, while ensuring that the environmental science is accurate and effective in interpreting and assessing environmental problems without inciting panic. This requires a carefully crafted communications plan and potential solutions for the environmental problems being faced that has consensus among stakeholders.

Therefore, although we study the environment, we need to follow a process that includes a literature review, a list of the problems that have been identified in the literature, new ideas, assess, and understand processes to determine whether the data collection, analyses, and interpretations are reasonable and feasible, and where improvement is needed. The most important thing to remember is that when we observe the environment, we interpret our observations from the societal norms, values, and constructs that we are a part of, while recognizing these elements are spatially and temporally variable, in addition to theoretical estimation and speculation. The research of EE must of course be based on theory and must be corroborated by sound science, practical analyses, and summarized and organized in a manner that the public at large can understand and when appropriate, contribute to developing solutions or the discussion. The following issues need to be emphasized when combing research methods (Estabrooks 2001):

3.1.1 Instrumental Research Utilization

Instrumental research is the application of specific research results and transforming them into materials that are suitable for EE.

3.1.2 Conceptual Research Utilization

It is said that research may change one's thinking, but it does not necessarily change one's behavior (Heimlich and Ardoin 2008). Heimlich and Ardoin (2008) declared that human behavior is grounded in rational thought. Why environmental education is not effective? How it could be more effective? In this case, it is necessary to inform decision makers of the research, what the results mean, and then let the decision makers ponder why EE is not effective.

3.1.3 Symbolic Research Utilization

Conceptually, environmental education research is often abstract and based on paradigms that fit the social and economic needs of human populations in space and in time (Stevenson 2007; Ardoin et al. 2013, 2020) The scientific and environmental concepts that are studied and ultimately presented to the public are explained or framed in a manner that pushes the science forward, but not too quickly or in a manner that society will lose interest or consider the results irrelevant to the problems society faces at that time (Ardoin et al. 2020). For example, the global change issues that we are concerned with today were not prevalent 100 years ago. Evolution was not accepted as a valid hypothesis until Darwin's (1859) seminal work that appeared in the *On the Origin of Species* (1859), and even today, there are sectors of society that do not accept this hypothesis that shapes all life on the planet. Environmental education is the persuasive summary of what we know to be true, which in turn regulates human environmental behavior from actionable knowledge (Mach et al. 2020). Therefore, the scientific community and stakeholders need to develop processes where existing and new environmental/scientific data are collated, made relevant to society so that educational pedagogies and policies can be developed for the benefit of humans and the environment.

Therefore, research on EE must first emphasize the importance of the relevant topics, including the value of assessing the state of the art (all that is known on a topic) (Freeman III 1986), the problem statement (uncertainties or gaps) (Zehr 1999), the research to fill the gaps, and the writing stage (dissertations and peer-reviewed publications). Building on what is known is the process that the scientific community has followed for hundreds of years. It follows a process that helps scientists recognize emerging and important issues discussed by the international academic community (Fig. 3.2). Second, EE research requires an in-depth analysis of disputes (Lucas 1980; Tilbury 1995; Van Weelie and Wals 2002). In order to conduct logical dialectics, we need to propose a preliminary analysis structure, think repeatedly, and demonstrate the ability to control the research object. Third, in research, the research goals should be as specific as possible to avoid vague descriptions. In addition, performance needs to be measured, and currently this is based on the quality of academic research results published, important personal contributions, talent cultivation, and the research team's academic community building and service experience (Estabrooks 2001). Although these elements are of the utmost importance, a dichotomy between science and the public is created (Eden 2010). The distillation and presentation of new ideas and relevance of the scientific data to the public at large is generally not considered important in academia. Environmental/science educators are then tasked with interpreting and translating the results generated in academia to the stakeholders, including and the public. This task is often difficult given that the goals of academia and society are generally not aligned.

When the physical environment refers to something other than human beings, it is necessary to define the environmental boundaries or what is the environment. We can consider brainstorming as a form of productive discussion. EE emphasizes the importance of close cooperation between professional teams, local people, and stakeholders in a face-to-face setting to establish lines of communication, the exchange of ideas opinions, and trust. As a result, these types of local engagements or personal relationships are considered more effective than written documents (Wang et al. 2021). In addition, EE advocates the need for appropriate planning to experience the natural environment and promote concepts of environmental protection based on novel ideas. In the education process, continuous improvement of teaching skills encourages understanding the rapid changes in the environment and environmental conditions are needed to respond to the development of pedagogies focused on teaching environmental aspects. Therefore, regardless of the interaction between the individual and the environment, whether the individual cares for the environment or not, we must grasp the "initial intention" (Bratman 1981)

Fig. 3.2 From the discussions of education scholars with all folks, we understand that education is an educational activity that develops theory from practice from brainstorming to face-to-face communication (Person standing on the right side: President Cheng-Chih Wu, National Taiwan Normal University, Taiwan, 2022) (Photo by Wei-Ta Fang)

that human beings generally benefit from the environment and the resources it provides.

According to the *Avatamsaka Sutra* 《華嚴經》 (V.17, in Chinese "Huayan," in the late third or the fourth century CE) (Gimello 2005): "Those who traverse the three times in the worlds, you never forget why you started,". In V. 19 it is said: "Like a bodhisattva's original heart, it is not the same as the latter heart." Later generations summarized their original intentions, explaining that "without forgetting their original intentions, they can always be obtained; their initial intentions are easy to obtain and always difficult to keep." Therefore, in the process of EE, it is inevitable to feel lonely, because this is a lonely job (Hart 2002). The hardest part of doing any work that is beneficial to environmental protection is persistence, hoping to communicate hard, and persevere; endure the torment, and be able to win.

Therefore, in a team setting we need to establish good and clear lines of communication and collaboration for the course development team. In the initial stage of curriculum development, when the needs of learners cannot be fully collected, it is necessary to periodically review and adjust the needs of the people and adjust the curriculum based on feedback (Bester et al. 2017), which is a focus on the practical process of truth-seeking (Ansori et al. 2020). Research can be monetarily and intellectually expensive and could drive or impede the research questions being asked. Therefore, in the process of conducting research, even if in a vacuum, it's important to stay on task and follow through on your original intentions and ideas. Building

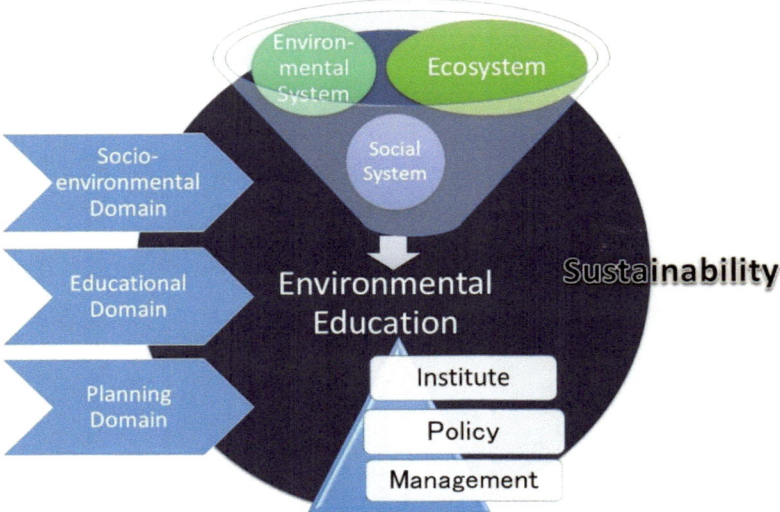

theory around hypotheses is challenging because changing paradigms not only requires the sound science to support new ideas, but acceptance of new ideas and in some cases changes in human behavior are needed. In addition to detailed observation, analysis, discussion, and constant self-criticism, it is also necessary to use seek the gift of feedback from one's peers to modify better shape the initial theoretical prototype. Environmental education is structured around a mutually coordinated framework to achieve the goal of sustainable development. For details, please refer to Fig. 3.3.

3.2 Types of Environmental Education Research

The term EE appeared in 1947. When we talk about EE research, we think we need to break the question down. That is, what exactly is EE research? The mission of EE research is to promote research and academic understanding of EE and education around the goals of sustainable development. These goals are achieved by publishing one's research findings in peer-reviewed journals. These journals were developed from EE programs and scientists across the globe. Many

are founded on outstanding schools of education thought and practice and have been able to adapt to philosophical changes in space and in time (Stewart 2020). Therefore, it is important to identify journal philosophies, practical EE experience, education for sustainable development (ESD) goals, and the government policies that helped create the ecosystem around the science of the research and high-quality innovative papers from a new social contract (Lubchenco 1998). It is important to determine whether the scientists and editors are driving the direction of the science or it's a question of dollars. Publishing houses businesses that like newspapers, are focused on generating revenue. The agencies that are providing the research funding are reacting to public pressure. The scientists then find themselves in a political, economic, and ethical mire because they need to follow the money and publish in the right type of journal. As the liaison between the academic community, stakeholders, and the public, environmental educators find themselves acting as sheriffs, lawyers, politicians, priests, and/or therapists in the process of distilling and disseminating the results of environmental study research and the meaning of these data to their students, stakeholders, and public.

3.2.1 International Periodicals

The Journal of Environmental Education (JEE) was founded in 1969 and Stapp (1969: 30–31) published an article within which EE was defined. The journal was dedicated to the research and development of environmental protection communications, highlighting the opportunity of the media to attract public attention to environmental conditions and issues. The mission of *JEE* was to provide a critical and constructive forum for research on the theory and practice of environmental and sustainable education. Today *JEE* publishes articles on EE experience and theoretical analysis, including critical, conceptual, or policy analyses articles on environmental or sustainability-related education edited by Alberto (Tico) Arenas, Editor in Chief, the faculty of the Environmental and Sustainability Education in the College of Education, University of Arizona,

USA. Arenas received his Ph.D. at the fields of Sociocultural Studies in Education at the University of California, Berkeley. Papers are interdisciplinary and cover research that ranges from early childhood to higher education, formal to informal approaches, literature reviews, and program evaluations.

A second influential journal, *Environmental Education Research (EER)* was founded in 1995 and the papers published in this journal are often literature reviews highlighting innovative empirical and theoretical research approaches, and papers within which key concepts and of EE and Sustainable Development Education (SDE) methods are analyzed (see Fig. 3.4). The *JEE* and *EER* provide readers with perspectives on EE theory and methodology. Their purpose is to improve research and practice on EE and SDE topics. The research published encourages examining methodological issues and challenges

Fig. 3.4 *Environmental Education Research (EER)* was founded in 1995 and the papers published in this journal are often literature reviews highlighting innovative empirical and theoretical research approaches. Left: Alan Reid (Faculty of Education, Monash University, Australia), the Editor of *EER*, inquired one of the Ph.D. students, Chia-Huan Hsu (Right), during the Taiwan's EE Conference, Jianshanpi Jiangnan Resort, Tainan, Taiwan in 2018 (Photo courtesy of Yi-Hsuan (Tim) Hsu, Middle, back to photographer) (Photo by Wei-Ta Fang)

to the existing theoretical dialogue. New articles provide in-depth connections between theory and practice and strengthen the conceptual framework across disciplines. Both journals welcome reader responses to published papers to attract ideas and generate academic dialogue that promote the theory and practice of EE and SDE research.

We reviewed research articles published in *EER* and *JEE* and all of the papers that we reviewed were based on international problems, such as climate change, global warming, sustainable development goals (SDGs), and empirical articles that included critical analyses and discussions of the research methods and findings. Conclusions and suggestions are based on policy practice derogation. As such, according to the papers published in the *JEE* and *EER,* EE and SDE have a wide range of research content, including critical articles and analyses on education policy, philosophy, theory, and history. In addition, articles within which qualitative analyses and verification of the reliability and validity of the analyses that were performed are predominant. Data on program evaluations show the progress of innovation in this field (Reid and Scott 2013), explains program goals, and record the background, processes, and research results. These results are based on the consistency and empirical nature of the arguments and can be extended to other educational and cultural backgrounds.

3.2.2 Local Periodicals

In the Sinophone society there are not many journals that are focused on EE research, but they are becoming more prominent. The *Journal of Environmental Education Research* published by the *Chinese Society for Environmental Education* (CSEE) in Taiwan was launched in 2003 and contains research papers that address a variety of topics, including educational discourses and environmental philosophical works. The main contents include:

- **Research Articles**: Reports on scientific research,

- **Academic Articles**: Reviews on EE research and practices; and
- **Essays and Analyses**: Discourses on the historical development, ideas, practices, and philosophy of EE.

Environmental education, as defined in the *Journal of Environmental Education Research (JEER)*, edited by Shin-Cheng Yeh at GIEE, NTNU, is extensive and includes "formal" and "informal" EE. Topics can cover a variety of environmentally related subject areas, such as: environmental ethics, environmental philosophy, sociology, psychology, commentary, communication, economics, planning and design, science and engineering, tourism, leisure and recreation, natural resource management, geography, culture and history, sustainable development, public health, food and agriculture (https://www.ipress. tw/J0088?pWebID=54) (in Chinese). Approaches to propose/test/assess theories and practices of EE policy, curriculum planning, or teaching and learning from multiple perspectives are the focus of the papers published in *JEER*.

3.2.3 Research on Environmental Education

In the aforementioned discussion the development of international EE is akin to a tree Joy Palmer (1951–) used to compare the content and development direction of EE. He believed that EE in the twenty-first century is based on the roots of trees, some shallow, some deep, but prevalent in the soil. On this basis, students should understand the environment, possess environmental knowledge, skills, and values, and develop the ability to take care of the environment (Palmer 1998).

3.2.3.1 Theme of Environmental Education
Project Environment was implemented in the UK in 1974 and mentioned three topics: "education about the environment," "education in/from the environment," and "education for the environment" (Tilbury 1995; Palmer 1998). However,

how can teachers use the spirit of suspicion, curiosity, and exploration in the development of EE to conduct ecological surveys under different hypothetical situations and attract students towards the field of scientific investigation in a tempting manner? In reality it comes down to what a person/teacher wants to achieve? Do they want to teach students how to perform outdoor teaching projects based on critical and creative thinking, integrate past experiences and ongoing course learning content, and/or use education and research to develop the process? At the present time, in the process of developing education and research teaching, the following three directions are mainly discussed through instructivism, constructivism, and deconstructionism.

- **Education about the Environment**

The aim of EE is to teach students about the environment, let them understand environmental concepts, and to allow them to criticize issues in a logical and constructive manner. This approach needs to generate critical thinking and problem-solving abilities by building on existing environmental knowledge and developing an awareness of the biotic and abiotic elements of the world around them. Therefore, EE is to seek and discover the essence of environmental research by developing and testing hypotheses. Learning to identify environmental problems that society faces today and then formulate relevant questions is a learned process. Under the guidance of teachers, colleagues, and peers, information about the environment is collected that will form the foundation of the research being performed. Therefore, education about the environment is a type of guided EE.

- **Education in or from the Environment**

The field of EE emphasizing outdoor education is important. Teachers teach students to use the natural environment for learning; nature is a classroom and we can use our natural curiosity for inquiry and discovery to strengthen the learning process (Chang and Ow 2022). In the learning process, we integrate environmental

awareness, research data, and personal experience to develop environmental awareness and solve environmental problems. Using this model, an awareness is built around our knowledge and experience with the environment.

- **Education for the Environment**

At this stage, we have the experience that the environment is the cause and human beings are the effect; or human beings are the cause and environment is the effect. From the deconstruction approach, teachers can encourage students to study the relationships between individuals and the environment. Through the screening and discussion of environmental issues, students can begin to understand the causes of environmental pollution and encouraged to integrate environmental responsibility and action into their behaviors.

- **Education about, in, and for the Environment**

The focus of EE still needs to be integrated into a unified system because most educational processes, from knowledge to competence and construction, are full of divides. That is to say, the curved lines in Fig. 3.5 are all EE content and can achieve the goal of sustainability through education. In other words, the concept of EE is high, but after a student receive EE instruction of EE, does it produce an ethical value that has for the environment, cultivate the correct knowledge, build the ability to improve the environment, integrate a sense of environmental responsibility, and generate actions resulting in stronger environmental awareness? This is the ideal scenario for "about, in, and for" EE. Integrated EE learning and research can cultivate actions oriented towards developing problem-solving skills and knowledge that form environmental attitudes and values that contribute to the formation of responsible environmental behaviors (Fig. 3.5). Therefore, integrated EE is a process that develops environmental attitudes, values, morals, and ethics and educates students develop an awareness and behaviors in people that care,

Fig. 3.5 Focus of EE
(Illustrated by Wei-Ta Fang)

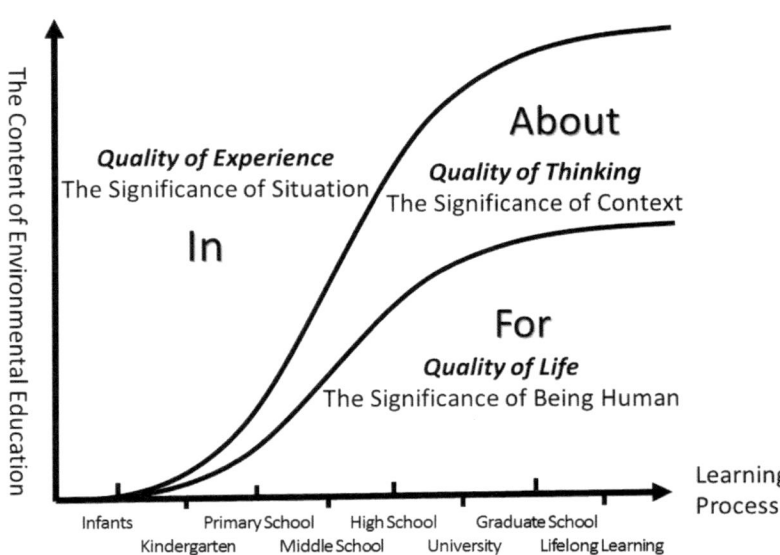

want to be close, and protect the environment through different learning processes.

3.2.3.2 Proposals for Environmental Education

Environmental education emphasizes self-learning and in academia, self-directed learning is expected. In-depth study is conducted based on the interaction of learning methods, situations, and free choice (Falk et al. 2009; Falk 2017). Learners are free to design personal education plans for their environment through autodidacticism. However, because people have the inclination to be liked, disliked, and work hard, all the previous situations are ideal learning methods for social environmental education. Therefore, how do we educate kindergarten, elementary school to middle school, and college students in the formal environmental setting? How do we integrate knowledge and experience into learning experiences and encourage learners understand the environment around them and the resources provided by these ecosystems? This is explained further.

- *Instructivism*

"Instructivism" is the approach of "guidance" that is advocated in the early stages of EE

courses. It is based on the premise that teachers establish/provide the necessary learning resources and restrictions for students, set learning goals, understand themselves the principles of EE, design teaching methods that will allow students to attain the program goals and objectives, and emphasize the importance of professional knowledge learning in the curriculum goals. Instructivism provides a learning approach where students learn about the environment based on stimuli that emphasizes EE is a discipline within the constructs of ecology. Teachers must present the learning content in the way that meets the goals and objectives of the curriculum and use the appropriate tests to measure whether the students grasped the concepts and further expanded their knowledge and abilities on the information presented.

Under the notion of guiding theory of media richness by Chao et al. (2020), communication can still take place, but the scope will expand in concert with that of the knowledge field, allowing scholars and graduate students to gradually understand their own capabilities and vision from computer-mediated communication (Chao et al. 2020). Based on guidance-based learning, it is emphasized that learning is a two-way contract learning between teachers and students, rather than one-way teachers exerting their teaching authority. The learning contract is a kind of autonomous learning (autodidacticism), allowing learners to accept two-way contracts on their own, and teachers and students accept each other. In the integration of instructionalism into learning, based on the connection between "stimulus-and-response" (S-R), scholars are encouraged to try mistakes, and teachers can correct mistakes, and through the process of guidance, perform wrong learning and correct answer problems.

• *Constructivism*

Constructivism is a philosophical idea derived from cognitionism and adopts a philosophical stand of "non-objectivism." Constructivists believe that the ability to generate knowledge needs to pass through the actual field. Although the ecological environment of the so-called EE field exists objectively, the human understanding of ecology and the meaning given to it are determined by individuals. Therefore, human beings construct the concept of "environment" with their own experience.

Constructivist learning is a self-learning theory that scholars should establish after accumulating the basic work of self-learning. While constructivism and instructivism help students to acquire knowledge, constructivism adopts an open-ended learning method and instructivism adopts problem-solving learning method (Edelson et al. 1996; Herrington and Standen 1999). Their educational concepts are different and Constructivists believe that the learning method of EE is closer to that in nature (Klein and Merritt 1994). Students can learn on their own and build knowledge from observing and interacting with the environment. Therefore, the study of natural ecological knowledge is an education in/from the environment. It is based on the experience and understanding of environmental situations.

Constructivists therefore believe that human beings choose for themselves and are responsible for those choices. This kind of thinking gives human-beings greater freedom, but they must also accept greater responsibility, which is close to the existentialist thinking mode. Existentialists believe that the meaning of human existence cannot be answered by rational thinking. Therefore, from the philosophical thinking of "non-objectivism," we understand that learning in/from the environment is personal, independent, and self-awareness is learned from subjective experience, which is not what teachers can do.

Constructivism encourages learners to actively experiment, experience, and take further actions in ecological experiences through the approach of "personal and direct participation in environment." This will complete the learning process of education. Knowledge of EE is "learning by observing nature," rather than relying on teachers to teach in the classroom what students should and should not do. Therefore, constructivism hopes that when scholars are confronting the conflict between theory and practice, that they form a sense of responsibility in the living environment and seek solutions on their own.

• *Deconstructivism*

Deconstructivism is a critical way of learning in the course advancement and growth education. We observe education about the environment and only teach what the environment is and cannot produce environmental actions. It is only a state of knowing and not doing. Education in/from the environment is integrated into the context of nature and the effects of nature connectedness, but whether knowledge and action can be generated as a result is still questioned by scholars (Tilbury 1995). We interpret the different perspectives and look for reasons for these conflicts. When dealing with classic narrative structures, it is best to interpret them in the existing context/setting (Gough and Price 2004). For example, from the deconstruction of any EE dissertation, the existence of a certain type of prototype is required.

However, even though the process of construction is perfect, it means that we have seen in our research the strong and powerful natural connection to nature and even reached the wonderful feeling of unity between heaven and man. However, from the perspective of critics there are deficiencies that need to be carried out by deconstructive criticism. Here, we quote the *Discussion of the Equality of Things, Zhuangzi* of ancient Zhuang Zhou (莊子; 莊周) (Zhuangzi, 369–286 BC), who said:

> There is nothing in the world greater than the tip of autumn down, and nothing in the world smaller than Mt. Tai (泰山). There is no one longer-lived than an infant died young, and no one shorter-lived than Ancestor Peng (彭祖), Heaven and earth and I are born together; the myriad things and I are one.

In other words, in the context of the two things, in the so-called environment, in terms of Zhuangzi's deconstruction, there is no absolute standard for all sizes; there is no absolute standard for the length of all time. From Zhuangzi's deconstruction method of time and space in the natural environment, in the process of deconstruction of "EE" textbooks and teaching methods, "instructivism" in the classroom and "constructivism" in the environment have always been in opposition and have a tense relationship.

"Instructivism" is described as "students memorizing the ecological knowledge to be a clear and unquestionable state forced by teachers"; however, "constructivism" is described as "both teachers and students do not know what to do. Teachers are only concerned with the psychological activities of students while learning knowledge, and not keen to test whether students really understand and memorized the knowledge required in the environment." Deconstructionists discuss the EE teaching model to guide questions and construct critical ideas and theories, so that students can further investigate and conduct research on controversial environmental issues and generate questions. However, deconstructionists sometimes find too many problems, have the ability to be critical, and cannot participate in the process of environmental improvement. They blame others for not understanding environmental protection, but all lack the ability to improve the environment and cannot be integrated into the actual mainstream society. Therefore, what we need is the fourth kind of doctrine, which is the integration doctrine of EE.

- **Integrationism**

The fourth doctrine refers to the integrationism of education about, in, and for the environment. The aforementioned methods of education do not actively fight against environmental, economic, and social injustices (Tilbury 1995). Even when criticism is made, it is limited to anonymous criticism and dares not to openly make constructive EE theory and practical contributions. Therefore, from the ontology of western scholarship, we need is an integrationism approach when it comes to the epistemology of the environment. "Nature" is not an absolute condition but relative in space and in time. In the dualistic structure of deconstruction, scholars criticize that although deconstruction can be used for academic criticism, it is difficult to understand its true definition and often belongs to political criticism. Therefore, we need a more rigorous academic and practical basis to explore the real complex interaction modes in the environment.

When investigating "deconstructivism," scholars should learn French philosopher Derrida's caring and self-reflection skills for the world. Through self-reflection, critical thinking, and group evaluation, he transforms from a virtual situation into an enlarged body of the real environment. Recognize that only through reflection and mirroring can we improve our prejudices and ideas, as well as strengthen the responsibility of the citizens of the earth through actions. In recent years, due to the emphasis on sustainable development and the promotion of education for sustainable development, the paradigm of educational research has shifted from an empirical paradigm to an ecological paradigm in the real world. Positivism has turned to critical theory and hermeneutics, so the connotation of EE is increasing. In other words, more social evidence, argumentation, criticism, interpretation, dialogue, and social participation are needed to respond to the changing trends of the times, see Table 3.1.

Table 3.1 The theoretical process of environmental education and research

Claim	Theme	Learning architecture	Learning process
Instructivism	Education about the environment	The instructor must specify the learning content in the way required by the course, and use appropriate tests to identify the concepts, knowledge, and abilities learned by the students	Use the link between "stimulus-and-response" (S-R) in your studies to encourage students to learn trial and error from their instructors
Constructivism	Education in or from the environment	Open and self-directed learning methods in outdoor environments to learn about ecology and the environment	Making observational experiences meaningful. Co-operation with one another to expand knowledge and develop new concepts and action strategies. In the face of theoretical and practical conflicts, the actual intervention from students is involved to learn to create value for other people, and it is up to the students to find solutions
Deconstruction	Education for the environment	The instructor guides questioning and constructs critical thoughts and theories, enabling students to conduct further research on topics, and generate more questions	Read and understand the "reading between the lines" in the text, and interpret the view of "dual opposition" of various conflicts, and look for the cause of the conflict of these views
Integrationism	Integrated EE	In recent years, due to the emphasis on sustainable development and the promotion of SDE, a shift in the EE paradigm has emerged, from a paradigm of pure experience to a paradigm of ecology in the real world, and from positivism to criticism and interpretation Therefore, the connotation of EE is increasing day by day	Through self-reflection and group evaluation and critical thinking, the virtual reality is transformed from concept of a virtual world into something into the expanded recognition of the real environment (Marín-Morales et al. 2018). Through reflection and mirroring, we can improve prejudices and concepts, to elicit and automatically recognize different emotional states and strengthen the responsibility of the citizens of the earth through actions

3.2.3.3 Research Directions of Environmental Education

- *Environmental Education Policy*

Environmental Education is based on the following concepts of "One Planet, Environmental Justice, and Sustainable Development (Magraw and Lynch 2006; Habib 2013);" therefore, how to improve the environmental literacy of the entire population and the practice of responsible environmental behavior is an important developmental direction for national environmental policies

and environmental governance (Fig. 3.6). At present, the national EE program is the basis for environmental literacy policies (Liu et al. 2015). It is formulated by the Environmental Protection Administration of the Executive Yuan (Environmental Protection Administration 2014), and consults with the Ministry of Education and other units, and reports to the Executive Yuan for approval of the Taiwan's environmental education program to carry out. This EE programs have been designed to enhance citizens' understanding and awareness of the world's environmental challenges, as well as to encourage active

Fig. 3.6 EE policy is based on the mechanism from local system to global system (Illustrated by Wei-Ta Fang)

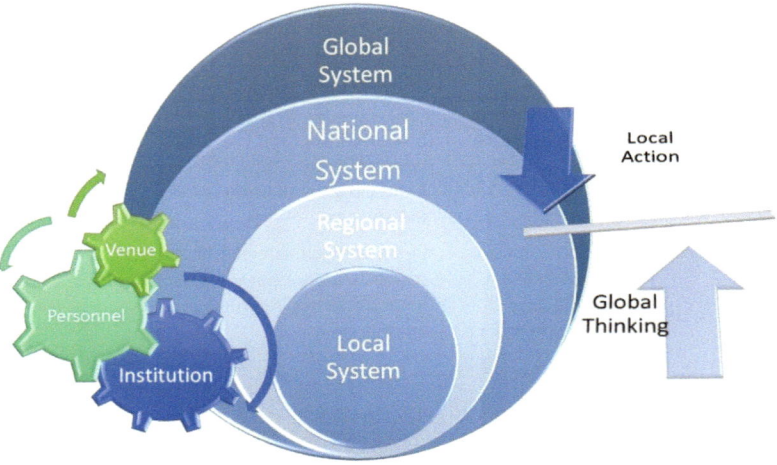

participation in environmental protection and, sustainable development (Huang et al. 2021).

- *Environmental Education in the Schools*

The school's EE program aims to strengthen the nation's establishment of environmental-related programs through the school system (from kindergartens, elementary schools, junior high schools, high schools to university undergraduates, graduate schools), classrooms, and outdoor environments. Environmental Education are built on a teachers' environmental literacy of knowledge, attitudes, skills, and values.

- *Corporate Environmental Education*

To promote corporate social responsibility, reduce environmental pollution and promote the recovery, regeneration or effective use of a producers' product(s), industry and government need to develop partnerships that promote environmental protection, improve employee environmental literacy, and environmental education.

- *Environmental Science Education*

To strengthen the disciplines involved in the environmental sciences (e.g., ecology, geology, geography, conservation biology, resources technology, environmental engineering,

environmental psychology, environmental politics, environmental society, environmental culture, environmental economy, and environmental engineering) science learning activities in the classroom, laboratory, and field must be organized in uncertain times (Wals et al. 2014; Kidman and Chang 2022). Environmental science education includes a good understanding of the living and physical aspects of the world around us.

- *In-Service Education*

EE at communities is a process of disseminating environmental knowledge and skills from in-service education in society (le Roux and Ferreira 2005). Disseminating environmental knowledge and learned skills from learning fields such as museums, social education centers, EE facilities, ecotourism, community tours, and the visits strengthened the connotation of environmental literacy of community residents and In-Service Education and Training (INSET) for teachers (le Roux and Ferreira 2005).

- *Environmental Philosophy*

Environmental philosophy explores the relationship between natural environmental values, human dignity, animal welfare, and the interactions between humans and nature. Environmental philosophy includes environmental

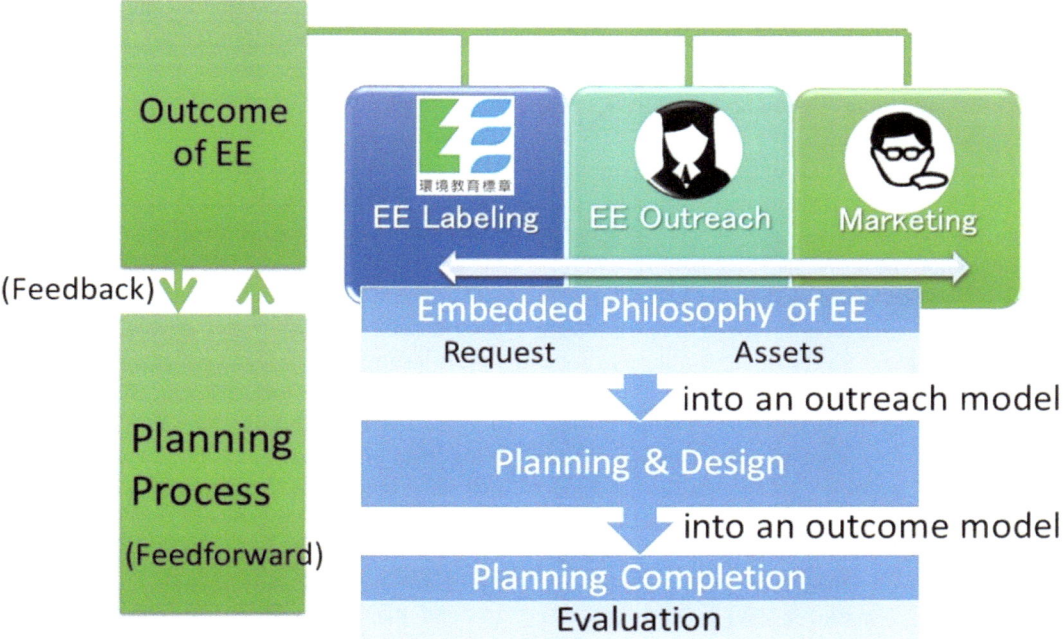

Fig. 3.7 The research directions of EE could be embedded in the environmental philosophy associated with project planning, design, and evaluation (Illustrated by Wei-Ta Fang)

ethics, land morality, and the meaning of sustainable development. Environmental philosophy studies the earth's ethics of earth resources, human depletion, environmental protection, and philosophical practice toward project planning, design, and evaluation (Fig. 3.7).

• *Environmental Interpretation*

Environmental interpretation is suitable for nonformal EE. Through the strategy of environmental outdoor fields and to explain the planning and implementation of outdoor ecological basics, ecotourism, ecological guides, and outdoor education methods are used to communicate knowledge and strengthen the human and natural environment. Therefore, an interactive opportunity can inspire learners to improve their knowledge, attitude, and activity skills of environmental ecology.

• *Environmental Communication*

Environmental communication is an activity that transmits environmental knowledge, methods, and thoughts through communication media, and cultivates environmental literacy for all. Environmental communication conveys the status and problems of environmental events and the creative process of multimedia forms such as text, sound, images, animations, and videos, that results in a tangible explanation of environmental protection. Environmental communication explores the symbols, discourses, and contextual relationships of environmental issues. The dissemination of environmental information through books, videos, media, and social networking platforms has aroused readers' interest in environmental knowledge. Furthermore, Artificial Intelligence (AI) and digital technology will be a new topic for revolutionary transformation for EE research in science (Yeh et al. 2021; Stagg et al. 2022).

3.3 Research on the History of Environmental Education

Historical studies of EE can be traced back to the emergence of the fields of formal education and educational research (Gough 2012). If we study the history of EE, we can then use curriculum history or genealogy.

Generally speaking, the historical research of EE is a type of research that must first look for objective knowledge in archives. Topics related to the subject can be found in the library, documentary library, and computer network searches. How to explore the context of archives through archival retrieval systems has become a key factor in thinking about archive/metadata research. In the archive, we understand that the reason why we need to involve in EE is because of the threat of global environmental change. International conferences have mobilized scientists to think about how to save the planet from the scourge of global change. Therefore, the history of EE illustrates our efforts to survive. From the historical trajectory, it we can observe the change of human collective behavior and use the "structuralist and "post-structuralist" approaches" research on using historical data.

The process of these meetings is long. In 1972, the United Nations Conference on the Human Environment advocated "the importance of education." The meeting stated that "Education in environmental matters, for the younger generation as well as adults, giving due consideration to the underprivileged, is essential in order to broaden the basis for an enlightened opinion and responsible conduct by individuals, enterprises and communities in protecting and improving the environment in its full human dimension." The Tbilisi Declaration in 1977 participants, discussed and formalized the field of EE (Knapp 1995). The Earth Summit in Rio de Janeiro, Brazil, in June 1992. The meeting participants provided education, public awareness, and training for the global action plan on Basic Principles tabled in Agenda 21. However, in the Conference on Environment and Development, EE was focused on promoting sustainable development

and improving people's ability to solve environmental and development problems. Later, EE was called "education for sustainable development." In 2009 the Bonn Declaration was developed and describes education for sustainable development and prescribes formal, non-formal, informal, vocational, and teacher education actions.

In many countries, the development of EE and the education for sustainable development (ESD) are different. Some scholars believe that "EE" has been diluted by "sustainable development education." Knapp (1995: 9) concluded that a name change was not in the best interests of EE. But overall, because EE has expanded into the economic and social fields, EE has been able to deepen its promotion effect in the world. European scholars of EE have incorporated EE research into their ESD programs. If we make inferences with the above-mentioned historical analysis methods, we can then understand that the budding and robustness of EE originated in the 1960s.

3.3.1 The Rise of Environmental Education

The field of EE originated in the 1960s. As the global environment deteriorated, it threatened human development. In the 1960s, scientists increasingly paid attention to the increasing scientific and ecological problems of the environment, and the public's need to understand these problems. These problems include the growing pollution of land, air, and water. In addition, the world's population is growing, and natural resources are continuously depleting. As explained by the Declaration of the United Nations Conference on the Human Environment, or Stockholm Declaration in 1972:

> We see around us growing evidence of man-made harm in many regions of the earth: dangerous levels of pollution in water, air, earth, and living beings; major and undesirable disturbances to the ecological balance of the biosphere; destruction and depletion of irreplaceable resources; and gross deficiencies, harmful to the physical, mental and social health of man, in the man-made environment, particularly in the living and working environment.

If education comes first in environmental improvement, educational research can stimulate effective ways of thinking and discussing human beings facing environmental problems (Carson 1962). American scholars Rachel Carson (1907–1964), Garrett J. Hardin (1915–2003), and Paul R. Ehrlich (1932–) yelled loudly, hoping to include education in the environmental agenda. However, EE is not just a social issue, but an educational issue. In addition, the relationship between science education and EE is implicit. In view of the seriousness of environmental problems, scholars in the 1970s hoped to solve environmental problems with science and technology. But a few scientists believe that science and technology alone are not enough. Therefore, through environmental chemistry, ecology, geology, geography, conservation biology, resource technology, and environmental engineering, the problems of environmental hazards that have been disturbed, could not be solved. Human ecologist Stephen Boyden said in 1970 (Boyden 1970: 18):

> The suggestion that all our problems will be solved through further scientific research is not only foolish, in fact dangerous. The environmental changes of our time have arisen out of the tremendous intensification of the interactions between cultural and natural processes. They can neither be considered left to the natural scientists nor as problems to be left to those concerned professionally with the phenomena of culture. All sectors of the community have a role to play, certain key groups have, at the present time, a special responsibility.

3.3.2 The Construction of Environmental Education

Since the 1970s, school education in western countries have incorporated ecological and environmental content into integrated teaching and incorporated it into the education curricula of schools at all levels. The Intergovernmental Biosphere Conferences in 1968 and 1970 both recommended EE should be incorporated into school curricula. Stapp (1969: 31) emphasized

the relationship of the environment using four goals EE (Stapp et al. 1969: 31):

- It is recognized that the human system is composed of human, cultural and ecological environments (Boyden 1970: 18). The biophysical environment is an integral part of the system, and humans can change the interrelationships of this system;
- A broad understanding of natural and manmade biophysical environments and their role in contemporary society;
- Basic understanding of the ecological and environmental problems facing human beings, how to solve them, and the responsibility of citizens and governments to work hard to solve problems; and
- Attitudes about the quality of the biophysical environment, which will encourage citizens to participate in the solution of ecological and environmental problems.

According to Stapp and his colleagues, this method of education is different from conservation education (Boyden 1970: 18). Conservation education focuses primarily on natural resources, not on the community and its related issues. Therefore, EE is not only concerned with the natural environment, but also with the working environment, as well as human well-being (Stapp et al. 1969: 30). In 1970, Stapp was invited to participate in the Australian Academy of Science Conference and proposed a curriculum in Australia. He emphasizes curriculum development procedures and administrative strategies rather than professional philosophical analysis. The experimental direction of Stapp led the practical development of EE.

The definition and goals of EE form the basis of some other concepts in the field. For example, in September 1970, the First International Working Meeting on EE in the School Curriculum, accepted the definition of EE as follows (UNESCO 1970):

> Environmental education is the process of recognizing values and clarifying concepts, in order to develop skills and attitudes necessary to understand and appreciate the interrelations among man,

his attitude and his bio-physical surroundings. Environmental education also entails practice in decision making and self-formation of a code of behavior about issues concerning environmental quality.

3.3.3 Ideology of the Declaration on Environmental Education

Environmental science education, usually in the form of ecological concepts, are incorporated into school curricula. However, true EE and learning have not been regarded as a priority for education, because in the West, EE is valued by scientists, environmentalists, and scholars, but not by governments.

In addition, the drafters of the EE declaration are all men. Although EE is based on novel ideas no attention was paid to gender equality. For example, 1975 was the International Women's Year and the United Nations issued non-sexist writing guidelines, hoping to use as much gender equality as possible in international declarations. For example, try to replace sex identification in the declaration with neutral words/descriptors. These instructions were followed writing the 1975 Belgrade Charter, but the 1977 Tbilisi Declaration didn't, which would make this document discriminatory on the basis of gender identification. Although some women consider human male activities to be a major factor in environmental degradation, it is important that all human beings be regulated by EE statements.

3.3.4 The Practical Power of Environmental Education

In the historical demonstration of EE, the practical power of EE needs to be evidence-based practice through empirical research. In science education, empirical research requires an experimental group and a control group to conduct evidence inference by employing intervention. However, EE is not education of knowledge, but education of practice. In other words, practical education requires behavioral changes, but this

behavioral change comes from a sincere change in the heart, not a short-term plan that can be manipulated in a classroom-based laboratory. Therefore, the experimental limits of EE have received considerable criticism for its research results (Blumstein and Saylan 2007). In recent years, EE research has been mainly discussed in terms of "positivism," "post-positivism," "structuralism," or "hermeneutics" and "critical theory."

The Frankfurt school Jürgen Habermas (1929–) has criticized the issue of instrumental rationality. His claim on epistemology holds that human knowledge can be divided into three types (Habermas 1971):

- Scientific research of experience-analysis: including the cognitive interest of technology;
- The scientific research of history-hermeneutics: including the cognitive interest of practice; and
- Critically oriented research: contains the cognitive purpose of liberation.

It could be argued that students who advocate the supremacy of techno-science, don't care about the environment. Therefore, if EE is an instrumental research utilization, which then means that through manipulation and intervention, one can change the way one thinks, which is empirical; but changing one's way of thinking does not necessarily change one's practices. Therefore, we need to adopt conceptual research, which is based on Habermas' "hermeneutics" and "critical theory." Conceptual research requires the power of social practice, and its purpose is to confirm the relationship between practice in a social context. In social practices, researchers emphasize the promise of change. There are two forms of this kind of commitment: one is activity, and the other is inquiry. In other words, social practice is usually applied in the context of human development and involves knowledge production and theoretical analysis. This knowledge is the knowledge generated after practice. Therefore, how to use research from the physical world to make sense requires research procedures. In other words, we need research as a persuasive tool, which requires time and money. Our research on how to continue to grow over time requires long-term observations, hypotheses of behavioral intent, and a correct measure of how others respond. In social practice, environmental literacy is regarded as a key factor in human growth (see Chap. 4). The practical factors occur that produce environmental literacy with the material, meaning, and procedures of the human world. The aforementioned research must confirm that ontology/epistemology/methodology have different assumptions, and the constructed worldview is also different.

3.3.5   Reflection on Environmental Education

If we say, the practical power of EE is action research. Action research is a type of self-criticism under collective action (Fig. 3.8). The purpose of understanding how practitioners deal with matters in a social context is to improve the public interest of the entire population, generate social justice, and understand the meaning of practice (Kemmis and McTaggart 1982). Kemmis and McTaggart (1982) adopted a self-reflective circle that can be divided into four elements: plan, action, observation, and reflection, and the extension of the plan will continue to be revised by plan, action, observation, and reflection (Fig. 3.8). A detailed research guide

and experimental design reference manual were designed based on the concept of the course of the circle of action research (Kemmis and McTaggart 1982). His plans, actions, observations, and reflections, through revising and improving the plan, produce a circle of re-action, forming the characteristics of the EE field to promote EE.

> **Box 3.1 Examples of Environmental Action Research in Taiwan in the 1990s**
>
> A self-reflective circle, which is divided the plan, action, observation, and reflection elements. This model is effectively a continuous improvement model where improvements/changes to the research and experimental design are made each time one moves around the circle. Therefore, according to the plan of the Environmental Protection Administration of the Executive Yuan (Taiwan EPA), how to promote campus ecological protection and community environmental protection, we cite the following case of the "National Little Environmental Planner" from Taiwan EPA.
>
> 1. **Planning Basis**
>
> First, to promote EE in schools and strengthen environmental protection education, the Environmental Protection Administration of the Executive Yuan (Taiwan EPA) began to hold a National Conference on Children's Environmental Protection in 1990, which was expanded in 1991 and renamed the National Conference of "Little Environmental Protection Administrator." From 1991 to 1997, a total of 4500 national elementary school students participated in the conference. Therefore, how to pay attention to environmental issues through the activities of the Little Administrator of Environmental Protection. How to protect the environment when problems occur in the campus environment or when problems are encountered

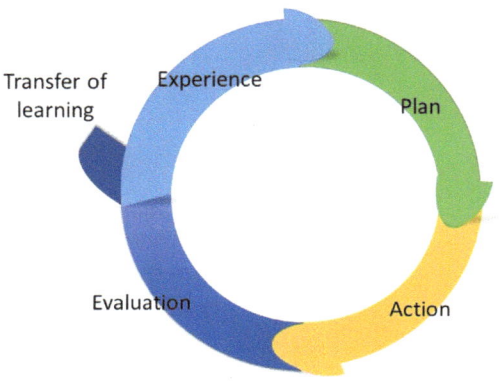

Fig. 3.8 Elements of research guidelines and experimental design (Illustrated by Wei-Ta Fang)

during the implementation of EE. For example: "How to deal with the fact that the ponds on the campus are seriously eutrophicated and become a breeding ground for mosquitoes?".

2. **Reference Methods**
i. **From European and American countries**: Environmental planning methods, planning theory, citizen participation, case studies, roundtable discussions, computer-aided mapping, report presentations, The Global Learning and Observations to Benefit the Environment Program (GLOBE).
ii. **Japan**: Town making plan, drawing of the environment map of amenity, outdoor visits, etc.
iii. **Taiwan, Republic of China (Taiwan, ROC)**: Local teaching materials and teaching methods, off-school teaching, national science exhibitions, etc.

3. **Research Strategy**

Think about the idea and solutions first. This project can be carried out from outside or inside the school. If it is outside the school, then start from the neighborhood, conduct environmental surveys, draw environmental maps, discuss urban and rural issues, and publish the research findings. When encountering environmental problems on campus, first perform the tasks and use scientific methods to collect information related to the issue, including expert interviews and detailed investigation. First, we can organize tasks and use scientific methods to collect information related to the issue, including expert interviews, data query, site surveys, questionnaires, etc. After the data are collected, we conduct preliminary consolidation (triage) and then use various democratic procedures, such as: conduct debates, discussions, and decisions to determine the

best issue resolution method to use as a yardstick for follow-up action or provide it to administrative units for reference.

4. **Take-Action**

Take-action based on the above-mentioned solution strategy, that is, "how do we manage to solve it by hand?" Therefore, under the promotion of the plan of the Little Environmental Planners nationwide in 1997, through the guidance of schools' teachers and the guidance of college student associations, the mothers and all residents of the community were affected and promoted (Fig. 3.9).

5. **Reflective Thinking**

Through "evaluation," the actual implementation results of the strategy are resolved, and the results are reflected on reflection. At the beginning of the promotion of the plan, in 1997, this project was praised as an "unprecedented national EE plan." Neighborhood neighborhoods started with a lot of interest, because planning in the United States started with towns and cities and communities.

6. **Government Action**

(1) In order to evaluate the performance of Little Environmental Protection Administrators of counties and cities in promoting environmental protection, the Taiwan EPA selected the first (the year of 1996) National Little Environmental Protection Administrators and arranged to meet with President Li Teng-Hui (1923–2020), Republic of China (Taiwan) at the Presidential Office and Record a TV show. In 1997, he promoted the "Little Environmental Planner" activity, using the four elements of planning, action, observation, and reflection. In response to the actual needs of

Fig. 3.9 Assembly photos. Take-action based on the solution strategy for natural conservation, that is, "how do we manage to solve all environmental issues by hand?" (Photo by Wei-Ta Fang)

social environment changes and EE, the "National Conference of Little Environmental Protection Administrators" organized by the Taiwan EPA. "The Little Environmental Protection Administrators Meeting" organized by the states focused on actually promoting environmental protection work as an educational focus, in line with the concept of "general transformation of the living environment, and discussed the format of holding. According to the Taiwan EPA and the Ministry of Education (MOE) compiling the "National Records of Excellent Little Environmental Protection Administrators," and "Environmental Protection Seeds" in 1998, it is good to look for local environmental issues and uses various scientific methods to collect relevant information. After collecting the data, draw an environmental map, write a small paper, and promote local environmental improvement, including river research, coastal protection research, campus noise decibel research. There are 79 environmental planner reports regarding to "The Little Environmental Protection Administrators Meeting" on environmental studies of the elementary school with fruitful results.

(2) After more than 30 years of education reform, the MOE has promoted twelve years of national education, and has made quality-oriented education a priority in EE. It is believed that the cooperation between school education and

community development promotes the planned *Curriculum Guidelines of 12-Year Basic Education* of the Ministry of Education in Taiwan, Republic of China (ROC). Community cooperation perspective. EE has undergone a generation of educational reforms, from the promotion of neighborhood education to the integration of cross-regional cultures, forming a new situation. Environmental education from parents, through the struggle of science and practice, has produced the next generation of EE from infusion education to transformation education. That is: The Sustainable Future Equation:

Sustainable future = (Awareness of issues + Knowing how change is detectinginourchangingworld) × Purposed action

3.4 Quantitative Research on Environmental Education

Quantitative research on EE is mainly based on "positivism." In the research, we attach importance to collecting evidence, conducting data analysis, using validity and reliability to strengthen the reliability of data, and using variable operations to control variables. Perform statistical analysis to describe the phenomena such as personnel, features, etc. to be discussed.

3.4.1 Preparation and Measurement of Questionnaires and Tests

The differences between the scale and the questionnaire and the compilation structure include the theoretical basis for the scale. The questionnaire is only required to meet the theme. Therefore, the compilation of the scale is based on the theory proposed by scholars to determine its content. The researcher compiles the questionnaire according to the following three steps: determining the subject, collecting information, and compiling the question. Once this information is collected the questionnaire can be developed.

To strengthen the validity of the questionnaire/scale, it is recommended to ask three EE scholars and experts to review the questionnaire, provide feedback, and discuss whether the questionnaire/scale topics need to be revised. The most common form of attitude measurement is the Likert five-point scale.

3.4.2 The Experimental Research Methods of Environmental Education

The experimental approaches can be repeated. Different experimenters can get the same results if the premises/assumptions are the same and the operation steps are the same. Experiments are usually published in the form of experimental reports. Due to the need for funding for the experiment, under the consideration of reducing the probability of experimental failure and reducing the cost of the experiment, the quantitative experiment should divide the experimental object into small phenomena; also, because the reality cannot be recognized by the experimenter, it needs to be cut into experiments one by one to analyze. Quantitative experimental research includes scientific experiments on EE, the methods of which are described below:

3.4.2.1 Observation and Formation of the Topic

After thinking about the EE topic of interest, researchers conduct research on the topic. The subject should not be selected randomly, but should be based on topics of interest. After selecting, you need to read many documents to fully understand what all the documents in this field include, in order to reduce information gaps. Therefore, the topic should be selected carefully, and the knowledge of that topic needs be connected.

- **Forming of Hypothesis**: Specify hypothetical relationships between two or more variables, test them, and make predictions;
- **Conceptual Definition**: Describe the concept and generates correlation with other concepts;
- **Operational Definition**: Define the parameter variables and how to measure and evaluate parameters in research;
- **Gathering of Data**: Include determining the size of the parent space, where the parent parameter is a statistical measurement and is unknown. The sample space is selected for parameter sampling distribution, and specific research instruments are used to gather information from these samples. Instruments used for information and data collection must be safe and reliable;
- **Analysis of Data**: Analyze the data and interpret the results to summarize the conclusions;
- **Data Interpretation**: Use tables, graphics, or photos to represent, and then describe in words; and
- **Testing and Revising of Hypothesis**: Make conclusions and repeat operations if necessary (i.e., conclusion, reiteration if necessary).

As mentioned previously, empirical research in "science education" requires the establishment of an experimental group and a control group to conduct evidence inference utilizing intervention. However, "EE" is not a knowledge-based education, but a practical education. That is to say, it is difficult for EE to conduct psychological experiments in a classroom-type laboratory through short-term plans to get the answers we want. Therefore, the "experimental results" of EE require careful examination of the "empirical results" and careful verification.

3.4.2.2   Quasi-experimental Research Method of Environmental Education

It is difficult for the social sciences to adopt an "experimental research" approach; therefore, most research is "quasi-experimental design." When researchers are unable to use random sampling methods to assign research subjects in educational situations and strictly control experimental situations, the ideal experimental design is to use "quasi-experimental design." For example, if EE researchers have compiled a new "environmental education textbook," they need to know whether this textbook is better than the traditional "EE" textbook. Researchers were unable to randomly select subjects from National Primary Schools and randomly assign them to experimental and control groups. However, when approaching the school, the researcher must use the quasi-experimental research method when the original class is used as the experimental object.

Therefore, the principle of "quasi-experimental research" design is "design of forward and backward measurement of unequal groups." The experimental and control groups are classified as follows:

- **Experimental Group**: pre-test (measurement before the experiment), test (experimental teaching, or new teaching of experimental "EE textbooks"), post-test (measurement after the experiment), delay test (in the measurements were taken three months after the experiment).
- **Control Group**: pre-test (measurement before the experiment), test (without experimental teaching, or traditional teaching of experimental "EE materials"), post-test (measurement after the experiment), delayed test (measured three months after the experiment).

Although the above quasi-experimental design cannot control all the factors that affect the internal validity of the experiment like the actual experimental design, it can control most of them, and can avoid the experimental situation of EE that is too artificial if missing. In education research, there are four most commonly used quasi-experimental designs: (1) Unequal control group design; (2) Equivalent time sample design; (3) Adversarial equilibrium design; and (4) Time series design.

3.4.3 Reliability and Validity Analysis

3.4.3.1 Reliability

The purpose of reliability analysis is mainly to analyze the consistency of test results. Reliability refers to the consistency or stability of the results obtained by the testing tools (questionnaires/scales), and an indicator that reflects the true degree of the tested features. There are four main methods for reliability analysis:

- *Retest Reliability Method*: The retest reliability method uses the same questionnaire and repeats the test at a certain interval for the same group of participants to calculate the correlation coefficient between the two test results. Because the retest reliability method needs to be tested twice for the same sample, the questionnaire survey is easily affected by events, activities, and subjects, and the interval is also limited, there are also certain difficulties in implementation.
- *Replica Reliability Method*: The replica reliability method allows the same group of participants to fill out two copies of the questionnaire at one time, and calculates the correlation coefficient between the two copies. The reliability of replicas hopes that the two replicas are completely the same in terms of content, format, difficulty, and direction of the corresponding items, in addition to the different expression methods. In actual surveys, the questionnaires meet this requirement, so this method a bit less.
- *Half-Reliability Method*: The half-reliability method is to divide the survey item into two parts, calculate the correlation coefficient of the scores of the two parts, and then estimate the reliability of the entire scale. Half-reliability is used for reliability analysis of attitude and opinion questionnaires. When performing a half-reliability analysis, if the scale contains reverse items, the scores of the reverse items should be reversed first to ensure the consistency of the scoring direction of each item, and then all items should be odd numbers. Or even, divide it into two parts that are as equal as possible, calculate the correlation coefficient between them, and finally get the reliability coefficient of the entire scale.
- *Alpha Reliability Coefficient Method*: Cronbach alpha reliability coefficient is the most used reliability coefficient at present. The alpha coefficient is the consistency between the scores of various items in the scale, which is an inherent consistency coefficient. This method is suitable for reliability analysis of EE attitudes, opinion questionnaires (or scales).

3.4.3.2 Content Validity

Content validity refers to the appropriateness of the content of the test subject to the sampling of the relevant content, that is, whether a certain measured value can represent all the partial content of an event. The higher the content validity measurement, the more able it is to measure the content of EE textbooks, and the more able it is to measure whether the teaching goals are consistent with the original plan. The content of content validity verification requires a detailed logical analysis of the content being tested, so it is also called logical validity.

3.4.3.3 Criterion-Related Validity

If we study the relationship between environmental attitudes and behaviors, the validity of the correlation criterion then is to test the relationship between the measurement score and the actual attitude and behavior. Because the validity of the criterion requires actual evidence, it is also called empirical validity.

3.4.3.4 Construct Validity

Construction validity refers to the degree to which measurement results, also known as construct validity, can be consistent with theoretical concepts. This kind of validity is mainly to measure the degree of construction of a certain theory of environmental psychology, also known as the conceptual validity of the theory.

3.4.3.5 Degree of Difficulty

The degree of difficulty of the questionnaire refers to the difficulty of the question. Generally speaking, the knowledge and ability test can explain the difficulty of the test, but for the test of EE motivation, attitude, and personality traits, the difficulty refers to the rate of whether to answer the question.

3.4.3.6 Discrimination

Discrimination refers to the test questions, mainly environmental knowledge questions, whether it can distinguish the level of participants' ability, adopt internal consistency, arrange the participants in order of the total score, and take the top 25% of the highest score as the high group. Take the last 25% of the lowest score as the low group, and then find the correct answer rate of each question in the high and low groups, expressed as PH and PL, and D = (PH − PL) as the item's discrimination index (item) discrimination index). The D value is between −1.00 and +1.00. The larger the D value, the greater the degree of discrimination; the smaller the D value, the smaller the degree of discrimination; the D value of 0, which means no discrimination.

3.4.4 Retrospective Research

Ex-post facto research uses ex-post facto research to find out possible relationships or effects. Comparing retrospective research methods with experimental research methods, these two research methods are both looking for the relationship between the self-variant term and the dependent variable term. However, the self-variables of the retrospective study must be determined in advance before collecting data to explore the relationship with the observed variables. An analysis is usually performed using statistical records, personal files, and mass media reports. Therefore, ex post facto research is also called explanatory observational studies, or causal-comparative research.

3.4.5 Relevance Research

Correlation research is defined as the relationship between two (or more) variables that collectively change values. Statistical correlation refers to the degree of relationship between two groups or populations, or co-occurrence and interaction between variables. In statistical methods, Pearson correlation technology can be used to calculate the strength and direction of the relationship between variables, and we use correlation coefficients. Positive correlation and negative causation, respectively, represent the situation that when a value increases, the value associated with it also increases or decreases.

3.4.6 Data Analysis, Interpretation and Application, Presentation of Research Results

In the research, we will formulate research hypotheses. Data analyses mainly applies statistical methods, calculation of the existing relationships between the data, and draws out statistical diagrams to explain the meaning of the data. The data analyses explain the most useful part of the data and through the presentation of the research results, the application of the research results are used to transform the value of the research results for EE promotion.

3.4.7 The Research Method of Delphi

Besides setting assumptions for quantification to converge, other objective methods can also be used to deal with the construction of environmental indicators, such as using the "Delphi expert research method, which is a structured decision support technology". During the information collection process, independent subjective judgments of experts are used to construct relatively objective opinions and suggestions, so the composition validity of the experts is important. The "Delphi Research Method" investigates in a way that experts do not meet each other until the opinions converge (Clark et al. 2020).

3.5 Qualitative Research on Environmental Education

The qualitative research of EE can be applied in the field of environmental social sciences. Qualitative research is a process of inquiry and construction of multiple realities. There are many kinds of qualitative research methods, and so far, there are still new methods, which are constantly being researched and explored. Qualitative research tools are mainly used by researchers to make long-term observations of research objects through the research area. Qualitative research requires interviews to understand the patterns of daily life of the participating researchers, analyze their social and cultural environment, and the impact of these environments on their thinking and behavior.

Therefore, the main purpose of qualitative research on EE is to understand the personal experience of the research object, the construction of its meaning, and "interpretive understanding" of the overall context in a certain environment. Researchers interpret their life stories and meaning construction through their own experience. Also, researchers need to reflect on whether they have research biases due to data limitations. In the actual research process, the researcher is a patchwork of social reality. If something happens only at a certain time and space, such patchwork will bias the data and subsequent analyses and interpretations. Therefore, the qualitative research results have a large subjective component, which is only applicable to specific situations and conditions, and cannot be inferred to the scope of the study area and the sample. That is, qualitative research focuses on understanding social and environmental events in a particular social context, rather than inferring

situations like that event. In EE research, qualitative research often uses interviews, observations, grounded theory, action research, ethnography, and content analysis.

Of course, the above methods are not independent; it means that interview methods, observation methods, and other methods are also used in grounded theoretical research. The methods of qualitative research are rich and diverse, and they affect each other and are inexhaustible.

3.5.1 Interview

Interviews in qualitative research are a process of dialogue, asking questions to interviewees, and leading out meaningful messages for research and questions raised. Interviewing is a type of research-oriented conversation. It is a research method in which the researcher collects first-hand information from the research through oral conversation. The interview method is usually performed by a trained researcher, who asks the interviewees a series of interactive answers. In phenomenological or ethnographic research, interviews are often used to reveal the meaning of the center of life from the perspective of the interviewee. Because social science research involves human thoughts and ideas, interviews have become a very common and useful research method in social science research associated with questions (please see energy literacy scales and conceptual logic maps) (Yeh et al. 2017). The following methods are commonly used in interview methods.

3.5.1.1 Non-structured Interview
No predetermined outline of the interview was proposed to remain as open and compliant as possible to the priorities of the interviewee. In the interviews, the researchers took a "let it go" approach.

3.5.1.2 Structured Interview
The purpose of this approach is to ensure that each interview presents the same questions in the same order. This allows the interview data to be

easily and reliably compiled and compared between different interviewees or between different survey dates.

3.5.1.3 Semi-structured Interview

Different from structured interviews, there is a rigorous and standardized interview outline that does not allow respondents to easily shift the focus; semi-structured interviews are open. Although there are still preliminary interviews, new questions are allowed during the interview process and ideas.

3.5.1.4 Focus Groups Interview

This is a qualitative form of research that can be divided into environmental expert interview methods and focus group clinical interview methods. Focus group interviews are groups of people who are asked about their views, opinions, or attitudes about something or something. In focus group interviews, participants are free to talk to each other or ask questions. In the process, researchers record what the participants mentioned in the conversation. Also, researchers should carefully select focus group interview members to obtain a more effective response. Focus group interviews have many advantages that individual interviews do not have, so they can play a more special role in research. These include: (1) "Interviews are themselves the object of research; (2) Collective inquiry into research issues; and (3) Collectively constructing knowledge.

3.5.2 Observation

Observation can be either quantitative or qualitative. Observation is a method of collecting data by observing people, events, or in the natural environment and recording their characteristics. Observation is a process in which human sense organs perceive things (see Fig. 3.10), and it is also a process in which the human brain actively thinks about things (see Figs. 3.11 and 3.12). In qualitative research, observation depends not only on the perception of things, but also on the perspective of observation. Observation can be a

straightforward method of qualitative research. Human beings are research tools, and first-hand exploration of the object being studied. The research question chosen by the observer, personal experience and assumptions, and the relationship with the observed things will all affect the implementation and results of the observation. Therefore, observation can be divided into:

3.5.2.1 Participatory Observation

The researcher becomes a participant in the culture or background being observed. Researchers need to be part of the context, organization, and cultural context that is being observed in order to make a successful observation. Researchers who want to use the participatory observation method must enter the field for observation at the beginning of the study, that is, with the consent of the observation subject. Researchers are the main tools for data collection and analysis. Therefore, the researcher must obtain the trust of the observed person and must maintain a friendly relationship with the observed person during the observation period. More importantly, researchers must be able to understand and reflect on the environment in which they are located in order to obtain a wealth of research data, so that the collected data can respond to research questions. Currently, visual participatory methods (VPMs) are one of the approaches to understand in EE research in regarding to conservation about diverse worldviews (Swanson and Ardoin 2021).

3.5.2.2 Direct Observation

Researchers must try to be unobtrusive so as not to bias the results of observations (see some observations approach from children's studies, i.e., young children's affective and cognitive growth (Ardoin and Bowers 2020); see Figs. 3.13 and 3.14). Making good use of technology is a good way, such as directly recording (Fig. 3.16), but with the consent of the interviewer (see Figs. 3.13, 3.14 and 3.15).

3.5.2.3 Indirect Observation

Observe the interactions between individuals, such as the results of processes, behaviors, carbon footprint, or soil compaction (Fig. 3.16). For

Fig. 3.10 Observation is a process in which human sense organs perceive things (Photo by Max Horng)

example, observe the food waste left by students in the school cafeteria to determine whether they are eating a moderate amount of food.

3.5.3   Grounded Theory

Grounded theory is a systematic methodology in the social sciences that builds theory through methodical data collection and analysis (Martin and Turner 1986). Grounded theory can be described as a research method, or it can be interpreted as a type of qualitative research (Strauss 1987). Before the research began, researchers did not have theories or hypotheses, but directly summarized the concepts and propositions from the original data, and then rose to the point of theory.

Therefore, the grounded theory is the opposite of the hypothetic-deductive method. The grounded theory is studied inductively. At the beginning of research using grounded theory, there may be a problem of awareness in the mind of the researcher, or only the preliminary qualitative data collected. As researchers review the collected data, after rethinking the idea, the concept or element will gradually become clear, and the code will be used to classify these concepts or elements. However, these codes are extracted from qualitative data. As more data are collected and re-examined, coding can organize concepts first and then categorize them. Therefore, the grounded theory is very different from the traditional research model. The traditional research model selects the existing theoretical framework, and then only collects data to show whether the

Fig. 3.11 Observation is a process in which the human brain actively thinks about things, like a drawing activities (Drawing at constructed wetlands of grades 3–6 students from direct practice of experiential learning (Gonguan Campus, National Taiwan Normal University) (Photo by Yi-Te Chiang)

Fig. 3.12 Observation is a process in which the human brain actively thinks about things, like a drawing activity (Rubbing practice of experiential learning (Gonguan Campus, National Taiwan Normal University) (Photo by Yi-Te Chiang)

Fig. 3.13 An observation for grouped children for data collection is a good way on campus (Gonguan Campus, National Taiwan Normal University) (Photo by Yi-Te Chiang)

theory is applicable to the phenomenon under study (Allan 2003).

Grounded theory is to prevent the stagnation of a theory to generate a new theory. In order to show the observation of the research fields based on the root of theoretical innovation, it lays a sound scientific foundation for theoretical development. Therefore, this method of grounded theory can generate new theories and get hypotheses and concepts, categories, and propositions from the data. Concepts are the basic unit of data analysis in grounded theory; categories are a higher level than concepts and abstract than concepts and are the basis of development theory; propositions are categories and concepts, or categories between concepts and concepts, It can be said that it comes from basic hypotheses, except that propositions focus on the relationship between concepts, and hypotheses focus on the relationship between measurement data. Grounded theory consists of five phases:

- **Research Design Stage**: include literature discussions, that is, selecting research samples;
- **Data Collection Stage**: develop methods for data collection and enter the field;
- **Data Compilation Stage**: arrange according to the sequence of events in the time and age;
- **Data Analysis Stage**: use open coding to convert data into concepts, categories, and propositions, and to write data memos; and
- **Data Comparison Stage**: compare the initially established theory with existing literature to find the same or different places, as the basis for revising the initial theory.

Constant change is a permanent feature of real social life. We need to explore the specific direction of change and the process of social interaction. Therefore, the grounded theory places special emphasis on generating theories from actions and constructing theories from the

Fig. 3.14 Making good use of observation aimed at parents and children for data collection is a good way learning in museum (Photo by Wei-Ta Fang)

perspective of actors. The theory of grounded theory must come from the data and have a close relationship with the data. Grounded theory plays a very important role in the development of social science research theories. Theories at all levels are indispensable for a deep understanding of social phenomena (Glaser 1978).

3.5.4 Action Research

Action research can be research that solves the problem at hand or it can be a team member or cooperate with others to lead the community of practice and reflect on the problem-solving process as a way to improve, solve, or deal with problems (Stringer 2013). Action research is based on the theoretical basis of the practical community and jointly conducts research and participation, that is, "researchers are participants

themselves and researchers." The focus of action research is to explore the process of group problem solving and how to solve it, and to reflect on the process of problem solving. Action research is a spiral process of collecting data to establish goals and actions, and to intervene in problems to evaluate goals and understand the results. The purpose of action research strategies is to solve specific problems and develop guidelines for effective practice (Denscombe 2010).

Action research usually involves conducting active research and changing the situation through existing organizations. Action research can be conducted in large organizations or institutions, assisted, or directed by professional researchers, with the aim of improving strategies, practices, and knowledge in their environment. Research designers, stakeholders, and researchers collaborate with each other to propose new action plans to help their communities improve

Fig. 3.15 Making good use of technology is a good way, such as directly recording by a smartphone (Photo by Wei-Ta Fang)

their work or practice content. Action research is an interactive investigative process that balances the resolution actions performed in a collaborative environment with data-driven collaborative analysis or research to understand the root causes that can lead to changes in individuals and organizations. For example, "action research" can be one teacher leading a class (the focus of action research is on students), or several teacher leaders leading an academic year, or the action research of several classmates (the teachers and students are also action research focus), or teachers can form an action research team (action research focuses on teachers). In terms of analysis, action research challenges traditional social science by creating reflective knowledge by transcending external sampling variables. In the process, you can actively conceptualize the theory according to step by step and collect data to understand the instantaneous changes that occur

in the structure (Figs. 3.17 and 3.18). Therefore, action research is a process of continuously discovering problems, solving problems, and then discovering new problems, and continuously generating loops. The impact of "Environmental Education Action Research and Teaching" on students' EE awareness is almost like traditional teaching methods; however, the environmental action curriculum has impacts on students' environmental attitudes and behaviors through planning, action, review, reflection, and re-action, the effects are significant. Therefore, the increase of knowledge is a continuum of action after action, which needs to be taken as the starting point from this perspective. Therefore, we question the knowledge of the social sciences as to how to develop truly wise action; not just to develop reflection on action.

It is not enough for researchers to just communicate knowledge. In action research, the

Fig. 3.16 You may use direct/indirect approaches to observe the interactions between individuals to groups to detect soil hardness from soil compaction of the changes caused by human trampling in the soils and ground for a field (Scenic view in the surrounding areas of Lake Tahoe, CA, USA) (Photo by Wei-Ta Fang)

findings are implicit. We need to learn how to use action research findings to promote scientific consensus in different practice and conceptual contexts. Participatory action research is therefore a form of question-based investigation of the practice by practitioners, and therefore, it is an empirical process. The ultimate goal of action research is to create and share social science knowledge.

3.5.5 Ethnography

The generation of ethnographic knowledge basically depends on the comparison of two cultural experiences. Ethnographic methodology emphasizes that researchers must be "deliberate ignorance." Process researchers in the field not only obtain research information through "question," but also live in the present and use their own senses, including vision, taste, hearing, and touch. Wait for multiple senses as a channel for research data collection. At the same time, the researcher must always be introspective in the process of research, and must be fully aware of the influence of his cultural background and researcher's identity on the research. In addition, ethnography is used as the research method. The acquisition of data is produced by the interaction between the researcher and the researched person (see Fig. 3.19). The researcher must be able to discover the social meaning and cultural value implied by the event or action.

In short, the ethnographic research method represents the researcher's entry into the researcher's daily life world, tries to understand the researcher's world, reverses the passive role of the researcher, and allows the researcher's "local perspective" to be heard or seen. From the past to the present, ethnographic research has been continuously enriched. In the past, ethnography emphasized observing the interactions between people in the community. For example, in the case of ethnography, the analysis of

Fig. 3.17 The environmental action curriculum has impacts on students' environmental attitudes and behaviors through planning, action, review, reflection, and reaction (Prof. LePage and graduate students are looking for all-day events and a committed group focusing on urban wetland projects, thinking about checking out group study results at Daan Park, Taipei, Taiwan, 2021) (Photo by Wei-Ta Fang)

important life experiences of environmental protests can be used as an example. For example, in Taiwan, the ethnographic analysis of the Binnan Industrial Zone, the Guanxi Industrial Zone, and the Green Oyster incident in the estuary of Keya Creek, Hsinchu City in the 1990s; the RCA incident, and the Mai Liao Industrial Zone in the 2000s, the Guoguang Petrochemical Park, and the demolition of the Dapu Industrial Zone in the 2010s on environmental events are all good subjects, which can deeply describe the field environment and the details of interaction between people. In addition, recent anthropological research has added non-human beings to the writing of fields, and developed multi-ethnic ethnography, emphasizing that the composition of society is not only human, but also the participation of many non-humans, such as cats and dogs, Insects, bacteria, machines, etc. (such as: *Insectopedia, The Mushroom at the End of the World*), are the works of this multi-species ethnography (Raffles 2010; Tsing 2015).

3.5.6 Content Analysis

Content analysis is a method of studying documents, files, or correspondence. Research materials may include various formats, such as pictures, audio files, text files, text, or images. One of the great benefits of content analysis is that it is a non-intrusive way to naturally study social phenomena that depend on a particular time and place in a file. The implementation and concepts of content analysis will vary from

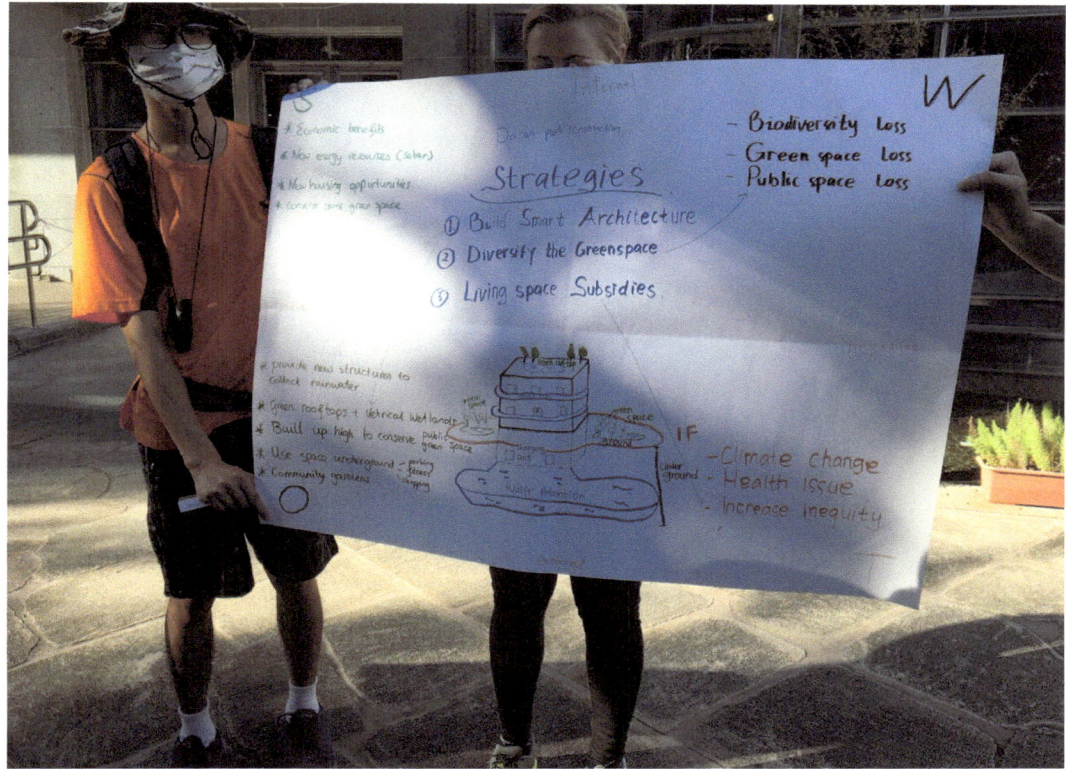

Fig. 3.18 Class exercises and presentation at Daan Park, Taipei, Taiwan, 2021 (Photo by Ben LePage)

discipline to discipline, but they all involve systematically reading or observing text content, and encoding on meaningful or interesting documents and archives. By systematically encoding a series of text content, researchers can use quantitative methods to analyze trends in big data content, or use qualitative methods to analyze text content.

3.6 Promotion of Environmental Education Theory

The EE department strengthens the development of human society by using practical technical knowledge, strengthening education, learning the environment, and continuously participating in and understanding activities to solve environmental problems. In this section we discuss the expansion and practice of EE theory, Darwinian

scientific integration, and comparison of learning methods.

3.6.1 Theoretical Expansion and Practice

The theoretical improvement of EE is not so much a transfer of technology as a linear 'top-down' approach, but rather a participatory 'bottom-up' approach (Black 2000). In the process of education, the above-mentioned formal education and non-formal education are conducted through bilateral one-to-one advice or information exchange, and in accordance with formal education in EE, using organized education and training methods—formal and non-formal education activities (Fig. 3.20).

Therefore, we conclude that a single model of EE is not feasible. Although we have criticized

the above-mentioned linear technology transfer model, we still need to rely on reliable scientific information and actively participate in the research and development process, from EE scholars and experts to front-line field teachers, through bilateral information exchange. On the student side, knowledge, attitudes, and behavioral models are enhanced through formal education and program training. In addition, new learning technologies will promote certain forms of education methods, training courses, and information exchange, and make up for the lack of application through promotion strategies.

3.6.2   Integration of Environmental Disciplines

We know that the discipline of EE crosses the traditional discipline boundaries, especially between the natural sciences and the social sciences. From the environmental sciences, the natural sciences, and the humanities, there is a tangled relationship, but we still need to be patient to go further integration. From the analysis, we can use Darwin's ecology to carry out the scientific integration of EE and the disciplines

Fig. 3.19 The acquisition of data is produced by the interaction between the researcher and the researched person, if you want to watch a child, be keeping a close to watch, and you may pay careful attention to a situation or a thing, so that you can deal with any physical/mind changes for a child in his/her living environment (Photo by Wei-Ta Fang)

Fig. 3.20 Environmental education is not so much a transfer of technology as a linear 'top-down' approach, but rather a participatory 'bottom-up' approaches (Illustrated by Wei-Ta Fang)

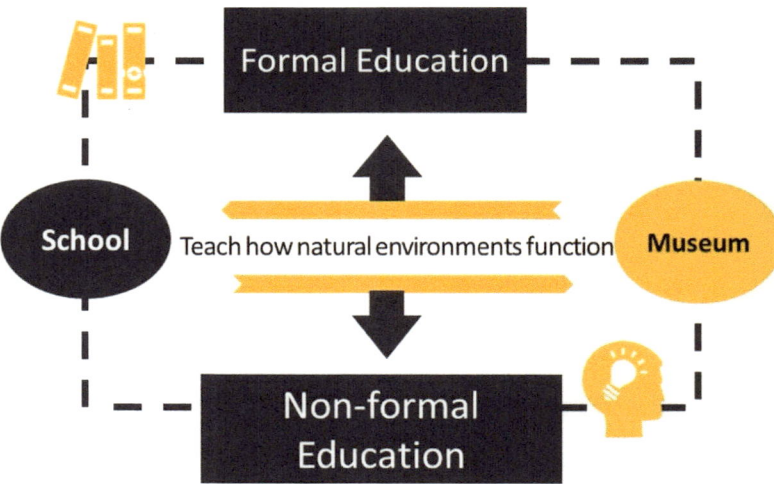

of ecology. Environmental thinkers generally think that the natural sciences and humanities are completely disconnected and there is little overlap between environmental sciences, natural sciences, and humanities, and even only limited to the methodologies of environmental and biomedical sciences, small overlap. However, today's subject areas are under the multiple social relationships that have come one after another. We observe emerging interdisciplinary areas, such as conservation biology, ecological economics, human behavioral ecology, and evolutionary psychology. We know that the clear barriers between natural sciences and social sciences are already being sold. Interdisciplinary scholars are helping to integrate the relationship between applied fields in the biological and human sciences. We need a new science, called human behavioral ecology or Darwin ecology, to complete this comprehensive doctrine.

3.6.3 Integration of Environmental Education Disciplines

The learning model of EE is mainly to borrow pedagogical methods to recognize and sense the environment and improve human behavior patterns. However, the cultivation of science education is based on brain science, life sciences, and cosmic sciences. It explains the nature and value of human learning science, and explores the ultimate thinking of philosophy, "What is human?" "Why am I here?" "What is the ultimate goal of the universe?".

Three major questions; however, EE involves the development of human behavior, which connects a very down to earth attitude toward learning, beyond the philosophical exploration of nihilism, to the practical thinking of the world.

What is the relationship between "EE" and the improvement of human cognition, the cultivation of mentality, and the formation of attitudes? We use "environmental learning" to improve human values and a sense of responsibility for the environment. Can environmental learning really achieve results? Since the 1950s, educators have considered the above issues through learning

theories; these issues need to be explored through educational psychology. We understand that there are three major learning theories in American academia that have a wide range of influences around the world. The views on the above points include the Bloom learning method, the Hungerford learning method, and the ABC Learning method of emotion theory (Fig. 3.21).

3.6.3.1 The Bloom-Style Learning Method

Bloom divides education goals into the Cognitive, Affective, and Psychomotor Domains (Bloom et al. 1956; Krathwohl et al. 1964).

- *Cognitive Domain*: Cognitive knowledge is aimed at knowledge, principles, applications, and problem-solving learning. The characteristics of cognition are the acquisition and application of knowledge;
- *Affective Domain*: Affection mainly refers to the positive or negative palpitations of external stimuli, such as emotional reactions such as hobbies and dislikes, which in turn affect the intentions adopted on maggots; and
- *Psychomotor Domain*: Action skills are a kind of energy generated by learning. The generated on this basis is the result of performance, and it is the precise expression of body movements. Therefore, after the teaching goals allow learners to learn through knowledge or skills, they should have the response they deserve.

3.6.3.2 The Hungerford-Style Learning Method

Hungerford divides EE goals into knowledge, attitude, and behavior areas (Hungerford and Volk 1990). He attaches great importance to EE curriculum planning and believes that "knowledge" affects "attitude" and "attitude" affects "behavior" theory. In other words, he believes that EE can ultimately affect human environmentally friendly behaviors and improve human-environmental literacy. Therefore, when human beings have knowledge, attitude, and skills, they can participate in solving environmental problems.

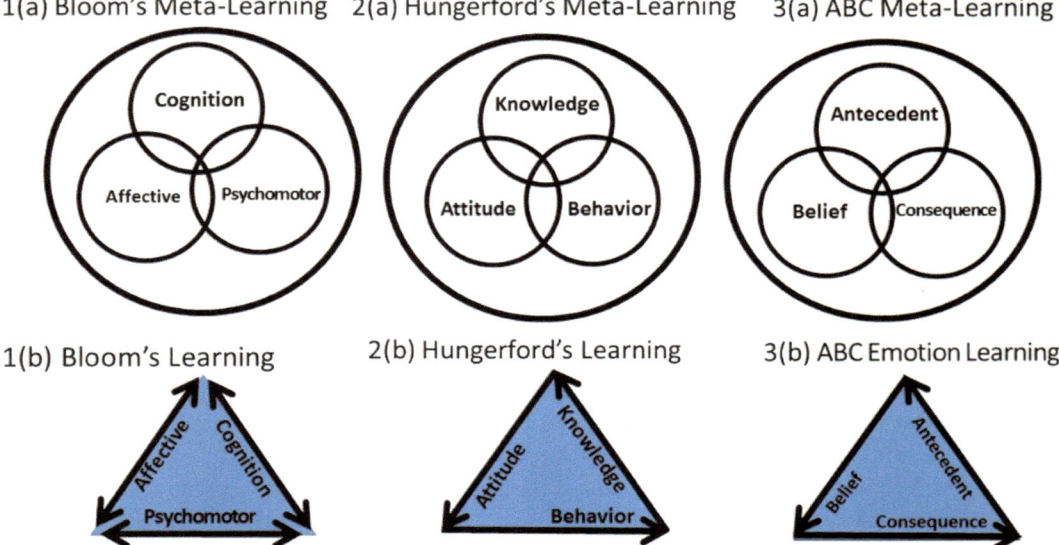

Fig. 3.21 Comparison of the use of the Bloom-style learning method, the Hungerford-style learning method, and the learning of ABC emotion theory. 1(a): Bloom-style meta-learning; 1(b): Bloom-style learning (adapted from Bloom et al. 1956; Krathwohl et al. 1964). 2(a): Hungerford-style meta-learning; 2(b): Hungerford-style learning (adapted from Hungerford and Volk 1990). 3(a): ABC meta-learning; 3(b): ABC emotional learning (adapted from Ellis 1957, 1962) (Illustrated and redrawn by Wei-Ta Fang)

- *Knowledge*: Knowledge can help us to establish the relationship between the object and the environment for the relevant information of the object we want to understand. This relationship needs to understand things through a cognitive schema;
- *Attitude*: Attitude is a person's psychological and neurological readiness, which refers to the judgment status of an individual on an object. This is an opinion organized through experience. When an individual's attitude affects, behavioral will intent through thoughtful decision-making processes and then through psychological responses; and
- *Behavior*: Human behavior refers to the spontaneous or passive behavior of human beings in the adaptation to a constantly changing and complex environment, or the interaction between the environment and other organisms or inorganic bodies physical response.

3.6.4 Learning Method of ABC Emotion Theory

The ABC Theory of Emotion was created by Ellis. A equates to an activating event; B is beliefs; and C triggers emotional and behavioral consequences (Ellis 1957, 1962).

- **A = Activating Event (Induced Event)**: the indirect cause of C is the inducing event A (activating event), and the direct cause of C is the individual's belief and evaluation of event A;
- **B = Belief**: Human emotions and behaviors (C) are not directly determined by life events (A), but by the cognitive processing and evaluation methods of these events. In other words, because the individual through this event incorrectly recognizes it, the error (B) is directly caused; and

- **C = Triggers Emotional and Behavioral Consequences**: Human negative emotional and behavioral disorder results (C) are not directly caused by an evoked event (A).

Activating events, beliefs, emotions, and behavioral results (consequences) are accompanied by people's thoughts, and emotional or psychological distress is caused by irrational and illogical thoughts. The basic idea behind the ABC emotion theory model is that activating events (A) will not cause the consequences of emotions (C); but beliefs (B), especially false beliefs, are also called the consequences of bad emotions (C) caused by irrational beliefs. Initially, Ellis considered his theory to be incompatible with religion, or at least incompatible with absolute religion, although he had accepted that certain types of religion were compatible with his theory (Ellis 2000). Specifically, according to Ellis, belief in loving God can lead to positive mental health outcomes; while belief in angry God can lead to negative mental health outcomes. The evolution of thought is especially true concerning religion.

If we take the tendency of human beings to have biology and sociology, human beings are then caught between rational reasoning of limited reason and irrationality. Under the emotions of fear and panic, human beings will produce unreasonable thought patterns. That is to say, in addition to understanding the impact of the ABC emotional theory model on the Hungerford model's "knowledge" and its impact on "attitude," the "attitude" affects the impact of the "behavior" theory. Looking at the behavioral model of meta, it seems that the "absolute cognition" of all things in the ternary model should be more transcendent. It is suggested that environmental educators should look at things objectively and have a more general and mature view of human environmental behavior (Ellis 2000; Hug 1977).

Therefore, we evaluate human environmental behavior, integrate the process of cognitive functions, learn to represent the intermediary role of the human intelligent system, perceive the "metacognition," understand the formation process of environmental protection significance and guide it. We have adapted three models of "meta-learning modes."

From the "metacognitive" learning model, we understand the individual's own cognitive process and can perform self-mastery, monitoring, evaluation, domination, etc. to comply with self-willed management, and at the same time, self-adjust to the goal to achieve control of unreasonable thoughts. In other words, EE has grown through mental growth and produced a mature mind. Through emotional growth, it has refined mature personal traits and a model of "responsible environmental behavior" and made "friendly earth" contributions and sublimation of ideas. Therefore, this chapter hopes that the study of EE can make theoretical corrections and adjustments to the EE in the process of forming the research significance, to achieve the real purpose of solving human-environmental problems.

3.7   Summary

From the Nevada Declaration, we learned that EE is to recognize values, clarify ideas, develop skills and attitudes, understand, appreciate, and thank individuals for their interaction with culture and the environment. And we learned how to enter the field to practice, in order to be aware of how good environmental attitudes, skills, care, decision making, and codes of conduct are generated. Therefore, EE research is a discussion of research methods that focus on attitudes, skills, care, decision-making, and behavior standards. We can discuss from three levels, including methodologies, research methods, or research methods, and discuss specific environmental improvement technologies and educational techniques. Therefore, the ways in which EE research questions are formed include understanding how to think and learn about the environment's awareness, and the ability of metacognition to cultivate environmental sensitivity so as to realize higher-level thinking ability. Therefore, for the development of research, we need to successfully judge whether our cognitive process has increased in order to judge whether the ability to change behavior is strengthened. Furthermore, the relationship between the researcher and the subject matter is very important. We care about the good and bad of the research results and have a good grasp of the nature of the research. In addition, through the interaction between the researcher and the research subject, we conduct in-depth and meticulous experiences, and then explain and clarify the literature, resolve the disputes in the literature, and stimulate thinking and initiate change, a new challenge that EE researchers and practitioners alike need to recognize.

References

Allan G (2003) A critique of using grounded theory as a research method. Electron J Bus Res Methods 2(1):1–10

Ansori AZ, Ibrahim M, Widodo W, Sutoyo S (2020) Development of truth-seeking learning model: validity and reliability. Int J Innov Sci Res Technol 5(1):385–389

Ardoin NM, Bowers AW (2020) Early childhood environmental education: a systematic review of the research literature. Educ Res Rev 31:100353. https://doi.org/10.1016/j.edurev.2020.100353

Ardoin NM, Clark C, Kelsey E (2013) An exploration of future trends in environmental education research. Environ Educ Res 19(4):499–520

Ardoin NM, Bowers AW, Gaillard E (2020) Environmental education outcomes for conservation: a systematic review. Biol Cons 241:108224. https://doi.org/10.1016/j.biocon.2019.108224

Bakhtin MM (1981) In: Holquist M (ed) The dialogic imagination: four essays. University of Texas Press

Bakhtin MM (1994) In: Morris P (ed) The Bakhtin reader. Oxford University Press

Bester L, Muller G, Munge B, Morse M, Meyers N (2017) Those who teach learn: near-peer teaching as outdoor environmental education curriculum and pedagogy. J Outdoor Environ Educ 20(1):35–46

Black AW (2000) Extension theory and practice: a review. Aust J Exp Agric 40(4):493–502

Bloom BS, Engelhart MD, Furst EJ, Hill WH, Krathwohl DR (1956) Taxonomy of educational objectives: the classification of educational goals. In: Handbook I: cognitive domain. David McKay Company, Philadelphia

Blumstein DT, Saylan C (2007) The failure of environmental education (and how we can fix it). PLoS Biol 5(5):e120. https://doi.org/10.1371/journal.pbio.0050120

Boyden SV (1970) Environmental change: perspectives and responsibilities. In: Evans J, Boyden S (eds) Education and the environmental crisis. Australian Academy of Science, Canberra, pp 9–22

Bratman M (1981) Intention and means-end reasoning. Philos Rev 90(2):252–265

Carson R (1962) Silent spring. Fawcett, Greenwich

Chang C-H, Ow P (2022) Inquiry-based fieldwork assessment for and as learning in geography. In: Bourke T, Mills R, Lane R (eds) Assessment in geographical education: an international perspective. Springer, Cham, pp 123–134

Chao S-H, Jiang J, Hsu C-H, Chiang Y-T, Ng E, Fang W-T (2020) Technology-enhanced learning for graduate students: exploring the correlation of media richness and creativity of computer-mediated communication and face-to-face communication. Appl Sci 10:1602. https://doi.org/10.3390/app10051602

Clark CR, Heimlich J, Ardoin NM, Braus J (2020) Using a Delphi study to clarify the landscape and core outcomes in environmental education. Environ Educ Res 26(3):381–399

Darwin CR (1859) On the origin of species by means of natural selection, or the preservation of favoured races in the struggle for life. John Murray, London

Denscombe M (2010) The good research guide: for small social research projects. McGraw-Hill House, New York

Edelson SC, Pea RD, Gomez L (1996) Constructivism in the collaboratory. In: Wilson BG (ed) Constructivist learning environments: case studies in instructional design. Educational Technology Publications, Englewood

Eden S (2010) NGOs, the science-lay dichotomy, and hybrid spaces of environmental knowledge. In: Geographies of science. Springer, Dordrecht, pp 217–230

Ellis A (1957) Rational psychotherapy and individual psychology. J Individ Psychol 13:38–44

Ellis A (1962) Reason and emotion in psychotherapy. Lyle Stuart, New York

Ellis A (2000) Can rational emotive behavior therapy (REBT) be effectively used with people who have devout beliefs in God and religion? Prof Psychol Res Pract 31(1):29–33

Environmental Protection Administration (2014) National environmental education action plan (2016–2019). EPA, Taipei, Taiwan (in Chinese). https://eeis.epa.gov.tw/front/resources/ResSearch/item.aspx?id=208. Accessed 6 Apr 2022

Estabrooks CA (2001) Research utilization and qualitative research. In: Morse JM, Swanson JM, Kuzel AJ (eds) The nature of qualitative evidence. Sage, Thousand Oaks

Falk JH (2017) Born to choose: evolution, self and well-being. Routledge

Falk JH, Heimlich JE, Foutz S (eds) (2009) Free-choice learning and the environment. AltaMira, Lanham

Freeman III AM (1986) On assessing the state of the arts of the contingent valuation method of valuing environmental changes. In: Cummings RG, Brookshire DS, Schulze WD (eds) Valuing environmental goods: an assessment of the contingent valuation method, pp 180–195

Gimello RM (2005) Huayan. In: Jones L (ed) Encyclopedia of religion, 2nd edn, vol 6. Macmillan, Detroit, pp 4145–4149

Glaser BG (1978) Theoretical sensitivity. Sociology, Mill Valley

Gough A (2012) The emergence of environmental education research: a 'history' of the field. In: Stevenson RB, Brody M, Dillon J, Wals AEJ (eds) International handbook of research on environmental education. Routledge, Oxfordshire

Gough N, Price L (2004) Rewording the World: poststructuralism, deconstruction and the 'real' in environmental education. South Afr J Environ Educ 21(2):23–36

Guez JM (2010) Heteroglossia. In: Western humanities review. University of Utah, Salt Lake City, pp 51–55

Habermas J (1971) Knowledge and human interests. Beacon, Boston

Habib A (2013) Sharing the Earth: sustainability and the currency of inter-generational environmental justice. Environ Values 22(6):751–764

Hart P (2002) Narrative, knowing, and emerging methodologies in environmental education research: issues of quality. Can J Environ Educ (CJEE) 7(2):140–165

Heimlich JE, Ardoin NM (2008) Understanding behavior to understand behavior change: a literature review. Environ Educ Res 14(3):215–237

Herrington J, Standen P (1999) Moving from an instructivist to a constructivist multimedia learning environment. In: Collis B, Oliver R (eds) Proceedings of ED-MEDIA 1999—World conference on educational multimedia, hypermedia & telecommunications. Association for the Advancement of Computing in Education (AACE), Seattle, pp 132–137

Huang YS, Asghar A, Nichols NE (2021) Implementing a national policy initiative to support education for sustainable development: lessons from Taiwan's Environmental Education Act. Educ Res Policy Pract 20:187–205

Hudson SJ (2001) Challenges for environmental education: issues and ideas for the 21st century. Bioscience 51(4):283–288

Hug J (1977) Two hats. In: Hungerford HR, Bluhm WJ, Volk TL, Ramsey JM (eds) Essential readings in environmental education. Stipes, Champaign, p 47

Humbetov S (2021) Data-intensive computing with map-reduce and Hadoop. In: 2012 6th International conference on application of information and communication technologies (AICT), pp 1–5. https://doi.org/10.1109/ICAICT.2012.6398489

Hungerford HR, Volk TL (1990) Changing learner behavior through environmental education. J Environ Educ 21:8–22

Kemmis S, McTaggart R (1982) The action research planner. Deakin University Press, Melbourne

Kidman G, Chang C-H (2022) Numbers and graphs—what sort of mathematical literacy do we need for geographical education in uncertain times. Int Res Geogr Environ Educ 31(1):1–4

Klein ES, Merritt E (1994) Environmental education as a model for constructivist teaching. J Environ Educ 25(3):14–21

Knapp D (1995) Twenty years after Tbilisi: UNESCO inter-regional workshop on re-orienting environmental education for sustainable development. Environ Commun 25(6):9

Krathwohl DR, Bloom BS, Masia BB (1964) Taxonomy of educational objectives: the classification of educational goals. In: Handbook II: affective domain. Allyn and Bacon, Boston

Le Roux C, Ferreira JG (2005) Enhancing environmental education teaching skills through in-service education and training. J Educ Teach 31(1):3–14

Liu S-Y, Yeh S-C, Liang S-W, Fang W-T, Tsai H-M (2015) A national investigation of teachers' environmental literacy as a reference for promoting environmental education in Taiwan. J Environ Educ 46(2):114–132

Lubchenco J (1998) Entering the century of the environment: a new social contract for science. Science 279(5350):491–497

Lucas AM (1980) The role of science education in education for the environment. J Environ Educ 12(2):33–37

Mach KJ, Lemos MC, Meadow AM, Wyborn C, Klenk N, Arnott JC et al (2020) Actionable knowledge and the art of engagement. Curr Opin Environ Sustain 42:30–37

Magraw D, Lynch O (2006) One species, one planet: environmental justice and sustainable development. In: The World Bank legal review, vol 2: Law, equity and development, pp 441–482

Marín-Morales J, Higuera-Trujillo JL, Greco A et al (2018) Affective computing in virtual reality: emotion recognition from brain and heartbeat dynamics using wearable sensors. Sci Rep 8:13657. https://doi.org/10.1038/s41598-018-32063-4

Martin PY, Turner BA (1986) Grounded theory and organizational research. J Appl Behav Sci 22(2):141–157

Palmer J (1998) Environmental education in the 21st century: theory, practice, progress and promise. Routledge, Oxfordshire

Raffles H (2010) Insectopedia. Vintage, New York

Reid A, Scott W (2013) Identifying needs in environmental education research. International handbook of research on environmental education. Routledge, Oxfordshire, pp 518–528

Roque NA, Ram N (2019) tsfeaturex: an R package for automating time series feature extraction. J Open Source Softw 4(37):1279. https://doi.org/10.21105/joss.01279

Stagg BC, Dillon J, Maddison J (2022) Expanding the field: using digital to diversify learning in outdoor science. Discip Interdiscip Sci Educ Res 4(1):1–17

Stapp W (1969) The concept of environmental education. Environ Educ 1(1):30–31

Stevenson RB (2007) Schooling and environmental education: contradictions in purpose and practice. Environ Educ Res 13(2):139–153

Stewart A (2020) Developing place-responsive pedagogy in outdoor environmental education: a rhizomatic curriculum autobiography. Spring Nature Switzerland, Cham

Strauss AL (1987) Qualitative analysis for social scientists. Cambridge University Press, Cambridge

Stringer ET (2013) Action research. Sage, Thousand Oaks

Swanson SS, Ardoin NM (2021) Communities behind the lens: a review and critical analysis of visual participatory methods in biodiversity conservation. Biol Cons 262:109293. https://doi.org/10.1016/j.biocon.2021.109293

Tilbury D (1995) Environmental education for sustainability: defining the new focus of environmental education in the 1990s. Environ Educ Res 1(2):195–212

Tsing AL (2015) The mushroom at the end of the world: on the possibility of life in capitalist ruins. Princeton University Press, Princeton

UNESCO (1970) International working meeting on environmental education in the school curriculum. Final report, at Foresta Institute, Carson City, Nevada. IUCN and UNESCO

Van Weelie D, Wals A (2002) Making biodiversity meaningful through environmental education. Int J Sci Educ 24(11):1143–1156

Wals AE, Brody M, Dillon J, Stevenson RB (2014) Convergence between science and environmental education. Science 344(6184):583–584

Wang YL, Derakhshan A, Zhang LJ (2021) Researching and practising positive psychology in second/foreign language learning and teaching: the past, current status and future directions. Front Psychol. https://doi.org/10.3389/fpsyg.2021.731721

Yeh S-C, Huang J-Y, Yu H-C (2017) Analysis of energy literacy and misconceptions of junior high students in Taiwan. Sustainability 9(3):423. https://doi.org/10.3390/su9030423

Yeh S-C, Wu A-W, Yu H-C, Wu HC, Kuo Y-P, Chen P-X (2021) Public perception of artificial intelligence and its connections to the sustainable development goals. Sustainability 13(16):9165. https://doi.org/10.3390/su13169165

Zehr SC (1999) Scientists' representations of uncertainty. In: Friedman SM, Dunwoody S, Rogers CL (eds) Communicating uncertainty: media coverage of new and controversial science, pp 3–21

Part II
Contextualization: The Ultimate End of Human Ecology

Environmental Literacy

4

Environmental education is aimed at producing a citizenry that is knowledgea-ble on the biological, physical, economic, and social issues that are created and/or associated with environmental problems and how the people can be moti-vated to minimize or mitigate for these issues by implementing environmentally friendly and sustainable solutions.

William B. Stapp, *The Concept of Environmental Education*, 1969

Abstract

Environmental literacy is an abstract concept and a subjective imagination. We see that this chapter discusses environmental education learning motivations, awareness and sensitivity, values and attitudes, mobilization skills, mobilization experience, environmental behavior, and aesthetic literacy in the cultivation of literacy. The above connotations of environmental literacy all need to construct the inherent goodness of human beings. We particularly hope that environmental literacy can be externalized to achieve changes in human-friendly environmental behavior. In other words, if the environmental literacy of the entire population can be strengthened, we can work together to form environmental cohesion, cultivate modern social citizens, generate environmental collective consciousness and awareness, and then based on the eternal belief in natural decision-making and environmental protection. This could promote a comfortable space and a clean home for sustainable development. Therefore, from the process that human beings can perceive and understand the environment, we have experienced the awareness of environmental changes. We need to improve environmental literacy to form the transformation of the collective human consciousness structure, so as to be aware of the external environment, that is the learning process. If, literacy is the overall effect of a learning process, then our final collective environmental consciousness will change from thought to proper behavior. These changes will affect the stage tasks of sustainable development. Then, based on empathy and awareness of all things, we should realize the sense of responsibility and eternal value as human beings, protect nature, and accept the challenges of future environmental changes.

W.-T. Fang et al., *The Living Environmental Education*, Sustainable Development Goals Series,
https://doi.org/10.1007/978-981-19-4234-1_4

4.1 Introduction

In Chap. 1, we iterated that the ultimate goal of Environmental Education (EE) is to protect the environment, improve the quality of human life, and use our natural resources sustainably as we strive towards the goal of sustainable development. Therefore, as part of the education process, awareness of our (human) activities and the use of our natural resources are essential for identifying gaps in our processes so that solutions developed can be developed. To change human behavior and move towards a more sustainable society, people need to know where these resources are from, how they develop, and how they are obtained. Understanding these elements helps the public develop solutions for environmental problems and fine-tunes the process of environmental education so that the people are presented with the correct information in order to move towards a sustainable society.

The International Union for Conservation of Nature (IUCN) uses the environmental education expert, Jan Cerovsky (1930 ~ 2017), in the definition of environmental education used in their *Handbook of Environmental Education with International Case Studies* (1976). Cerovsky's definition is as follows:

> Environmental Education is the process of recognizing values and clarifying concepts, in order to develop the skills and attitudes necessary to understand and appreciate the interrelationship between humans, cultures and their biological and physical environments.

Therefore, environmental education is an introspection effort and a decision of free will. We should want to live in a healthy environment because this is what we want, not because this is what we are being told. For this reason, we explore environmental literacy in much greater depth and define environmental literacy as the process that shapes or develops a person's environmental values and ability to solve environmental problems. This, however, implies that the process of environmental education is globally unified and not geopolitically motivated. We can't have people from one region/country incorporating sustainability practices into their day-to-daily lives while its neighbors are following different practices.

To begin, we should all know agree on the concept of environmental literacy and what it means. The word "Literacy" was coined in the early fifteenth century and refers to the person who can read and write. The meaning is broad and includes personal education and common skills, but does not involve moral or value judgments. However, environmental literacy is associated with value judgments and environmental ethics. The term "Environmental Literacy" was coined by Charles E. "Chuck" Roth (1934 ~ 2016). Charles E. Roth cited in this paper, *Environmental Literacy: Its roots, evolution, and directions in the 1990s*, for all processes he detected (Roth 1968, 1984, 1991, 1992), confirmed this opinion. As a teacher and a senior research/development associate at the Education Development Center, Newton, Massachusetts, in New England during 1992. Charles E. Roth has been served as the Director of Education for the Massachusetts Audubon Society from 1968 to 1988, also actively working in Drumlin Farm (Scopinich 2016).

Later, the 1990 *National Environmental Education Act* (Public Law 101–619) in USA has brought environmental education back to the attention of many educators and environmentalists (Marcinkowski 1990–91). Meanwhile, Disinger confirmed Roth's idea, published in one of their papers, *Environmental Education Research News*, represented their conveying information about environmental literacy (EL) (Disinger and Roth 1992).

In Roth's previous discussion (Roth 1968, 1984, 1991, 1992), he provided a detailed explanation of the concept, but the definition is quickly clouded by consideration of the cognitive process, societal norms and values, policy, culture, psychology, geographic, and economic elements that comprise a person's daily life (Roth 1968). Nonetheless he ultimately defines environmental literacy as "*A person's environmental sensitivity, knowledge, skills, attitudes, values, personal investment, responsibility, and active involvement*". He refines the term later in the paper by subsuming environmental sensitivity, attitudes, and values under the term "affect". We believe that this modification is important because attitudes and values are shaped by societal norms and values, so what society deems to be important at any point in time is really a reflection of the people's perception of environmental issues, their sensitivity towards these issues, and how they should be managed. Roth (1992, p. 10) believed much of the environmental degradation that has occurred in the past and continues today, is the result of the failure of our society and is educational systems to provide citizens with the basic understanding and skills needed to make informed choices about people-environment interactions and inter-relationships. Since the term first coined by Roth (1968), its been re-defined (e.g., Hungerford and Peyton 1976) to meet the needs of the world's population, spatially and temporally, but conceptually, the concept and its associated terms and components remain applicable in the modern world. Hungerford and Peyton (1976) said environment literacy is "*reflected by human beings who have the knowledge and ability to communicate the need for environmental action strategies and be willing to use these skills and knowledge to develop and implement strategies to remediate and/or address environmental issues.*" The big levels of the environmental literacy are listed as follow (see Fig. 4.1).

- Level I: Ecological Foundation;
- Level II: Conceptual Awareness of Issues and Values;
- Level III: The Investigation and Evaluation of Issues and Solutions; and
- Level IV: Citizenship Action and Participation.

How then are the "cornerstone, keystone, stepping stone, and capstone" elements illustrated in the model presented in Fig. 4.1 related?

Level I: Ecological Foundation and Level II: Conceptual Awareness of Issues and Values are much like cornerstones, serving as first stones laid in the constructing of a masonry foundation. Level I and II have been considered the most important elements for building environmental literacy because all the other stones that follow are laid in relation to Level I and II. Levels II and III are the keystones for the model. Without these elements the model would be meaningless.

In Level III is focused on Investigation and Evaluation of Issues and Solutions. Level III provides the philosophy, process, business proposition, or principles to support citizenship action and participation before moving to Level IV. The four-level model proposed by Hungerford et al. (1980) was later divided into a more complex Environmental Literacy Model (Hungerford and Tomera 1985). In this model

Fig. 4.1 The level of environmental literacy, which is one of the goals of environmental education (originally adapted from Hungerford et al. (1980); pp. 42–44, revised and illustrated by Wei-Ta Fang)

citizens with responsible environmental behaviors to be citizens possessing environmental literacy was considered (Hungerford and Tomera 1985).

At this time environmental literacy models were the subject of research and new models were proposed. Wilke (1985) described competencies in Environmental Education should include the following elements: Level 1: Ecological Foundations, Level II: Conceptual Awareness, Level III: Investigation and Evaluation, and Level IV: Environmental Action Skills. The model centers on cognition, affection, and skills in taxonomy such as Bloom (Bloom et al. 1956), and belongs to the field of cognition with "environmental knowledge to the problem," "ecological concepts (cognition)," "environmental sensitivity." The areas of affection include "attitude," "values," "beliefs" and "control concept"; those in the field of skills adopt "environmental action strategies (skills)." The above seven variables are interconnected. Here, he moved from four levels to seven variables in Fig. 4.2.

The environmental literacy model in Fig. 4.2 is based on the study of environmental behavior and a retrospective of environmental education as it relates to today's issues (Hungerford and Tomera 1985; Hungerford and Volk 1990; Hungerford et al. 1990) (please also see Box 4.1: Tbilisi Declaration). How relevant will these models be in 10, 50, and 100 years? We suspect that we will see impacts in the next 20 years when seawater levels start creating major flooding problems in our coastal cities. Are our models and environmental education process and we deliver such messages equipped/prepared for such change? For example, people living in coastal areas are going to care a lot about sea level rise, while people living inland aren't really going to care unless it impacts their supply chain/way of life.

Since then, environmental literacy has integrated the fields of cognition (knowledge), affection (attitude), and skills (behavior). In 1990, the United Nations designated the "Environmental Literacy Year" calling for "human environmental literacy" to strengthen basic knowledge, skills, and motivation for learning to strengthen sustainable development. This is a good point from UN's declaration, but this is developed some decade ago. Is it relevant today and/or in another 20 years? We therefore, developed a new model shown in Fig. 4.3.

Fig. 4.2 Variables affecting the environmental literacy. Each variable interacts with one another (adapted and modified from Hungerford and Tomera (1985); Hungerford and Volk (1990); Hungerford et al. (1990)

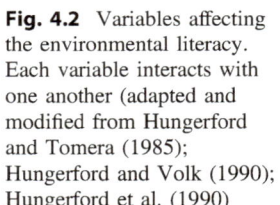

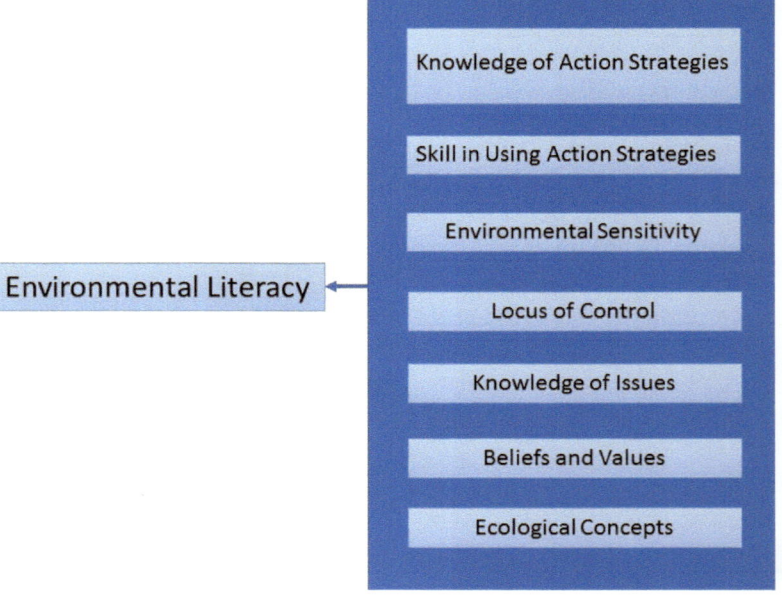

Fig. 4.3 In the study of environmental behavior, the scale of environmental knowledge and environmental attitudes needs to be considered. Modified from Hungerford and Tomera (1985), Hungerford and Volk (1990), Hungerford et al. (1990), Liu et al. (2015), Liang et al. (2018), Cherdymova et al. (2018) (Illustrated by Wei-Ta Fang)

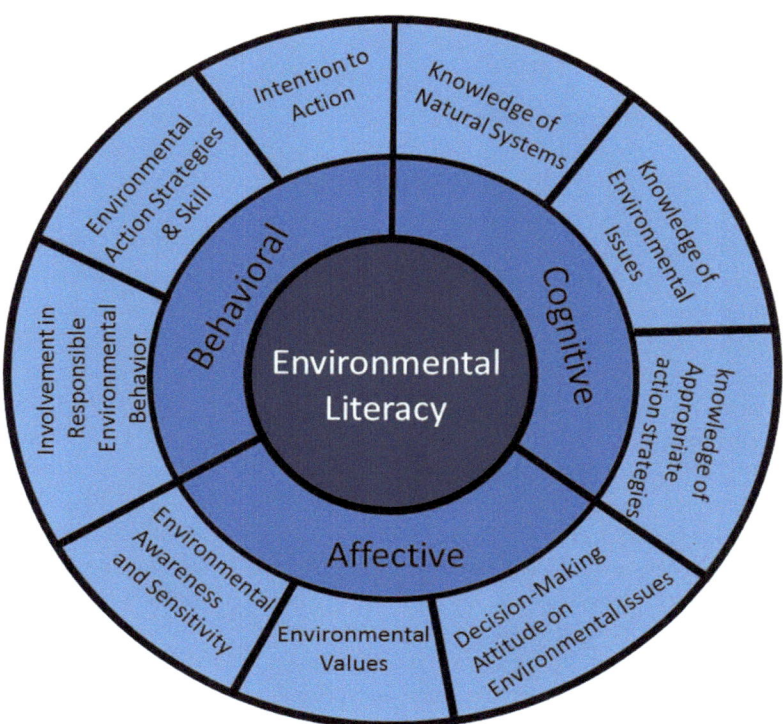

Box 4.1: Tbilisi Declaration: Another Literacy Cited?

According to Hungerford and Peyton (1976), it is believed that environmental literacy includes three components: (1) cognitive knowledge, (2) cognitive process, and (3) affective processes. Environmental literacy means that a person has the ability, knowledge, and skills on environmental issues and is able to teach others. In 1977, UNESCO proposed five characteristics of "environmental literacy," include: (1) awareness and sensitivity to the overall environment; (2) understanding and experience of environmental issues; (3) possessing values and feelings for the environment; (4) possessing the skills needed to identify and solve environmental problems; and (5) the ability and willingness to participate at all levels to solve environmental problems.

When people emphasize sustainable development and generational justice, a correct attitude towards the environment, a concept of control, and a sense of personal responsibility have produced a strong intention of environmentally friendly behaviors, which involves environmental education and learning motivation, awareness, sensitivity, values, skills, and experience. Unfortunately, people will not react until it's too late to save energies. Environmental Education (EE), however, can and should play a large role in managing environmental issues and sustainability, but the practice of sustainability is akin to masturbation. We feel good about what we're doing, but it really isn't changing much.

4.2 Motivation for Environmental Education

4.2.1 Pedagogic Reasons

EE requires pedagogic processes through which the concepts, skills, attitudes, ethics, and values of the environment and society can be used by the entire population to understand the sustainable use of resources, maintain environmental quality, and achieve ecological balance. In recent years, because the global environment is threatened by human development, the carrying capacity of the environment is limited (Sepkoski 1993). Sepkoski at University of Chicago who developed an algorithm for species longevity (Sepkoski 1993, 1997, 1998). He declared that the fossil record shows large variation. Rates of origination have declined through the Phanerozoic. This appears to have been largely a function of sorting among higher taxa (especially classes), which exhibit characteristic rates of speciation and extinction that differ by nearly an order of magnitude. However, a somber sub-question has been added: How many species are dying every year? We believe species longevity is only a few million years. This then puts adaptation into context because humanity will not be able to survive forever.

4.2.2 Social Learning Reasons

From a scientific point of view, we have an intrinsic motivation to learn about environmental issues. After all, we live on a planet where biotic and abiotic change is expected, which as an intelligent species provides us with the motivation to learn more about the world around us. Understanding the world around us allows us to promote additional learning activities, albeit incrementally, and identify the gaps in our knowledge base. The most recent pandemic is an excellent example of a gap that we missed or ignored given that there have been at least 17

pandemics since the 1300 s. It is also the direct cause and internal motivation of human behavior. Therefore, the learning motivation of environmental education is mainly composed of two factors: internal driving force and external incentive. These factors, driven by the "self-course," have become a driving force for human beings to continuously seek knowledge and advance. Albert Bandura (1925 ~ 2021), a professor of Department of Psychology, Stanford University, is dedicated to exploring the "self-course," that is focused on the study of personal goals, self-evaluation, and the thinking process of self-expression ability and belief, treating people as their "agents" That is, we can influence our own development, and not only analyze individuals, but also emphasize social impacts, such as how socioeconomic conditions affect human beings' belief that they can change things.

In the 1950s, Bandura recognized the motivation of human learning with "social learning," which originated from "behavior/learning theory." He began to shift his attention from animals in the laboratory to human behavior. Is this a journey of continuous interaction between individuals and the social environment? If so, then, most human behaviors develop through the learning process; since birth, individuals have been learning others' behaviors all the time. With the increase of age and experience, under the urging force of external environmental factors, human actions, thoughts, and feelings will gradually mature and become socially acceptable by the family and society. From Fig. 4.4, this series of learning activities involves social stimuli, so it is called social learning, and this type of learning is the main way that individuals learn social protocols. Finally, the name "social cognition," emphasized on two important characteristics. First, the human thinking process should play a core role in personality analysis and second, the thinking process must take place in the context of society. That is, humans understand their relationship with the surrounding environment through human interaction.

Fig. 4.4 Social learning motivation (Bandura 1977) (Revised and illustrated by Wei-Ta Fang)

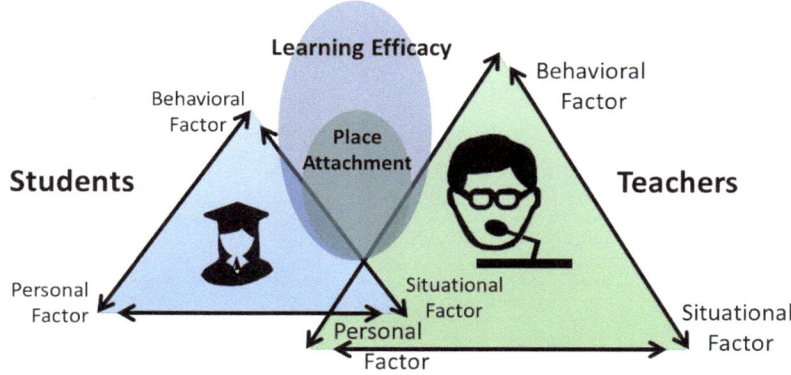

4.2.3 Better Environmental Protection Behaviors

In the behavior research of scholars at home and abroad, elementary school students have the best environmental protection behavior, followed by middle school students, and college students become worse after they reach adulthood. Among college students, the environmental protection behaviors of girls participating in colleges are better than the environmental behaviors of boys not participating in colleges (Liang et al. 2018). Female undergraduate students exhibited a more satisfactory fit in environmental literacy than did male undergraduate students (Wongchantra and Nuangchalerm 2011; Cincera et al.

2013). The higher level of environmental literacy attained by females could be explained by their social status and norms expectation in the Eastern society. For example, females in Taiwan have traditionally been taught to love (Fig. 4.5) maintain cleanliness in their homes, saving energies, and an animal caring role (Fig. 4.6) (Liang et al. 2018).

In addition, the environmental behavior of civil servants in central government agencies is better than that of local civil servants because the local civil servants do not feel responsible for the environmental crises. These factors appear to be related to the age and qualifications of civil servants (Fang et al. 2019). The results indicate people's attitudes towards environmental issues

Fig. 4.5 The higher level of environmental literacy attained by females could be explained by their social status and norms expectation in the Eastern society (Photo by Wei-Ta Fang)

Fig. 4.6 Females in Taiwan
have traditionally been taught
to love, maintain cleanliness
in their homes, and an animal
caring role (Photo by Wei-Ta
Fang)

is more important than their level or amount of
knowledge that they possess. Therefore, environmental education, like any other aspect of
education, is part of life's education. The delivery of the material on environmental issues plays
a much more important role in environmental
literacy as opposed to memorizing the
information.

In terms of teaching mode or delivery, environmental education should adopt an integrated
teaching method, that strengthens the teaching
mode from motivation theory, awareness,
knowledge, attitude, and action skills to increase
problem solving competencies. In addition,
environmental education needs to use teaching
strategies to focus on students as a teaching
center, emphasize students' active learning,
strengthen their sense of connection with the
world around them, and use students' actual environmental action experiences to establish an
interactive relationship between teachers and
students (Bandura 1977).

We need to mention that students could learn
in different ways to present information (i.e.,
visual, tactile, audio, or some combination
thereof) and the delivery needs to take these
learning styles into consideration so everyone
gets to learn (Fig. 4.7). Use the relationship

between learning behavior, student personality,
and learning environment as a causal relationship
to practice what Bandura calls "social epistemology." If we look at the development point of
view of students' growth, the methods for students to acquire knowledge and skills need to
learn, observe, and control themselves through
their own observations, and regulate personal
behavior and emotions.

Therefore, in the theory of learning motivation, it is necessary to observe the process of
students learning on their own. Motivation is
strengthened by educational methods, and from
this we can begin to understand environmental
awareness and sensitivity, values, and attitudes,
see Box 4.2.

**Box 4.2: Similar Environmental Nouns
in Environmental Literacy**

In environmental literacy, we often see
similar nouns, including attitude, awareness, awareness, consciousness, emotion,
perception, and emotion.

1. **Attitude**

Attitude is the way a person feels about
others, things, or situations, or generates
behavioral intentions. It is also an

Fig. 4.7 We need to mention that students learn in different ways, so we need to consider these learning styles, so everyone gets to learn (Photo by Wei-Ta Fang)

established way of thinking or feeling. Attitude includes mindset, viewpoint, beliefs, norms, affective dimension, and desire dimension. We logically assume that there is a positive correlation between cognitive and emotional attitudes. For example, in terms of environmental science, humans recognize that acid deposition will destroy the forest, but if humans themselves do not care about forest damage, then they will not care about the fate of the forest. In other words, environmental knowledge does not form environmental attitudes, because attitudes involve emotional evaluation factors. If you are interested on more details about scales, you can also see **Appendix: Environmental Attitude Scales** at the end of this book.

• **Positive Attitude**

There are two components to a positive attitude towards environmental change. A way of thinking about the threat posed by changes in the environment and actively acquiring a defense model for adaptation. Another positive attitude emphasizes human health and wellness. No matter how bad the natural environment is, we need to face the environmental challenges with a positive attitude in order to solve environmental problems.

- **Negative Attitude**

There are two components to a negative attitude to environmental change. One is to completely negate and adopt the ostrich mentality to face the environmental issues of climate change. In addition, another attitude is a negative attitude that we should avoid. As for environmental change, people generally evade passively and deny the challenges brought about by climate change. They cannot get out of their predicament, find solutions to problems, and have no way to solve environmental problems. Since people are saturated with negative opinions about the environment. We hear about climate change every day and anything bad in the news circles back to the environment. It also erodes the public's faith in science, so people will follow anyone who "promises" they have an answer. Therefore, as the proverb is saying: "people are like cows, except cows are smarter most times."

- **Neutral Attitude**

A neutral attitude to environmental change is another common attitude. There is no doubt that people who maintain a neutral attitude will not hold any hope to solve environmental problems. Humans often tend to ignore the problems in their living environment and maintain a peaceful attitude of "nature straight from boat to bridgehead." They wait for others to solve the environmental problems they face. They don't care about the improvement of living environment, they don't consider the complicated life phenomenon, and they don't care about others with the slightest emotion. They are lazy in heart and never feel the need to change themselves; they believe that they can live a simple self-lifestyle.

2. Awareness

Awareness is a kind of consciousness. It is also the perception or feeling of the surroundings, or the awareness of events, objects, or states or abilities that can be perceived. At this level of consciousness perceptual experience is confirmed by the observer, without necessarily implying understanding. More broadly, awareness is the state or quality of knowing. Environmental awareness is often used for public environmental knowledge or understanding environmental social or political issues. Environmental awareness is a synonym for whether the public is willing to participate in environmental movements, or whether they agree to jointly advocate environmental protection.

3. Concern

Environmental concern is the degree to which human beings are aware of environmental problems and are willing to support the solution of these problems, or to the extent that human beings are willing to make individual efforts to solve these problems.

4. Consciousness

Consciousness is a state in which humans perceive their existence. Consciousness has characteristics of sentiment, subjectivity, and the ability to experience or feel.

5. Emotion

Emotion is the mental state of a person's physical and psychological experience. Emotion can be a combination of multiple sensations, which intersects with mood, temperament, personality, disposition, and motivation, resulting in a psychological and physiological state.

6. Perception

Perception is a person's interpretation of the overall information of the external environment stimuli received. These interpretations can be influenced by the recipient's experiences and ability to recall them. Perception reflects the overall sense composed of the attributes and relationships of objects.

7. **Sentiments**

Emotion is a collective representation of a person's emotions and feelings.

4.3 Environmental Awareness and Sensitivity

The goal of environmental education curriculum is to improve a student's environmental awareness by training or improving their sensory awareness, which strengthens their skills in observation, classification, ordering, spatial relationships, measurement, inference, prediction, analysis, and interpretation (Gagne 1968; McTear 2002; Peuquet 2002; Auer 2008). Such skills can cultivate a students' situational awareness of environmental damage and pollution. It increases the degree of appreciation and sensitivity to the environment. In this section we discuss the implications of environmental awareness and environmental sensitivity.

4.3.1 Environmental Awareness

Once an individual understands how ecosystems function and the associated environmental issues, the fragility and importance of protecting our environments become apparent. So why is environmental awareness an important element of environmental education and the world around us? For people that are in constant contact with nature (Braus and Milligan-Toffler 2018), establishing and maintaining environmental connections are as easy as taking a walk in the woods. But for people who have lived in cities since childhood, opportunities to interact with nature are not easy or even available. If we haven't ever interacted with nature or since we were young, then we've lost the ability or skills to interact with nature. This phenomenon is called extinction of experience.

In the history of environmental education, environmental awareness began to attach importance to the nature in the second half of the twentieth century. Environmental education scholars believe that outdoor games are important for the healthy development of young children (Fig. 4.8).

Humans have become more aware of the environment because of their interest in biology and because children are naturally curious. Children are fascinated with nature and look for patterns or movement and insects are generally targeted because they can be easy to find. Interactions with nature help develop the plasticity of

Fig. 4.8 Strengthening outdoor games is important for the healthy development of young children (Photo by Wei-Ta Fang)

neural networks and strengthen environmental protection awareness (Li and Song 2019). Children playing outdoors feel more comfortable interacting and being a part of nature because they do not consider themselves isolated from nature and perform better in environmental behavior (Fang et al. 2017; Omidvar et al. 2019). As our environmental awareness develops or becomes sharper, we become aware of all types of environmental information.

As children grow older, they develop an understanding of how the world works and form a "mindset." We become more rational and build a framework around our perception of the world around us that is shaped by intrinsic and extrinsic factors/experiences. We ultimately develop a definition of this world and create models and develop ideas of nature that support our beliefs/concepts, which inevitably limit our ability to understand or interact with nature effectively. We consciously decide what is real, choose environmental signals that fit our paradigm, and filter out the parts that we either don't understand, care about, or fit our model.

Understanding of the living environment depends on human experience. In fact, when humans are close to the environment, they feel related to nature and the world around them. It can be said that humans create environmental awareness in order to find the right affinities to their close relationships to nature. Is this a situation like a runner who runs because they experience the dopamine high? Or is it an alignment of their beliefs with their knowledge base? Are we reducing contradictions? Perhaps this sense of affinity with nature strengthens our neural network systems as they relate to perception. Human beings want to understand nature and gain more of a sense of security (Selye 1976; Chawla 2020). This is also a primal response because we know we are part of the environment and if we screw it up, we've screwed ourselves. When human beings live in the environment, they feel more satisfied with their surroundings and care more about the environment.

The interventions for the environment from humans are needed to increase and improve green space, which can deliver positive health, social, and environmental outcomes for all persons, especially in urban environments (World Health Organization 2016).

Empirically, life determines the sculptural elements of nature; however, intellectually, the school's curriculum is based on books. These are theoretical explanations of the environmental experience. But, in the end, we rely more on life experiences to initiate environmental action, remedial measures to practice sustainable agriculture, repair community ecosystems, and develop ideas or concepts to protect our fragile ecosystems that help shape our lives. Charles Percy Snow (1905 ∼ 1980) talked about the dissatisfaction between the natural and social sciences and was saddened by the gulf between scientists and cultural intellectuals.

In 1959, he delivered a speech entitled *The Two Cultures and Scientific Revolution*, which sparked widespread and heated debate. He then used *The Two Cultures and Scientific Revolution* to explore the breakdown of communication between the sciences and humanities. Snow believed that the quality of education worldwide is declining. He wrote:

> A good many times I have been present at gatherings of people who, by the standards of the traditional culture, are thought highly educated and who have with considerable gusto been expressing their incredulity at the illiteracy of scientists. Once or twice, I have been provoked and have asked the company how many of them could describe the Second Law of Thermodynamics. The response was cold: it was also negative. "Yet I was asking something, which is the scientific equivalent of: Have you read a work of Shakespeare. I now believe that if I had asked an even simpler question – such as, What do you mean by mass, or acceleration, which is the scientific equivalent of saying, Can you read? – not more than one in ten of the highly educated would have felt that I was speaking the same language.

Snow mocked political leaders:

> So the great edifice of modern physics goes up, and the majority of the cleverest people in the western world have about as much insight into it as their Neolithic ancestors would have had.

Since being fired in 1959, his lectures condemned the British education system because the

Victorian era over-rewarded the humanities (especially Latin and Greek) at the expense of science. In his opinion, the practice of these deprived British elites (especially political, administrative, and industrial) is well prepared to manage the modern scientific world. Snow believed that:

> In contrast, schools in Germany and the United States are trying to promote science and humanities in an equal way and prepare for the way civil society thinks. Better scientific teaching methods make these countries. Rulers can compete more effectively in the age of science.

Later Snow's discussion of *The Two Cultures* tended to focus on understanding the differences between national competition systems in British school education and social class systems. If Snow was worried about the basic science of physics, then the understanding of modern politicians on global climate change, such as the speech of US President Donald Trump (cognitive bias), might look like the breakdown in communication between the sciences and the humanities in the modern world. This was again illustrated when Trump iterated *"The kidney has a very special place in the heart"* at the Advancing American Kidney Health (AAKH) event on July 10, 2019, which again illustrates the complete failure of American health, science education, and environmental education systems. US President Trump signed an Executive Order, AAKH, that could significantly change the administering of kidney care to double the number of organs available for transplantation by 2030 (Knight 2019).

Understanding environmental issues can promote environmental sensitivity and cultivate the desired environmental ethics and values. Environmental awareness is the foundation of environmental education because it provides students with the ability to become environmentally aware on many levels including:

- Improvement of sensory awareness (observation, classification, ranking, spatial relationship, measurement, inference, prediction, analysis, and interpretation);

- Appreciation and sensitivity of the natural environment;
- Awareness of environmental damage and pollution;
- Situational awareness including the natural and social environments;
- Awareness of the impact of our behavior on the natural and social environments;
- Awareness that human beings are closely related to the environment, natural resources, society; and culture;
- Awareness of the environmental responsibilities that we should bear; and
- Awareness that the everyday operation of society is reliant on natural resources.

Kuppusamy and Mari (2017) investigated the correlation between environmental awareness and environmental knowledge using "AKASA" Model, which is based on awareness, knowledge, attitude, skills, and action (Fig. 4.9). The study was based on responses from 234 s- and third-year students of architecture from five private universities in Malaysia. The UNESCO-UNEP commitment on the evolution of environmental education identified five essential components and from these, the objectives of environmental education were targeted. However, Kuppusamy and Mari (2017) only looked at environmental awareness and environmental knowledge. The purpose of the study was to investigate the relationship of between environmental awareness and environmental knowledge. The results show that the relationship between these variables is positive and strongly related (Kuppusamy and Mari 2017).

4.3.2   Environmental Sensitivity

Environmental sensitivity describes the ability of an individual to perceive and process information about their environment. A classification could be explained the definition from the environmental sensitivity which providing a concise summary as follows.

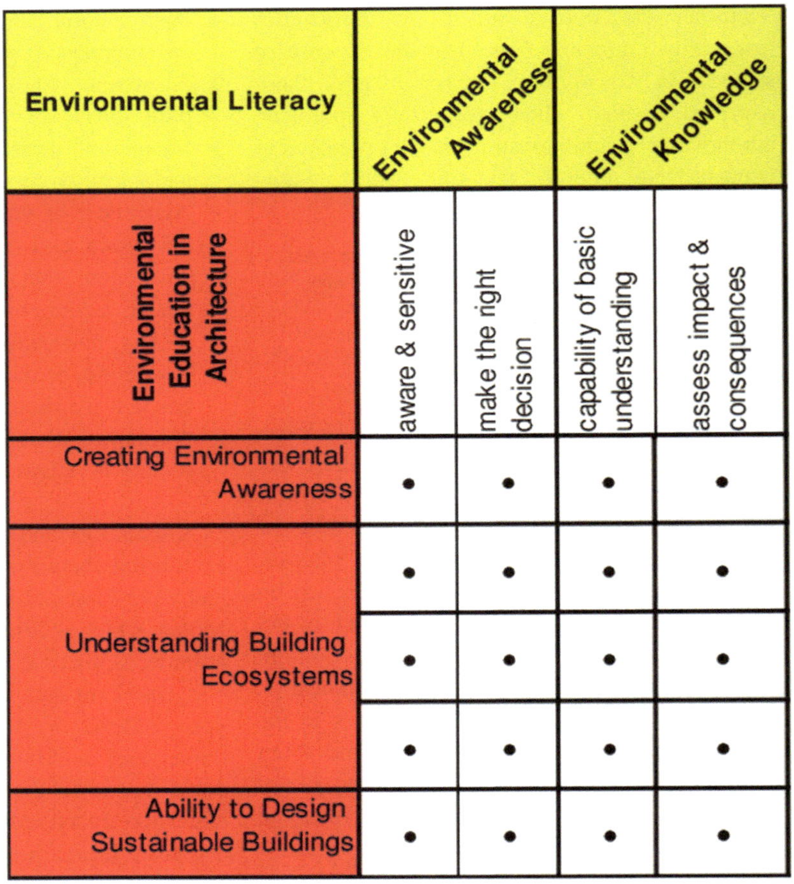

Fig. 4.9 Environmental awareness and environmental knowledge with questionnaire matrix, the dots = yes for each item (Kuppusamy and Mari 2017) (2020 University of Reading's institutional repository, CentAUR; Illustrated and redrawn by Arba'at Hassan)

Environmental Literacy	Environmental Education in Architecture	Environmental Awareness		Environmental Knowledge	
		aware & sensitive	make the right decision	capability of basic understanding	assess impact & consequences
Creating Environmental Awareness		●	●	●	●
		●	●	●	●
Understanding Building Ecosystems		●	●	●	●
		●	●	●	●
Ability to Design Sustainable Buildings		●	●	●	●

4.3.2.1 Orienting or Directional Sensitivity

Orientation sensitivity includes changes in how a person's perceptions and thoughts change with subtle changes in their environment. A sensitive person can be aware of low-intensity environmental change or detect the emotional stimuli of others. These relationships come from feelings and consciousness. People possessing strong orientation sensitivity can perceive subtle environmental factors and trigger association and response.

4.3.2.2 Chemical Sensitivity

Chemical sensitivity includes a variety of sensitivities to products such as food, paint pets, plants, fuel, molds, pesticides, detergents, fossil products, electromagnetic radiation, cigarette smoke, fragrance products, and cleaning products, which can trigger allergic reactions. Chemical sensitivity varies between individuals and is now an important workplace concern.

4.3.2.3 Aesthetic Sensitivity

Humans respond to the external environment and the mode of emotional response or overall impression is used to judge their aesthetic taste. The importance of the concept of aesthetics is difficult to assess, measure and discuss because aesthetic sensitivity is a non-verbal manifestation, that can only be understood from an individual's point of view. We have different views on aesthetics and while these views vary within a population, they can change in an individual or society over time. It is not uncommon to despise an artists' work as a young adult and later become enamored with their work as one matures.

These sensitivities contribute to an individual's sense of belonging within their respective environments, as well developing care and respect for the environment. When we have emotional concerns about the local environment and try to resolve problems, we feel that we are part of the community and made the effort to care for its resources. In other words, we tend to respond spontaneously and develop environmentally friendly behaviors because our actions negatively impacted the environment, and in some cases human health and well-being.

Louise Chawla (1949 ∼), emeritus professor of environmental psychology at the University of Colorado, believed that environmental sensitivity and environmental awareness are the reasons that we take responsibility for our activities in an effort to minimize impacts to the environment, and help children and youth to plan their sustainable communities (Chawla 1998; Derr and Chawla 2018). When we have a sense of environmental sensitivity, society has higher requirements for environmental protection. It is hoped that the government will take more remedial measures in response to environmental problems. Therefore, improving human environmental awareness and sensitivity is a way for humans to properly participate in environmental steward and create a better future. Through environmental education, we can arouse, or better yet, ignite our responsibilities and obligations to respect nature, understand the fragility of ecosystems, and develop solutions to solve the environmental threats and problems we created. At the same time, through discussion and communication, more people can reach a consensus on the environment, and at the same time instill hope for the future to inspire people.

4.4   Environmental Values and Status

In environmental literacy, environmental values and decision-making attitudes on environmental issues are also an important part of environmental attitudes. In the field of environmental philosophy, value is an ethical concept, a belief formed by the concepts, systems, laws, and symbols shared by social groups. If we talk about environmental values, then it is then a mentality under human consciousness, and a product of the interweaving of the natural ecology and human culture against the environment, which results in the judgment of the environment and behavior in individuals and societies (standard).

4.4.1   Environmental Values

Environmental values are the criteria by which humans judge the worth of an environment. Western philosophy has endowed the internal value dispute of environmental ethics and environmental moral judgment; because environmental ethics is not to explore the validity of intrinsic values, but to observe all of the values we have. Currently, we bring together the contributions of philosophy, economics, political science, sociology, geography, anthropology, ecology, and other disciplines that we attempt to monetize the value of past, present, and future environments om an anthropogenic point of view. This provides us with a sense of responsibility and justification for our actions, good or bad. In this process, the clarification of environmental values, through the verification of basic disciplines, deals with the relationship between the conversion of the environment into currency and the basic principles of public accountability. Therefore, with the development of social economy, environmental values have established a new "man-land relationship."

Stephen R. Kellert (1944 ∼), Honorary Professor of Social Ecology at Yale University, explores biology from the perspective of social ecology in his book *The Value of Life: Biological Diversity and Human Society* as the practical importance of diversity to human society. He divided the important values of nature and wild animals and plants to us into ten basic types, which are used to evaluate the universal existence of human beings on the ecological environment (Kellert 1996). Kellert believes that humans are not the only creatures in nature and judging nature cannot be based solely on human needs. At the same time, we should consider the laws of natural development and these laws do not necessarily depend on human development. There is a crucial relationship between survival and development, given the importance of life. However, human decision-making has had many serious environmental and social impacts that are based on economic or social environmental development. He developed the basic values of life and described these biologically based values and based on human culture, learning and experience.

Kellert (1996) summarized how we can assess natural differences in these values by gender, age, race, occupation, and geographical location. How does human activity in the ecological environment affect changes in values between species? How to show the significance in different cultural policies and ecological management? Kellert argues that the protection of biodiversity is fundamentally closely related to human well-being. He clarified the importance of biodiversity in human sociocultural and ecological psychology, and proposed his ideas as follows:

4.4.1.1 Aesthetic Value

Nature has aesthetic value, because the beauty of nature is everywhere (e.g., mountain peaks and moons, scorching flames, lush forests, lakes and mountains, autumn sky, choppy waves, roaring deer, lonely sunsets). It is a naturally rich mysterious aesthetic experience because each person has a point of view that is refined by personal taste and experience. One person's most beautiful sunrise/sunset is just another sunrise/sunset to another person.

4.4.1.2 The Value of Domination

Since ancient times, we've have wanted to dominate nature. We use nature to grasp and control the world's diversity of plant and animal species. We hope that "man will prevail in the sky" and have the right to dominate the natural world. It is given to mankind by the Bible, such as the Bible. "The Garden of Eden - Genesis 1:26" and God said, "Let us make man in our image, according to our likeness; and let them have dominion over the fish of the sea, and over the birds of the air, and over all the wild animals of the earth, and over every creeping thing that creeps upon the earth." The term "Dominion" is controversial. Many biblical scholars believe that the true meaning of the Bible is to require humans to take care of the earth and its beings, and that God has the so-called obligation of human beings to care for all beings. Therefore, because of the challenges of dominance and care, people face the modern society with decreasing biodiversity. They need to break away from the man-made method of mankind and control the narrow values of nature.

This concept is highly anthropomorphized because our species, *Homo sapiens* (300,000 years) is a recent addition is a relatively recent addition to the planet's inventory of plant and animal species that have a fossil record that can be traced back about 3.7 billion years.

4.4.1.3 The Value of Ecological Science

This observation is based on the ecological structure, function, and time series of nature. Science reveals the position of humans and all other species in their respective environments. Ecological value is the interdependent relationship between species. Ecologically speaking, we are just another species in a long list of species that inhabited this planet. In some cases, we happen to be smarter. But as a species, our time on Earth is measured.

4.4.1.4 The Value of Human Nature

Values depend on our culture and experiences. Culture preserves our values, including fairness and justice, compassion and charity, obligations and rights, and the protection of species for their survival, and human well-being. When choosing environmental values, we tend to consider the comfort and convenience of the economic or social environment as opposes to out own beliefs, which is also a part of human nature.

4.4.1.5 The Moral Value

Value is a moral code based on behavior. The fact that we have a relationship with nature is related to our responsibility as good citizens. Moral values underline the sense of loyalty and belonging in a group. However, when faced with environmental ethic issues, we are usually reluctant to take expensive actions, but resort to convenient, fast, trouble-free, and selfish actions. Therefore, the throbbing of ecological protection has a positive meaning because it is backed by a firm moral claim and needs to sacrifice human convenience. This new theory of moral behavior shows that if individuals do not act in accordance with their own interests and make the environment worse, and individuals will not benefit from it, they need to encourage everyone to put on moral conscience, advocate altruism, and further choose to take effective action.

4.4.1.6 The Value of Naturalism

Value can be framed in economic terms or its importance to a person. From an ecological perspective our respect for nature comes from caring for nature. The value of naturalism is to add people's feelings for nature. Therefore, when appreciating nature, our senses can play a large role in what we value.

4.4.1.7 The Negative Value

Nature causes negative emotions such as hatred and fear. The development of economic and social conditions may cause many of us to ignore nature and the damage to its environmental value. On the other hand, the lack of understanding of the environmental impact negates the value of nature, leading to cumulative negative problems in the environment and consequently acts that damage the environment.

4.4.1.8 The Spiritual Value

The spiritual value of nature is where we experience the spiritual connection with nature. Zhuangzi's *Theory of Qiwu* (《莊子·內篇》齊物論)(Theory of Equality) points out a concept of "satisfaction with oneself," which is also the state of *qiwu* (齊物)(equality or egalitarianism). There is no difference between human beings and all things. With the spiritual power of nature, he explained:

> The heaven and earth live with me, and everything is one with me.

The basic structure of human relations is the ultimate goal of adapting to nature in the course of human evolution. From the subjective consciousness of nature, the establishment of an integrated spiritual realm that connects me with the world is the foundation of all ecological cognition, judgment, and evaluation, based on the spiritual beliefs of nature and spirituality.

4.4.1.9 The Symbolic Value

When nature is an identity of a primary bearer of value and, secondarily, a bearer of value itself, then it has symbolic value (Davis 2019). The meaning of the symbols in nature leads to the metaphorical symbols in our experience. The symbolic value of nature refers to the semantic and cultural universe linked to it. Maybe we use this simple notation and often take it for granted. But symbolism in nature is quite subtle. We can find deeper meaning from the symbols hidden in nature. But we must take the time to find and promise to accept the symbolic information of nature. When we borrow everything in nature to express our thoughts and emotions, that is, in our hearts, we regard nature as the transformation and sublimation of symbolic meaning. It is, even representing an important a symbol as a marker (Davis 2019), although it has little practical effect. A place or a public good, therefore, can have both economic and symbolic value. Scientists are trying to make is clear how to answer the questions "What is the nature of meaning from her value" (Mills et al. 2010).

4.4.1.10 The Value of Pragmatism

We obtain tangible substances from the biological world. The basis of these values seems to be the biologically practical nature of human beings. These values are influenced by human learning and experience, and if they are not developed through a connection with nature, they may undermine human sustainable development goals. Therefore, we should recognize and respect this natural and practical value of the environment and change some traditional values that are not conducive to the environment. For example, we believe that some environmental resources are inexhaustible and this is wrong.

A case in point related to Table 4.1, Covid-19 is nothing but a strand of RNA and it doesn't think, have a family, go to work, or pay taxes, but it has brought the human species to its knees. That is, which species has mastery and dominance over another? This is a real "redneck value" of what we want to say. We should be concerned with living in harmony with all other species rather than being overlords.

4.4.2 Environmental Attitude

Environmental attitudes are the psychological tendencies of a person that manifest themselves as preferences for the natural environment or an evaluation response to dissatisfaction. Environmental attitudes are a potential construct of human beings, so we cannot observe them directly. What does this mean? Is this an evolutionary response to assess the safety or suitability of an environment for our survival? In other words, is the selection of a suitable environment hard wired into our genetic code? This is a relatively new ideas for our attitude towards the environment. This also presents considerable challenges for the study of human health. That is, environment can impact phenotype and vice versa (Ralston et al. 2008). Environmental factors and genes form a fully interactive system at all levels (Schneider 2007). Behavioral genetics become increasingly important as the full complexity of gene–environment relations. Behavioral analysis, therefore, both contributes to and gains of how nature and nurture really work.

Humans can infer the likes and dislikes of environmental attitudes based on survey responses. For surveys of environmental attitudes, we can use direct self-report methods for measurement or use hidden measurement techniques, such as observation methods (see appendix).

4.4.2.1 Navigating Environmental Attitudes

There are several methods that can be used to measure environmental attitudes (Cotgrove 1982; Wiseman and Bogner 2003). Opposition of

Table 4.1 Ten ecological values in Stephen R. Kellert's *The Value of Life*

Value	Definition
Aesthetic value	Appreciate the charm and beauty of nature
Dominant value	Mastery, physical control, and dominance of nature
The value of ecological science	Appreciate the structures, functions, and relationships in nature
Human value	Humanistic; strong emotional attachment and love for nature
Moral value	Moralistic; caring for Nature with the Principle of Land Ethics
The value of naturalism	Naturalistic; enjoy immersed in nature
Negativistic value	Negative value, such as fear, aversion, and alienation
Spiritual value	Spiritual transcendent feelings; respect for nature
Symbolic value	A symbolic revelation of nature based on human language and thought
The value of pragmatism	Utilitarian; Benefit from practical use and material use of nature

Source Kellert 1996

environmental attitudes however, is the process where positive and negative aspects comprise our attitudes about the environment and associated concerns and problems. For example, environmental conservation implies a certain level of environmental protection and people possessing this attitude focus on protecting nature from abusive practices that contribute to environmental change. On the other hand, utilization or what has been called resource management is an attitude that reflects or justifies the use of our natural resources for our benefit and people and the belief that nature requires a level of management in order to be healthy. This attitude or belief justifies our use, transformation, and management of natural resources is an appropriate approach to protect, conserve, and sustainably use these resources. It's a version of Ramsar's wise use approach to wetlands, but the concept of "wise use" can be applied to management system so that we can justify how we use our natural resources without guilt. It's a delicate balancing act between the real need for these resources in order to survive and abuse of these resources for profit or personal gain.

Thomas Heberlein believed that values and attitudes are a yearning for certain things or an outcome regardless of whether it is right or wrong; individual, collective, or both. Therefore, solving environmental problems requires a good understanding of the attitudes of the general public on scientific issues and in particular where an individual lands on the scale compared to the community. Heberlein (2012) tried to explain human attitudes, how humans change attitudes, and influence behavior and said, "*Solving environmental problems requires a scientific understanding of public attitudes.*" He explained that the concept of attitude has neither is qualitative and difficult to measure. He believed promoting environmental protection is not an attempt to change society's attitudes on this matter, but rather, it's about designing solutions that strengthen environmental policies. As such, he said, "*Attitudes measured at very general levels should not be expected to be associated with a specific behavior.*"

He elaborated on these views on attitude by using Aldo Leopold, a well-known environmentalist, as an example. In his classic book *A Sand County Almanac*, Leopold (1949), went from using a shotgun to hunt wolves to protecting them. Although Leopold changed his view about wolves, he did not repeal the wolf hunting bill illustrating the point that human attitudes can, and often contrast with our behaviors. Observing the attitudes of a person or group of people is not difficult, but determining the impact of the social, economic, ecological factors that shape behavior is challenging, and changing individual and collective attitudes is even more challenging.

4.4.2.2   Transformation of Environmental Attitude

Environmental attitude is the sensibility and the rationality of interactive thinking, perception, and dialectical perspective. Perception manifests emotional responses via sensory stimuli. If these experiences generate emotional responses through subjective cognition alone, then they are all perceptual experiences. Rationality is the ability of human beings to reason. From a rationalist point of view, environmental education should be rational education. We talked about the "existence of the environment" and "the nature of the environment" in Chap. 1, and in Chaps. 2–5, we used concept theory to talk about the "concept system" of environmental psychology. The concept system is abstract, which is not easy to understand, and is not of interest to environmental scientists. From the point of Immanuel Kant's (1724 ~ 1804) *Critique of Practical Reason*, they all want to derive knowledge through "independence from experience." Even in pure logic, one obtains a rational essence that is independent of time.

Of course, environmental education beyond the environment is a kind of rational thinking, and the effect achieved is a "quantitative, emotionless" education. However, education is not "teaching the truth"; true "environmental truth" or education that is transmitted only through one-way transmission of teachers and students can achieve a real state of learning that fully understands the truth. From the empiricism of David Hume (1711 ~ 1776), environmental education is closely related to empirical theory. Empiricism hopes to establish theories from evidence through modern scientific methods, not to obtain answers through purely logical reasoning, nor to obtain answers from Kant's "independent of experience." We should explore what Hume's *An Enquiry Concerning Human Understanding* talks about, using rationality to pursue intellectual "relation of ideas." This kind of practice is deductive. In addition, the "matters of fact" depend on empirical evidence. This is achieved through observation, induction, and understanding of nature. We don't know if we will learn the uniformity of nature of the universe in our lifetime. This is of course the goal of scientists, but not the goal of educators. Because in addition to environmental education, we must learn the logical concepts and mathematics of abstract rationality, and need to understand the real world with a perceptual understanding as much as possible.

Therefore, the particularity of environmental education lies in the unity after contradiction. We understand that we have a continuous, rational, and chaotic personal attitude to the "environment, economy, and society." This self-struggling self-struggle is just the beginning of self-growth. When economic development and environmental protection are fighting, when human interests invade species, and when human left-brain rational thinking and right-brain emotional thinking are in conflict, we must fight against our own emotions, not against others. Fighting with yourself is seeking self-growth; fighting with others, social struggles, and national struggles will only produce chaos that can never be unified.

As the saying goes, "There is nothing in the world, and mediocre people disturb themselves." This is the truth. The Hindu Vedanta philosopher Sankara (686 ~ 718) believes that his own subjective thinking, that is, imposing "I" on the truth, forms a hallucination of the human world, and at the same time produces pain.

Hume once said that human beings usually assume that the "I" is the same as the "I" five years ago. However, people's attitudes change. "I" has been circulating; "I" is not a fixed form. In the process of introspection, he suddenly realized that the Song Dynasty poet Xin Qiji (辛棄疾) (1140 ~ 1207) said in the *Sapphire Case*: "He looked back, but the man was in the dark." If you seek him out, then you are pursuing your life's self-worth. However, looking back suddenly, the ego was not asking him, but was near. Hume believes that human beings have always been in the flow and suddenly felt the ego, and that ego is a collection of many different feelings accumulated. "Looking back, I feel like I'm changing and extending at a fast flow speed." In other words, attitudes are always changing, and the process of changing long-term attitudes changes through thinking, and at the same time, feelings also change. With the habits that have been cultivated for a long time, the way of thinking has also changed, and new ideas have emerged. When we understand that everything is transmuted, everything is stretched. There is no need to adhere to rigid environmental attitudes, but to keep pace with the times, producing "spirit and flesh," "rational/perceptual," "deduction/induction," "ideal/reality," "logical/empirical," "peace/emotion"—adjustments and conversions. There is no absolute, no uniformity, only the colorful and brilliant, only the brilliant and splendid, and only the Russian literary critic theorist Mikhail Bakhtin (1895 ~ 1975) "heteroglossia," which can completely present the diversity of the environmental role dialogue (Bakhtin 1981, 1994; Guez 2010).

4.5 Cognitive, Affective, and Psychomotor Skills

In the first three sections, we talked about the motivation of environmental education, environmental awareness and sensitivity, and environmental values and attitudes. All the above are implicit environmental driving forces. Explicit learning has become the mainstream of environmental education. Environmental education consists of consciously solving problems and actively working to learn skills and produce a clear learning process. In the implicit stage of education, Benjamin Bloom (1913 ~ 1999) described the education of the "Cognitive Domain" and "Affective Domain," David R. Krathwohl (1921 ~ 2016) and Bloom's description of the "sphere of affection" in 1964 was revised in 2001 by Krathwohl. From the three main areas of learning by Bloom and Krathwohl for curriculum construction, there are related classifications from cognitive thinking to emotional feelings to technical (physical/ kinesthetic) activities. According to Bloom et al. (1956) book *Taxonomy of Educational Objectives, Handbook 1: Cognitive Domain*, the cognitive domain is divided into six levels (Bloom et al. 1956). By 1964, the area of affection, *Taxonomy of Educational Objectives: The Classification of Educational Goals, Handbook II: Affective Domain*, began to be classified (Krathwohl et al. 1964). But the field of skills was not fully described until the 1970s. The following describes the content of cognition, affection, and skills.

4.5.1 Cognitive Domain

4.5.1.1 Knowledge
It includes memory and knowledge, and can recall important nouns, facts, methods, standards, principles, and principles.

4.5.1.2 Comprehension
The meaning of important concepts can be grasped and can be translated and explained.

4.5.1.3 Application
Can apply the learned abstract knowledge, including knowledge concepts, methods, steps, principles, and general principles to practical application in special or specific situations. For example, after learning about resource recovery, you can know how to classify resources.

4.5.1.4 Analysis
Information that can be used to communicate. It contains ingredients, elements, relationships, and

organizational principles, and is analyzed and explained so that others can better understand the meaning and can further explain the organizational principles of these messages and the effects of communication.

4.5.1.5 Synthesis
Refers to the ability to integrate the piecemeal knowledge learned to form a self-complete knowledge system, or to show the relationship among them.

4.5.1.6 Evaluation
After learning, you can give a value judgment on the knowledge or methods learned based on your personal point of view.

4.5.2 Affective Domain

4.5.2.1 Receiving
The attitude of voluntarily accepting and paying attention to the learning activities they engage in while studying or after studying. Acceptance includes the existence of a conscious situation, a willingness to accept it, and conscious attention.

4.5.2.2 Responding
Actively participate in learning activities and get satisfied from participating activities or work. For example, in resource recovery, you can work silently. These reactions include voluntary and contented responses.

4.5.2.3 Valuing
It expresses positive affirmation in attitudes and beliefs about the content learned in environmental education. These contents include value acceptance, value affirmation, and value practice.

4.5.2.4 Organizing
After conceptualizing the content of learning, it is incorporated into the personality traits of individuals and becomes individual values. Form value conceptualization and constitute personal value system.

4.5.2.5 Characterizing by value Set

Synthesizing the connotation of learning by individuals, after receiving, reacting, evaluating, organizing and other internalization processes, the knowledge or ideas obtained from a personal character. This is the environment. The ultimate practice of education character formation attitude.

4.5.3 Psychomotor Domain

After the connotations of cognition and affection were discussed, discussions in the field of skills only began in the 1970s, including discussions on scoping (Dave 1970; Simpson 1972). Anita Harrow proposed the connotation of the Psychomotor Domain in 1972, as follows:

4.5.3.1 Reflex Movements

Reflex motion involves spine motion and muscle contraction.

4.5.3.2 Basic Fundamental Movement

Basic movements include walking, running, jumping, pushing, pulling, and manipulating related skills, actions, or behaviors. The simple basic actions of human beings are combined parts that form complex actions.

4.5.3.3 Perceptual

Perceived ability involves the functions of the body, including related skills such as vision, hearing, touch, or muscle coordination. These skills draw information from the environment and react.

4.5.3.4 Physical Activities

Physical action is related to endurance, flexibility, agility, strength, and reaction time.

4.5.3.5 Skilled Movements

Skills and movements learned in games, sports, dance, performance, or art.

4.5.3.6 Non Discursive Communication

Creative gesture expression through gestures, gestures, facial expressions. These actions are to understand how the brain strengthens memory through learning, through physical movements, and helps to assist positive learning with embodied learning.

4.5.4 Modify the Cognitive Domain

In 2001, Bloom's students, Lorin W. Anderson (1945 ~) and David R. Krathwohl (1921 ~ 2016), revised the Cognitive Domain. In the original version, the list of functions from simple to the most complex was ordered by knowledge, understanding, application, analysis, synthesis, and evaluation. In the 2001 version, steps were changed to verbs and ranked according to recall, understanding, application, analysis, evaluation, creation (Anderson et al. 2001). Knowledge is related to learning and retention, and the remaining five are related to learning transfer.

4.5.4.1 Remember

Extract relevant knowledge from long-term memory.

4.5.4.2 Understand

Create meaning from learning messages. Build new knowledge and connect with old experience.

4.5.4.3 Apply

After using procedures and steps, perform tasks or solve problems, and closely integrate with program knowledge.

4.5.4.4 Analysis (Analyze)

Involves the decomposition of materials into local materials, pointing out the correlation between local and overall structure.

4.5.4.5 Evaluation

Evaluate according to criteria and standards.

4.5.4.6 Create

Combines various elements to form a complete creative idea, cost, or plan.

From the modified Cognitive Domain of Anderson and Krathwohl, different learning models in the learning pyramid were mapped (Lalley and Miller 2007). The learning pyramid emphasizes positive forms and is more effective for long-term learning. If we say that humans remember 10% of what they read, about 20% of what they watch and hear, up to 90% of human beings teach others to teach them to protect the environment in order to understand knowledge (Fig. 4.10). Of course, some people are better at learning than others. Although in most cases the construction of the learning pyramid makes sense, it still faces criticism.

4.5.5   Correction of Skill Areas

The behavioral goals in the field of corrective skills proposed by Elizabeth Simpson in 1972 (Simpson 1972), in addition to motor skills, also include procedures and advanced actions before actions are generated, which may be more Educational skills needed for education:

4.5.5.1   Perception

The individual uses the senses to obtain clues for the required motor skills. Can be used to stimulate discrimination, make clue choices, and learn movement conversions.

4.5.5.2   Set

Before the study of motor skills, psychological preparation has been completed. This stage is preparation for action that belongs to

psychological tendency, movement tendency, and emotional tendency.

4.5.5.3   Guided Response

Acting under the guidance of the demonstrator to respond. This stage is followed by imitation and trial and error.

4.5.5.4   Mechanics

This stage refers to the attainment of a considerable degree of skill learning and the coordination of hand and eye movements in order to achieve a habitual degree.

4.5.5.5   Complex Overt Response

In complex reactions, learning a variety of motor skills can already reach the point of learning familiarity. At this stage, learn about motion positioning and automatic work.

4.5.5.6   Adaptation

After learning skills have reached a level of proficiency, you can change the skills at any time to meet the needs of the situation to solve the problem.

4.5.5.7   Origination

From the knowledge of innovation performance, further use of skills to surpass personal experience and achieve the effect of innovative design.

Whether it is the different skill areas proposed by Anita Harrow or Elizabeth Simpson in 1972, they want learners to reach the level of mastery learning. That is, through learning, to become proficient, and to enter the "creation" that Anderson and Simpson jointly praise. The learning context varies according to the learner's age, type of environmental education, learning method, and learning process. Although scholars have criticized the learning pyramid, the learning pyramid still exists in academia, and there is currently no more appropriate theory to replace it. Therefore, in the process of environmental literacy learning, we should recognize that environmental education and learning is a continuous process, and it does not exclude the use of more direct methods for further learning.

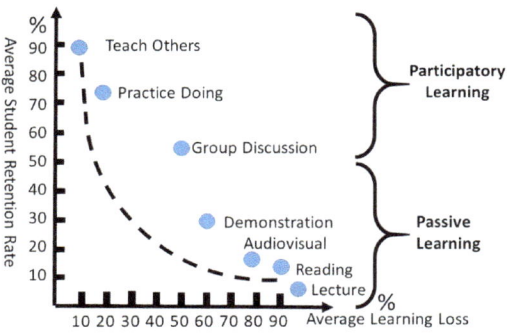

Fig. 4.10 Residual rates of environmental education methods (Illustrated by Wei-Ta Fang)

4.6 Environmental Action Experience and Pro-Environmental Behavior

In Sect. 4.5, we discussed about psychomotor skills. This section discusses environmental mobilization experience and pro-environmental behavior in environmental literacy. It explores the differences between action and behavior, and explains why humans take pro-environmental behavior.

4.6.1 Actions and Behaviors

Generally speaking, there is no obvious difference between behavior and action. In fact, the difference between behavior and action is an animal event that is distinguished by modern behavior. In the past, humans interpreted all behaviors, even physical objects, as intentional.

4.6.1.1 Action

Social action is the process that considers individual actions and reactions. Tom Campbell (1981:178) explained that action is an intentional activity that requires awareness or consciousness of the actor. But Alfred Schütz explained that individual actions are different. Therefore, action is a flow of movements and a process.

The action described by Max Weber (1864–1920) is different from what Schutz said. He explained the difference between "action" and "social action." He believes that when actors consider the actions of others and are therefore guided by their actions, this "action" is a "social" activity. Therefore, Weber found that action is an interesting and important concept in sociology. He explained "the meaning of the word action, which is a process of motives and feelings experienced by human beings. It is also an activity involving personal awareness. This activity has its purpose and is Activities that act in a certain way.

4.6.1.2 Behavior

Weber believes that "behavior" is a purely mechanical body movement. The behavior has no intention and has no special meaning to the individual. This is a distinction between behavior and action that automatically responds to specific impulses. In the last century, "behavior" was considered not the exclusive name of human beings (Ingold 1990; Kelso et al. 2006; Taylor 2021). Behavior is the system or living organism's response to various environments. According to Tom Campbell, behavior is just a kind of "reflection" and a response to what happened, so Campbell believes that objects need behavior to produce behavior (Campbell 1981:173).

4.6.2 Research on Actions and Behaviors

We sort out the differences between Weber, Campbell, and Schutz on action and behavior:

"Action" is a conscious activity. Action has a subjective meaning or goal for the environment and human beings involved.

If we say that a boy recycles plastic bottles correctly before a recycling bin, this is then an action.

If we say, a boy is holding a can of a drink, it is then a behavior.

When behaviors are the result of an unconscious response, we believe that these behaviors are meaningless and uncountable. Here, we refer to "behavior" as an uninterpreted or minimally interpreted description of an event.

Review related research on environmental protection, and rarely discuss the difference between action and behavior. Behavior defines how individuals act; act is anything an individual does; and the psychological idea of how they do these things is related to motivation. In contrast to socially normative human behavior, "action" is an event that is completed in order to achieve a goal.

4.6.3 Pro-environmental Behavior

When sociologists make a semantic distinction between behavior and action, the word usage of behavior plus "environmental behavior" becomes an adverb and even forms "Pro-environmental behavior". These are changes of unconscious behaviors in response to the needs of the times, and become conscious environmental behaviors.

After the 1970s, scholars analyzed the causes of environmental problems from different perspectives, and tried to verify human factors that may affect environmental problems. Although care may be an inducer of environmental behavior, this relationship is not linear. When its feasibility, importance, and necessity are fairly certain, humans will generate actions. Therefore, the development of a positive environmental attitude is a precursor to effective environmental action. From environmental attitudes to environmental actions, this is defined as a psychological disposition.

Arbuthnot et al. proposed pro-environmental behavior (Arbuthnot et al. 1976, 1977). They tried to use the "foot-in-the-door" technique to induce the pro-environmental behavior of recycling. Stern later proposed pro-environmental behavior again (Stern 1978). He used psychological experiments to learn about increasing recycling centers and reducing the use of household heating fuels to promote environmentally friendly pro-environmental behavior. Later, Schoenfeld et al. proposed the concept of pro-environmental trends (Schoenfeld et al. 1979). However, pro-environmental behavior has only gradually gained importance since the 1980s (Dunlap et al. 1983). According to the definition of pro-environmental behavior, "People are intuitively looked for an action with minimal negative impact on the natural and man-made environment" (Kollmuss and Agyeman 2002). In the definition of pro-environmental behavior, it is found that the definitions of the two concepts of "behavior" and "action" are no longer distinct, but can be used interchangeably. But "behavior" is closely related to individual

actions in environmental improvement. In other words, "environmental behavior" is "direct environmental action" (Jensen 2002).

The Belgrade Charter states that the goal of environmental action is to improve all ecological relationships, including the relationship between human with nature, and the relationship between man and man. Therefore, in light of national culture and environmental differences, basic concepts such as quality of life and human happiness need to be defined. We need to identify actions that will improve human potential and develop social and individual well-being in harmony with the natural and human environment.

According to the research of scholars Hungerford and Peyton (1976, 1977), Hungerford et al. (1980, 1983, 1985, 1990); Hungerford and Volk (1990), it is believed that the modes of environmental actions can be roughly summarized into five categories. They are (1) eco-management, (2) consumer/economic action, (3) persuasion, (4) political action, (5) legal action, or/and any combinations of these five (Table 4.3) (Hungerford and Peyton 1977; Hungerford et al. 1999; Hassan 1992). Their summaries are as follow:

4.6.3.1 Ecological Management (Eco-Management)

Ecological management (eco-management) are the actions taken by individuals or groups to maintain or promote existing ecosystems. Usually the work that the environment can do in person from garbage collection to forest conservation is all ecological management. The purpose is to maintain good quality

4.6.3.2 Consumer or Economic Action

Consumer or economic action refers to the economic threats made by individuals or groups to changes in certain business or industrial behavior. This is the action taken by consumerism.

4.6.3.3 Persuasion

Persuasion refers to interpersonal communication actions for environmental issues.

Table 4.3 Mode of Environmental Actions

Action	Description
Ecological management (eco-management)	Any physical action taken with respect to the environment. It is the actual actions taken by individuals or groups to maintain or promote existing ecosystems. Usually, the work that the environment can do in person from garbage collection to forest conservation is all ecological management. The purpose is to maintain good quality of the environment or the disadvantages of improving the environment
Consumer or economic action	Consumer or economic action refers to economic threats made by individuals or groups to change in certain business or industrial behavior. This is the action taken by consumerism. Consumer action is effective when a group of people get together to take action
Persuasion	Persuasion refers to interpersonal communication actions for environmental issues. Persuasion is used when a person or a group of people try to convince others that a certain action is correct
Political action	This mode refers of any action that can bring pressure on political or governmental agencies and/or individuals to take positive environmental action. It should change the government's decision-making through the strategic actions of individuals or groups. For example, referendums, civic processions, lobbying governments and public opinion organizations, and writing to public opinion representatives in jurisdictions such as legislators and city councilor
Legal actions	Usually, this action is appropriate to adults. It includes complaints, cautions, court injunctions, complaints, cautions, court injunctions, and so on. This is done through the jurisdiction of court procedures and processes

Sources Hungerford and Peyton 1977; Hassan 1992; Hungerford et al. 1999; Hungerford and Volk 2003

4.6.3.4 Political Action

Political action should change the government's decision-making through the strategic actions of individuals or groups. For example, referendums, civic processions, lobbying governments and public opinion organizations, and writing to public opinion representatives in jurisdictions such as legislators and city councilors.

4.6.3.5 Legal Actions

Legal actions include complaints, cautions, court injunctions, and so on. This is done through the jurisdiction of court procedures and processes.

Pan et al. (2018) believed that the environmental actions of college students in Taiwan are not active; they often implement ecological management and consumerist actions, and rarely do persuasion and citizen actions. In addition, the level of environmental literacy is moderate, and in terms of affection, knowledge, and skills, the affective change group scores higher. This college student study shows that environmental

hopes, action intentions, and civic engagement skills are important predictors of environmental action.

4.6.4 Responsible Environmental Behavior

Responsible environmental behavior has always been an important goal of environmental education. When citizens have the knowledge, attitudes, and skills on environmental issues, they will actively participate in approaches to solve current and future environmental problems (Hungerford and Peyton 1977).

At the broader social level, environmental sociologists have studied the relationship between environmental action and political action (Dunlap 1975), norms and values (Heberlein 1972; Heberlein and Shelby 1977), and other demographic factors, Examples include age, race, and socioeconomic status (Van Liere and

Dunlap 1980). Hudspeth (1983) analyzes environmental planning and decision-making participation, and argues that citizen participation addresses environmental issues. Early research has shown that among young, well-educated people, environmental issues are more important. But the dynamic process of attitude development at the individual level cannot be presented.

The reason is that in environmental education, it is necessary to cultivate citizens who are responsible for the environment. When citizens have the knowledge, attitude, and skills, they can participate in the solution of various problems, and then develop the premise of protecting the environment and developing Sense of responsibility to perform responsible environmental behavior.

Lipsey (1977) summarized ecologically responsible behavior and became the outpost of responsible behavior research. Borden and Schettino (1979) summarize the factors of responsible environmental behavior based on the basis of social psychology. They construct according to affective, cognitive, and behavioral attitudes to solve environmental problems.

Promoting responsible environmental behavior is a form of civic participation (Hines et al. 1986/1987). Civic awareness means that it is necessary to cultivate citizens with environmental behaviors, that is, to educate citizens with environmental literacy through environmental education. In order to achieve this goal, a curriculum structure needs to be established to plan and cultivate students with responsible environmental behavior. Hungerford and others first proposed an environmental literacy model (Hungerford and Tomera 1985).

Therefore, how to convert environmental literacy into human's daily environmental behavior is based on the environmental behavior model suggested by Jody M. Hines (Hines et al. 1986/1987), as shown in Fig. 4.11. Hines made use of the meta-analysis method. The Model of Responsible Environmental Behavior proposed in 1986/1987 has been widely used in environmental education.

Hines et al. (1986/1987) analyzed 128 articles published in journals and books on environmental behavior related research or unpublished dissertations since 1971, and proposed a responsible environmental behavior model. The main factor that produces environmental behavior is that the individual has an intention to take action, and this intention is affected by the individual's personality, skills, knowledge of action strategies, and knowledge of problems. This means that before people have the intention to act, they must first recognize the problem and strengthen their attitude, control, and sense of responsibility.

4.6.4.1 Attitude

Attitude is often defined as the persistent positive or negative feelings that humans have about other objects or problems, and thus, attitudes affect responsible environmental behavior (Hines et al. 1986/1987). Attitude has a positive impact on environmental intentions.

4.6.4.2 The Concept of Control

The concept of control is to the mode of operation of one's beliefs. When a person's personality traits "internal control tendency" tells the self to strengthen a certain behavior, it is more likely to produce that behavior; and this behavior will continue to strengthen the internal locus of control. On the other hand, if a person has an "outside control tendency" and does not believe that his actions can affect it, then the person may not do it. Many studies in the past have shown a link between control concepts and environmental behaviors (Hines et al. 1986/1987); control concepts are also factors that influence environmental behaviors; that is, control concepts have a positive effect on environmental action intentions to influence.

4.6.4.3 Personal Responsibility

Personal responsibility is also a factor that drives environmental action. Personal responsibility can have a positive effect on action intentions.

Since the 1980s, many paths and frameworks for environmental behavior have been proposed, involving responsible environmental behavior. In addition, there are included theories of planned behavior (Ajzen 1985, 1991), the "value-belief-

Fig. 4.11 Responsible
environmental behavior
patterns (modified after Hines
et al. 1986/1987; Illustrated
by Wei-Ta Fang)

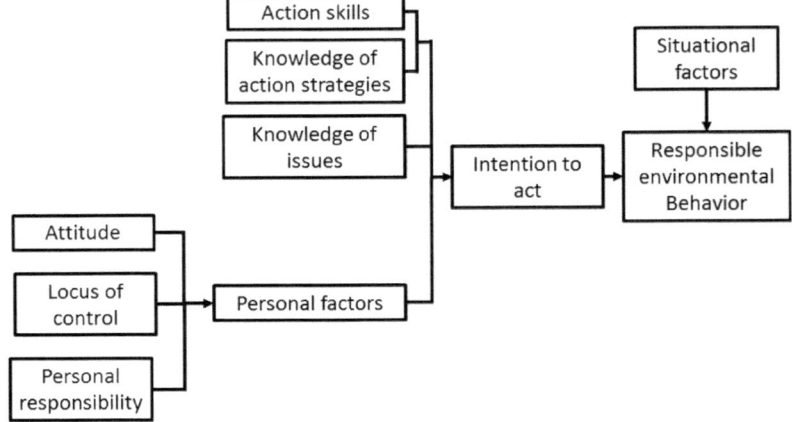

normative" theoretical model (Stern et al. 1999; Stern 2000), and the entry-level variables (entry-level variables), ownership variables, and models of empowerment variables (Hungerford and Volk 1990).

Among them, Chao (2012) analyzed structural models of Responsible Environmental Behavior and the Theory of Planned Behavior (Hines et al. 1986/1987; Ajzen 1985, 1991). Finally, it was found that the predictiveness of the Responsible Environmental Behavior model in the field of environmental education is more significant than the planned behavior theory model; and Chao's research also pointed out that the personality factor has the highest deduction and happens to represent a person's personality traits can have a significant impact on their responsible environmental behavior. We discuss personality traits in Chap. 5.

4.7 Environmental Aesthetics

In Sects. 4.3 through 4.6, we discussed many abstract concepts of environmental literacy, such as environmental awareness and sensitivity, environmental values and attitudes, environmental mobilization experience, pro-environmental behavior, and responsible environmental behavior. At the end of this chapter, we discuss environmental aesthetics.

Environmental aesthetics is an emerging discipline. From the perspective of artistic philosophy of classical aesthetics, environmental aesthetics originates from a reflection that places too much emphasis on artistic media and expression. Environmental aesthetics is the pursuit of appreciation of the value of the natural environment. However, this subjective aesthetic experience includes both the natural environment and the social environment affected by humans (see Fig. 4.12). At the same time, environmental aesthetics began to consider examining the environment, emphasizing the aesthetics of daily life. These aesthetic connotations involve not only objects, but also daily activities. Therefore, in the early twenty-first century, environmental aesthetics accepted the study of almost all environmental protection matters except art, which included aesthetic significance. Therefore, the category of environmental aesthetics, from natural environment to man-made environment, is an aesthetic connotation that can be explored.

Environmental aesthetics is related to the coordination of nature, whether it is hue, color system, balance, and comfortable wind, water, light, and rhyme. If we say that environmental aesthetics is pursuing the harmonious characteristics of human heart, then this is another reflection on modern art. Because of the art of "modernism" or "modernism," since the twentieth century, from the avant-garde and avant-

Fig. 4.12 This subjective aesthetic experience includes both the natural environment and the social environment affected by humans (Heping Island, the doorway to Keelung Harbor, Keelung, Taiwan. It was one of the settlement of Ketagalan Tribe at an early time)(*Island of Love*, Watercolor on Arches, 1986, 42 cm × 58 cm) (Illustrated and painted by Wei-Ta Fang)

garde colors, there have been many violent artistic thoughts and schools that break the tradition and resist nature. The above-mentioned modern art is formed by the combination of various types of visual styles, based on scientific and rational foundations. The main schools include Fauvism, Prophet, Cubism, Surrealism, and Expressionism, Abstract Expressionism, Constructivism, Futurism, Dadaism, Style, Bauhaus, Surrealism, Abstractism, Popular Art (Pop Art), Optical Art (Op Art), Conceptual Art, etc. After the 1960s, landscape art reexamined the relationship between human life and artistic landscape, and restored the art of man and nature.

However, modern art is stimulated by media that are represented by multimedia combinations such as sound, light, and video. Various artists are established through mutual subjective relationships with the audience. From the appreciation of environmental art, to return to the embrace of nature, this is a reflection on the intervention of human life through natural art, and also a response to the control of human society by modern artificial intelligence and criticism of the reality of science and technology on the depression of human nature. The following is a brief introduction to the artistic genre that provides reflection on the real environment under the turbulent environment of the twentieth century.

4.7.1 Dadaism

Dadaism was influenced by anarchism in 1916, and through anti-war leaders, it protested capitalist values with an anti-aesthetic work. The characteristics of Dadaism include the pursuit of sober irrationality, while refusing to agree on artistic standards. Therefore, this is an opposition to fait accompli, but also an opposing force to war. The Dada movement later influenced Pop Art in 1955. But the irony is that Pop Art is pursuing fashion and accepting the facts of convention.

4.7.2 New Expressionism

Neo-Expressionism was a genre that emerged in Germany in the late 1970s. Neo-Expressionism reflects on Expressionism of 1911 and does not emphasize simple "repetition of nature," nor is it mechanical imitation "unnatural." Neo-

Fig. 4.13 Action Act:
Exploring Womb (Acrylic on
card, 1987, 104 cm
× 76 cm)(Illustrated and
painted by Wei-Ta Fang)

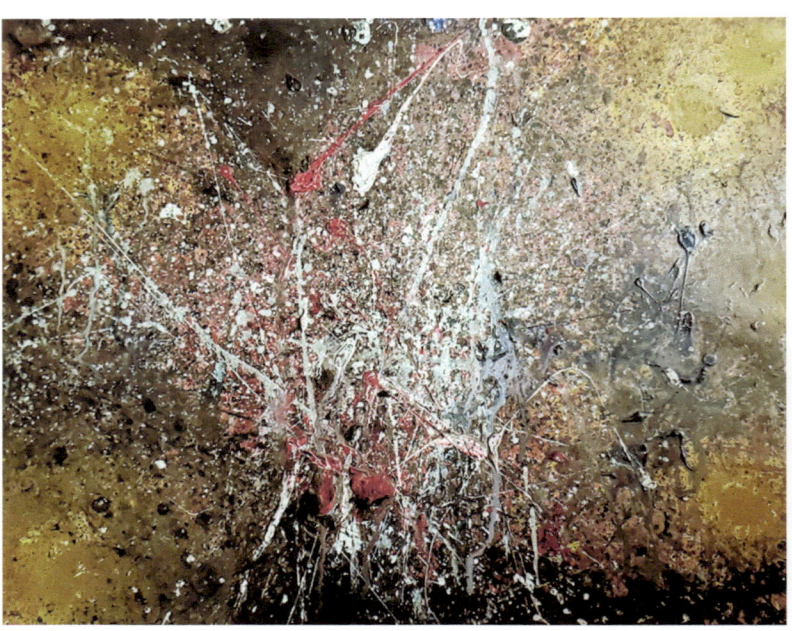

Expressionism is a reaction to Pop Art, emphasizing self-expression, showing the returning
tendency to Expressionism in terms of pictures,
brushwork, and mood. Neo-Expressionism
accepts the philosophy of existentialism, learns
the artistic tradition of abstract expressionism in
practice, pays attention to the emotional bursts
and improvisations of the painting process, pursues primitivism, and strives for the true colors.
Therefore, they are in the works, the ugly phenomenon of society, or self-mockery.

4.7.3 Action Art

Action art refers to modern art that originated in
Europe in the 1950s. It refers to an art composed
of individual or group behavior at a specific time
and place. Performance art is different from action art. It is an artistic process in which the
artists personally plan and promote and form a
group participation. Performance art must
include the shaping of the space between the
performance artist's body and the communication
with the audience at a specific time and place.
This art form is different from the art forms
composed of specific things, such as painting and

sculpture, but a unified expression of environment, body and space, see Fig. 4.13.

4.7.4 Landscape Art

Landscape art, also known as earth art, is an
evolution from environmental art, which is a kind
of environmental art. Landscape art originated in
the United States in 1960. After 1970, many
painters and sculptors showed landscape works
outdoors. Although the art of landscape is a kind
of display that combines with nature, it will
retouch nature and rethink the relationship
between human and nature. The audience visits
the landscape art in nature, and gets an artistic
perception of being integrated with nature.

4.7.5 Installation Art

Installation art used to be called "environmental
art," which means that the artist displays the
consumed objects in human daily life in a
specific space-time environment. The earliest
environmental art was exhibited in France by
Marcel Duchamp (1887 ~ 1968). He showed

the work of Fontaine in the shape of a urinal. In 1917, Duchamp purchased a ceramic urinal at a chain store on Fifth Avenue in New York, turned it 90 degrees, and left the signature "R. Mutt 1917." The first environmental artwork was born. Environmental art involves artists choosing, using, transforming, and combining materials, and then re-selecting spaces, using media's visual and auditory sensory experiences to produce artistic effects. Installation art welcomes the audience to get involved, so as to gain a new experience for the audience.

4.8 Summary

Although the process of globalization, pollution and environmental problems have become increasingly more pervasive, and environmental issues are getting more attention because they are now substantially impacting our way of life and the cost of doing business. Hungerford (1975:21) recognized that environmental education should not be based on "overgeneralizations, questionable logic, myths, pedagogical excesses, simple untruths, and at least a modicum of hypocrisy" and defined environmental education as the process of educating citizens about ecologically-related issues and the impact of these issues on our social fabric. This behavior focuses on the development of responsible /acceptable responses of the public regarding such issues. This definition is based on the environmental and social issues that have been created by humans and predicated on the notion that society itself determines the people's behavior or attitudes that are related to existing and emerging environmental issues and that there is an inherent expectation that the people who created the problems have the potential or should be held responsible to resolve the issues. We, therefore, all know that the basic spirit of environmental education lies in the "education process," "societal values," and our collective "knowledge, attitude, and skills," and ability to solve "problems.' In short, the desired effect of environmental education is whether we can improve the environmental literacy of the people and

formulate and implement corrective actions to resolve environmental problems without creating additional problems. For example, phytoremediation technologies that are based on plants and microbes can be used to remediate (clean) a wide variety of soil and water contaminants with H_2O and CO_2 typically being the end products of these novel technologies (Paz-Ferreiro et al. 2014). Despite the simplicity of such technologies, educating the public on the complex biological, chemical, and physical processes that occur and the risk of creating compounds that are more harmful to the environment and human health compared to the original pollutants is a challenge. In this Chapter we explore the processes of environmental education, including learning motivation, environmental awareness and sensitivity, environmental values, attitudes, experience, and pro-environmental behaviors, with the goal of providing our readers with an appreciation for nature,,the role nature plays in our way of life, and the knowledge and skills that citizens need to start taking action to correct some of the environmental issues that we are facing through environmental aesthetics (Saito 2010) and sound science. However, the development of environmental literacy for an entire population cannot be achieved overnight. Such a process requires time and a group of individuals that consciously care for the environment and work individually and collaboratively, typically for generations. Through educational channels, environmentally consciousness people, accumulate knowledge that develops community and cultural pro-environmental behaviors, that ultimately form a collective social force that is powerful and advocates for environmental rights.

References

Ajzen I (1985) From intentions to actions: a theory of planned behavior. In: Kuhl J, Beckmann J (eds) Action control: from cognition to behavior. Springer, Berlin Heidelberg, pp 11–39

Ajzen I (1991) The theory of planned behavior. Organ Behav Hum Decis Process 50(2):179–211

Anderson LW, Krathwohl DR et al (eds) (2001) A taxonomy for learning, teaching, and assessing: a revision of bloom's taxonomy of educational objectives. Allyn & Bacon, Boston

Arbuthnot J, Tedeschi R, Wayner M, Turner J, Kressel S, Rush R (1976) Self-perception theory and compliance: the foot-in-the-door technique applied to sustained pro-environmental behavior. Paper presented at Eastrn Psychological Association

Arbuthnot J, Tedeschi R, Wayner M, Turner J, Kressel S, Rush R (1977) The induction of sustained recycling behavior through the foot-in-the-door technique. J Environ Syst 6(4):355–368

Auer MR (2008) Sensory perception, rationalism and outdoor environmental education. Int Res Geogr Environ Educ 17(1):6–12

Bakhtin MM (1981) The dialogic imagination: four essays. In: Holquist M (eds). University of Texas Press, Austin

Bakhtin MM (1994) The Bakhtin Reader. In: Morris P (ed). Oxford University Press, Oxford

Bandura A (1977) Social learning theory. Prentice Hall, Hoboken

Bandura A (1986) Social foundations of thought and action: a social cognitive theory. Prentice Hall, Hoboken

Bloom BS, Engelhart MD, Furst EJ, Hill WH, Krathwohl DR (1956) Taxonomy of educational objectives: the classification of educational goals. In: Handbook I: cognitive domain. David McKay Company, Philadelphia

Borden RJ, Schettino AR (1979) Determinants of environmentally responsible behavior. J Environ Educ 10 (4):35–39

Braus J, Milligan-Toffler S (2018) The children and nature connection: why it matters. Ecopsychology 10 (4):193–194

Campbell T (1981) Seven theories of human Society. Clarendon Press, Oxford

Chao Y-L (2012) Predicting people's environmental behaviour: theory of planned behaviour and model of responsible environmental behavior. Environ Educ Res 18(4):1–25

Chawla L (1998) Significant life experiences revisited: a review of research on sources of environmental sensitivity. J Environ Educ 29(3):11–21

Chawla L (2020) Childhood nature connection and constructive hope: a review of research on connecting with nature and coping with environmental loss. People Nat 2(3):619–642

Cherdymova EI, Ukolova LI, Gribkova OV, Kabkova EP, Tararina LI, Kurbanov RA et al (2018) Projective techniques for student environmental attitudes study. Ekoloji 106:541–546

Cincera J, Krajhanzl J (2013) Eco-achools: What factors influence pupils' action competence for pro-environmental behaviour? J Clean Prod 2013 (61):117–121

Cotgrove SF (1982) Catastrophe or cornucopia: the environment, politics, and the future. Wiley, Hoboken, p p166

Dave RH (1970) Psychomotor levels in developing and writing behavioral objectives. In: Armstrong RJ (ed) Educational innovators. Tucson, pp 20–21

Davis R (2019) Symbolic values. J Am Philos Assoc 5 (4):449–467

Derr V, Chawla L (2018) Placemaking with children and youth: participatory practices for planning sustainable communities. New Village Press, New York

Disinger JF, Roth CE (1992) Environmental education research news. Environmentalist 12(3):165–168. https://doi.org/10.1007/BF01267599

Dunlap RE (1975) The impact of political orientation on environmental attitude and action. Environ Behav 7 (4):428–454

Dunlap RE, Van Liere KD (1978) The "new environmental paradigm." J Environ Educ 9(4):10–19

Dunlap RE, Grieneeks JK, Rokeach MM (1983) Human values and pro-environmental behavior. In: Conn WD (ed) Energy and material resources: attitudes, values, and public policy. Boulder, Nova Scotia, pp 145–168

Dunlap RE, Van Liere KD, Mertig AG, Jones RE (2000) Measuring endorsement of the new ecological paradigm: a revised NEP Scale. J Soc Issues 56(3):425–442

Fang W-T, Ng E, Chang M-C (2017) Physical outdoor activity versus indoor activity: their influence on environmental behaviors. Int J Environ Res Public Health 14(7):797. https://doi.org/10.3390/ijerph14070797

Fang W-T, Chiang Y-T, Ng E, Lo J-C (2019) Using the norm activation model to predict the pro-environmental behaviors of public servants at the central and local governments in Taiwan. Sustainability 2019(11):3712.https://doi.org/10.3390/su11133712

Gagne RM (1968) Contributions of learning to human development. Psychol Rev 75(3):177–191

Guez JM (2010) Heteroglossia. Western humanities review. University of Utah, Salt Lake City, pp 51–55

Hassan A (1992) The Perceptions of Grade 6 School Teachers Regarding the Status of Alam dan Manusia (Man and the Environment) in Primary Schools, Sabah, Malaysia. Unpublished doctoral dissertation. Southern Illinois University at Carbondale, USA

Heberlein TA (1972) The land ethic realized. J Soc Issues 4:79–87

Heberlein TA (2012) Navigating environmental attitudes. Oxford University Press, Oxford

Herberlein TA, Shelby B (1977) Carrying capacity, values, and the satisfaction model. J Leis Res 9:142–148

Hines JM, Hungerford HR, Tomera AN (1986/87) Analysis and synthesis of research on responsible environmental behavior: a meta-analysis. J Environ Educ 18(2):1–8

Hudspeth TR (1983) Citizen Participation in environmental and natural resource planning, decision making and policy formulation. Environ Educ Environ Stud 1 (8):23–36

Hungerford HR (1975) Myths of environmental education. J Environ Educ 7(2):21–26

Hungerford HR, Peyton RB (1976) Teaching environmental education. J Weston Walch, Portland

Hungerford HR, Peyton RB (1977) A paradigm of environmental action. ERIC Document Reproduction Services (No. ED137116). ERIC, Washington, DC

Hungerford HR, Peyton RB, Wilke RJ (1980) Goals for curriculum development in environmental education. J Environ Educ 11(3):43–44

Hungerford HR, Peyton RB, Wilke RJ (1983) Yes, Environmental education does have definition and structure. J Environ Educ 14(3):1–2

Hungerford HR (1985) Investigating and evaluating environmental issues and actions: skill development modules. A curriculum development project designed to teach students how to investigate and evaluate science-related social Issues. Modules I-VI: ERIC. ERIC, Washington, DC

Hungerford HR, Tomera AN (1985) Science methods for the elementary school. Stipes Publishing, Champaign

Hungerford HR, Litherland RA, Peyton RB, Ramsey JM, Volk TL (1990) Investigating and evaluating environmental issues and actions: skill development program. Stipes Publishing, Champaign

Hungerford HR, Volk TL (1990) Changing learner behavior through environmental education. J Environ Educ 21:8–22

Hungerford HR, Hagengruber D, Bluhm WJ (1999) Threatened and endangered animals: an extended case study for the investigation and evaluation of issues surrounding threatened and endangered animals of the United States. Stipes Publishing, Champaign

Hungerford HR, Volk TL (2003) Notes from Harold Hungerford and Trudi Volk. J Environ Educ 34(2):4–6

Ingold T (1990) An anthropologist looks at biology. Man 25(2):208–229

Jensen BB (2002) Knowledge, action and pro-environmental behaviour. Environ Educ Res 8(3):325–334

Kellert SR (1996) The value of life: biological diversity and human society. Island Press, Washington, DC

Kelso JS, Engstrom DA, Engstrom D (2006) The complementary nature. MIT Press, Cambridge

Knight R (2019) A patient's perspective on advancing American kidney health initiative. Clin J Am Soc Nephrol 14(12):1795–1797

Kollmuss A, Agyeman J (2002) Mind the Gap: why do people act environmentally and what are the barriers to pro-environmental behavior? Environ Educ Res 8 (3):239–260

Krathwohl DR, Bloom BS, Masia BB (1964). Taxonomy of educational objectives: the classification of educational goals. In Handbook II: affective domain. Allyn and Bacon, Boston

Kuppusamy S, Mari TS (2017) Relationship between environmental awareness and environmental knowledge using "AKASA" model among architecture students in private universities, Klang Valley, Malaysia. In: 2nd International conference on knowledge engineering and application, 21 October 2017, London, United Kingdom

Lalley J, Miller R (2007) The learning pyramid: does it point teachers in the right direction? Education 128 (1):64–79

Leopold A (1949) A sand county Almanac. Oxford University Press, Oxford

Li T, Song J (2019) Research on promotion methods of positive mental health of college students under the model of ecological sports teaching. Ekoloji 28 (107):1861–1868

Liang S-W, Fang W-T, Yeh S-C, Liu S-Y, Tsai H-M, Chou J-Y, Ng E (2018) A nationwide survey evaluating the environmental literacy of undergraduate students in Taiwan. Sustainability 10:1730. https://doi.org/10.3390/su10061730

Lipsey MW (1977) Personal antecedents and consequences of ecologically responsible behavior. A review. Catalog Sel Doc Psychol 7:70

Liu S-Y, Yeh S-C, Liang S-W, Fang W-T, Tsai H-M (2015) A national investigation of teachers' environmental literacy as a reference for promoting environmental education in Taiwan. J Environ Educ 46 (2):114–132

Malaysian Qualifications Agency (MQA) (2015) Malaysian qualifications register. Retrieved 30 April 2022 from https://www2.mqa.gov.my/mqr/

Malone MP, Ward MP (1973) Ecology: let's hear from the people: an objective scale for the measurement of ecological attitudes and knowledge. Am Psychol 28 (7):583–586

Maloney MP, Ward MP, Braucht GNŽ (1975) Psychology in action: a revised scale for the measurement of ecological attitudes and knowledge. Am Psychol 30:787–790

Marcinkowski TJ (1990–91) The new national environmental education act: a renewal of commitment. J Environ Educ 22(2):7–10

McTear MF (2002) Spoken dialogue technology: enabling the conversational user interface. ACM Comput Surv (CSUR) 34(1):90–169

Mills AJ, Durepos G, Wiebe E (2010) Symbolic value. In: Encyclopedia of case study research, vols 1–0. Sage, Thousand Oaks

Omidvar N, Wright T, Beazley K, Seguin D (2019) Investigating nature-related routines and preschool children's affinity to nature at Halifax children's centers. Int J Early Child Environ Educ 6(2):42–58

Pan S-L, Chou J, Morrison AM, Huang W-S, Lin M-C (2018) Will the future be greener? The environmental behavioral intentions of university tourism students. Sustainability 2018(10):634. https://doi.org/10.3390/su10030634

Paz-Ferreiro J, Lu H, Fu S, Mendez A, Gascoet G (2014) Use of phytoremediation and biochar to remediate heavy metal polluted soils: a review. Solid Earth 5:65–75

Peuquet DJ (2002) Representations of space and time. Guilford Press, New York

Ralston A, Shaw K (2008) Environment controls gene expression: sex determination and the onset of genetic disorders. Nat Educ 1(1):203

Roth CE (1968) On the road to conservation. Massachusetts Audubon, pp 38–41

Roth CE (1984) Elements of a workable strategy for developing and maintaining nationwide environmental literacy. Nat Study 37(3–4):46–48

Roth CE (1991) Does your curriculum foster environmental literacy? Paper Presented at the National Conference of the Association for Supervision an Curriculum Development, San

Roth CE (1992) Environmental literacy: its roots, evolution and directions in the 1990s. Education Development Center, ERIC Clearinghouse for Science, Mathematics, and Environmental Education, Columbus

Schoenfeld AC, Meier RF, Griffin RJ (1979) Constructing a social problem: the press and the environment. Soc Probl 27(1):38–61

Scopinich K (2016) Losing a legend in environmental education. Mass Audubon. Retrieved 30 April 2022 at https://blogs.massaudubon.org/yourgreatoutdoors/losing-a-legend-in-environmental-education/

Selye H (1976) Stress without distress. In: Serban G (ed) Psychopathology of human adaptation. Springer, Boston, pp 137–146

Simpson EJ (1972) The classification of educational objectives in the psychomotor domain. Gryphon House, Vancouver

Snow CP (1959/2001). The two cultures. Cambridge University, Cambridge

Stapp WB (1969) The concept of environmental education. Environ Educ 1(1):30–31

Stern PC (1978) The limits to growth and the limits of psychology. Am Psychol 33(7):701–703

Stern P (2000) Toward a coherent theory of environmentally significant behavior. J Soc Issues 56(3):407–424

Stern PC, Dietz T, Abel TD, Guagnano GA, Kalof L (1999) A value-belief-norm theory of support for social movements: the case of environmentalism. Hum Ecol Rev 6(2):81–97

Saito Y (2010) Future directions for environmental aesthetics. Environ Values 19(3):373–391

Sepkoski J (1993) Limits to randomness in paleobiologic models: the case of Phanerozoic species diversity. Acta Palaeontol Pol 38:175–198

Sepkoski J (1997) Biodiversity: past, present, and future. J Paleontol 71(4):533–539

Sepkoski J (1998) Rates of speciation in the fossil record. Phil Trans R Soc Lond B353315–326

Schneider SM (2007) The tangled tale of genes and environment: Moore's the dependent gene: the fallacy of "nature vs. nurture". Behav Anal 30(1):91–105

Taylor C (2021) The explanation of behaviour. Routledge, Oxfordshire

Van Liere KD, Dunlap RE (1980) The social bases of environmental concern: a review of hypotheses, explanations and empirical evidence. Public Opin Q 44:181–197

Weigel RH, Weigel J (1978) Environmental concern: the development of a measure. Environ Behav 10(1):3–15

White L (1967) The historical roots of our ecological crisis. Science 155:1203–1207

Wilke RJ (1985) Mandating preservice environmental education teacher training: the Wisconsin experience. J Environ Educ 17(1):1–8

Wiseman M, Bogner FX (2003) A higher-order model of ecological values and its relationship to personality. Pers Individ Differ 34:783. https://doi.org/10.1016/S0191-8869(02)00071-5

Wongchantra P, Nuangchalerm P (2011) Effects of environmental ethics infusion instruction on knowledge and ethics of undergraduate students. Res J Environ Sci 2011(5):77–81

World Health Organization (2016) Urban green spaces and health—a review of evidence. https://www.euro.who.int/en/health-topics/environment-and-health/urban-health/publications/2016/urban-green-spaces-and-health-a-review-of-evidence-2016

The trees come up to my window like the yearning voice of the dumb earth. The fish in the water is silent, the animals on earth are noisy, and the birds in the air are singing. But man has in him the silence of the sea, the noise of the earth and the music of the air.

Rabindranath Tagore, Stray Birds. (1916)

Abstract

We explore environmental cognition, personality traits, social norms, environmental stress, and the healing environment. Cognition is the learning process of identifying the light, sound, smell, and feel of the space around us and then forming concepts of what we sense and then create visual images in our minds of what we are perceiving. This then allows us to respond appropriately to stimuli and what we believe to be true. Therefore, we review the cognitive theory of environmental learning and then move to an exploration of the social theory associated with environmental learning and our understanding of nature. The use of different epistemological methods gradually unlocks the influencing factors of environmental behaviors, such as personality traits and social norms with the objective of explaining social behavior. Finally, by understanding environmental stress, it becomes apparent that humanity needs redemption and healing through the power of nature, including nourishment of phytoncide, vitamin D, and the exposure to of outdoor environments such as forests and oceans to reduce stress. This then helps restore our physical and mental health and strengthens our thinking and decision-making processes.

5.1 Environmental Cognition

We learned about the concept of environmental literacy in Chap. 4, which is comprised of psychological attributes such as environmental knowledge, skills, and attitudes. These attributes cultivate an individuals' environmental literacy and practical capability in achieving professional and social goals in personal environmental education. However, environmental education is not so simple. As a process it requires us to incorporate and balance our professional goals with societal norms and values and our personal views and beliefs on environmental education. The focus on the social goals of environmental education seems odd and why is this important? Environmental psychology is a discussion about social issues that are shaped by learning, feeling, perceiving, attention, memory, thinking, imagination, emotion, will, capability, and behavior (Fig. 5.1).

© The Author(s) 2023
W.-T. Fang et al., *The Living Environmental Education*, Sustainable Development Goals Series,
https://doi.org/10.1007/978-981-19-4234-1_5

First, we discuss the concept of environmental cognition, which is a way of thinking that is beneficial to a person's survival and social well-being. How can we extend this approach from the perspective of personal psychology to self, other, local communities, and global communities?

Cognition is the process through which the brain recognizes the world around us through our senses and then forms concepts, perceptions of what we sense and then provides us with an image or allows us to visualize what we sense. The act of visualizing and/or remembering allows us to develop strategies to respond to what we sense, which may be further modified by similar or past experiences and what society expects or considers acceptable at the time. This entire process is a psychological function that each human being produces through information. Cognitive processes can be conscious and even unconscious mental models that are based on perception, attention, morphological identification, input, registration, and output that are produce mental models through propositions, imagination, log-in access, and thinking. Therefore, from the perspective of cognitive science, when humans learn in the environment, the intrinsic and/or extrinsic changes that occur are interpreted as "cognitive processes." This kind of view regards a human being's understanding of the environment as a part of learning, so it is called "cognitive theory." Environmental education scholars are epistemologists that are different from "stimulus–response" theorists. By repeating the practice of stimulus responses, we can achieve a view of environmental education and learning. If students do not understand the secrets of the environment, even if they practice many times, then there is no way to achieve the effect of learning. For example, in the study of climate, if you do not understand the principles of the atmospheric circulation, then one doesn't really

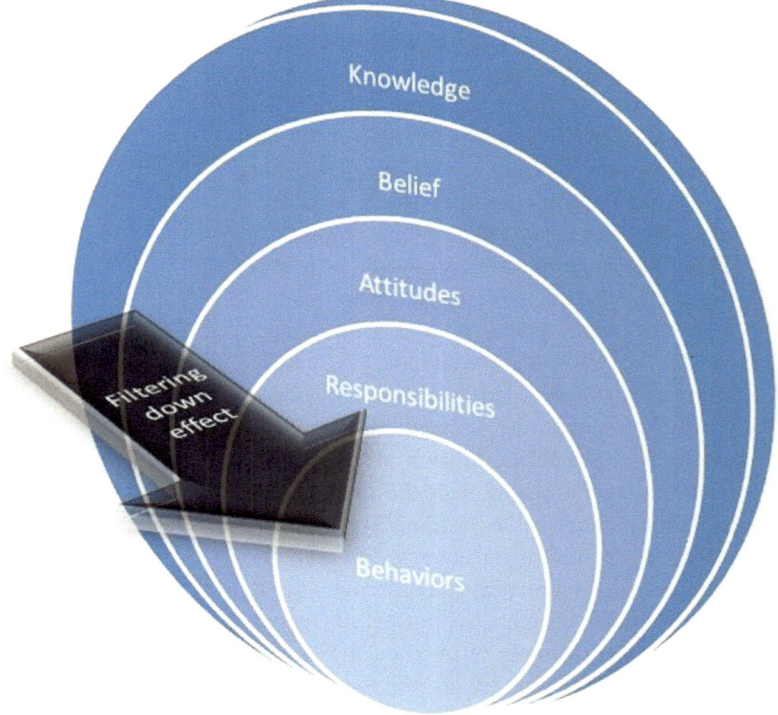

Fig. 5.1 Environmental psychology is a theory of the development of the processes associated with learning feeling, perception, attention, memory, thinking, imagination, emotion, will, capability, and behavior. The filtering down effect model is a socially, contextually, and culturally-bound goal-driven theory. If environmental knowledge is nothing more than information generated using sound science, then then this is considered real knowledge—especially "actionable" knowledge—by having desires and curiosity, through plotting and play in the real world (Illustrated by Wei-Ta Fang)

"understand the environment," and one cannot make circulation predictions. In the study of biodiversity, you cannot describe the anatomy, morphology, diversity of a species without "understanding the environment."

5.1.1 Cognitive Theory of Environmental Learning

Cognitive theory was derived from Gestalt Psychology in the early twentieth century, which values the wholeness of perception. In other words, when humans are in the environment, most people will not pay special attention to the irritation caused by the colorless, and odorless molecules or PM (particulate matter) 2.5 fine suspended particles in the air entering the nasal mucosa. Instead, most people focus on the stimulus provided by the entirety of the environment. Elements such as the plants, animals, sounds, and smells form the mind-body relationship. Environmental epistemology advocates whether human beings have a good learning effect in the face of learning situations and the following conditions need to be considered:

5.1.1.1 Situational Conflict

The first point that we need to consider is whether a new situation is consistent with an old experience. When human beings face new learning situations in the environment, they usually need to consider existing or similar situations that they've experienced and compare them with the new learning content that they are being presented with. When the learning objectives are clear and well understood, this kind of learning situation is more in line with the learning experience that the learner already has, so it's easier to enter a new situation and learn/ assimilate new data. However, if the new data conflict with a person's original or preconceived ideas, then the new data may not be assimilated or considered questionable.

There are three possible cognitive outcomes in climate change research (Tilbury 1995). The first is scientific consensus; the second is denial; the third is somewhere in between. Scientific consensus on any matter globally is rare, but that doesn't mean scientists should abandon discussions or solutions on global issues. We forget and need to be reminded periodically (e.g., Covid-19 pandemic) that we are just one, and not necessarily the most important species that inhabits this planet. Remarkably, this devastating virus has brought our species to its knees. The virus doesn't discriminate based on sex, race, color, or economic status. Our collective actions and behaviors impact all forms of life.

Evolutionary theory states that a species has few options to respond when faced with change: to evolve, migrate, adapt, or go extinct. Despite our jaded views on the outcome of this pandemic and others that are likely to come, environmental literacy will be essential, unified, and integrated into environmental education programs if humanity wants to continue to exist. As environmental educators, it is now our responsibility to inform and educate the public on the options available to survive the issues we are experiencing and will face in the future.

For example, the earth is getting hotter. July is typically the world's warmest month of the year, but July 2021 outdid itself as the hottest July and month ever recorded, said Rick Spinrad, the Administrator of National Oceanic and Atmospheric Administration (NOAA). July 2021 global surface temperatures were 0.93 °C above average, 0.01 degrees above the previous record set in 2016, 2019, and 2020.

In Taipei, Taiwan, in 2018, there were 60 days above 35 °C. In the summer of 2019, the dry and hot air brought from North Africa was transported north into Europe. The heat wave in Gallargues-le-Montueux southern France reached temperatures as high as 45.9 °C, which caused forest fires. Over the last decade the fire season in California is starting earlier and temperatures during the heat waves have been causing devastating fires with the loss of life and property throughout the state. Climate conditions are not expected to change soon, and agricultural practices and water use throughout the state remain unchanged. From a social and root cause standpoint, the people of California (or the world for that matter) either don't understand how climate change works or they don't want to know.

Environmental education programs may mitigate some of the more extreme changes, but our survival will require people to understand the world around them and adapt as needed. The temperature in Delhi, India is recorded to rise to 42.4 °C on April 9, 2022. In the summer of 2019, the highest temperature in Churu, Rajasthan, India was 50.3 °C and monkeys in the forest lost their lives. There were 21 heat waves in India in 2010 and in 2018, that number increased to 484 and during that time more than 5,500 people died. If this trend continues, India may be uninhabitable in the future.

5.1.1.2 Situational Reorganization

When a learning situation changes, a combination of old and new experiences arises, and the reorganization of previous experiences takes place. This kind of learning is not considered to be a fragmented learning experience, but a re-learned experience based on the old experience, injection of the new experience, and the form of the new experience. The cognitive response in this contextual reorganization includes the re-modulation of knowledge, meaning, and belief. This is a systematic view produced by the overall assessment of human cognitive processes, environmental situations, and individual emotions, feelings, and moods.

Vela and Ortegon-Cortazar (2019) assumed that there is a human subject and the external environment, and that the two systems interact. If the input relationship from the situation to the human subject leads to a psychological reaction, then the output from the human subject to the situation is called a human action. Under the influence of the natural environment, this kind of action has the following effects: cognitive response, affective response, and behavioral intention.

The components of the cognitive image include environmental elements that attract human subjects, which may affect human emotional responses and produce positive emotional systemic effects. This is an emotional representation that also has a positive effect on the behavioral intentions of others. Therefore, the emotional response should have a greater impact than the cognitive response (Vela and Ortegon-Cortazar 2019). Through measurement and analysis, we understand the environmental behavior intentions that affect human subjects. That is, in the study of environmental perception and environmental cognition, the existence or absence of environmental motivations, environmental goals, and attitudes to alternatives to environmental action are usually taken for granted. Moreover, the psychological response or process between human subjects and environmental actions is the focus of environmental psychology. These processes include obtaining information from the environment, the content of the information, the representation of environmental perception and cognition, and the judgment, decision making, and selection of the information represented. Knowledge of perceptual and cognitive processes can improve the quality of human environmental life through environmental decision-making, environmental planning, and design.

5.1.2 The Exploratory Theory of Environmental Learning

Environmental education is a process of learning and exploration, which needs to include cognitive, affective, and technical and participative domains (Tilbury 1995). In the process of constructing environmental competencies, we need to acquire ecological knowledge and interact with society. In addition, the people need to participate in actions as citizens through a sense of responsibility. If the learner is dependent on a place during the learning process, then a natural connection is created. Environment-based learning promotes the relationship between humans and the natural environment, building deep environmental knowledge as well as understanding of the world around the learner. Therefore, incorporation of local knowledge into environmental education activities is critical.

5.1.3 The Social Theory of Environmental Learning

In 1960s, the "Stimulus–Response Theory" developed by Burrhus Frederic Skinner (1904–1990) reached the peak of operant conditioning (Skinner 1967). But Noam Chomsky (1928–) criticized Skinner's classic conditioned reflexes and operating conditioned reflexes, which had a serious impact on human-driven psychoanalysis. In 1959, he challenged Skinner's *Verbal Behavior*, arguing that pure "stimulus–response theory" could not produce a learning response. So, what mechanism and under what circumstances can effective learning be achieved?

Against this background, Canadian psychologist Albert Bandura (1925–2021), in "Social Learning Theory," believes that learning is a cognitive process that occurs naturally in a social environment (Fig. 5.2). Teachers can observe or directly instruct the learning that takes place. In addition to observing student learning behaviors, teachers can also provide rewards and punishments, which is called alternative reinforcement. Bandura's Theory extends the traditional theory of behavior, in which behavior is strengthened and controlled, emphasizing the learning of various internal processes, which is important for human learning. Bandura studied the learning process that takes place in interpersonal relationships but didn't use the Theory of Operant Conditioning to explain it. He thinks that the weakness of conditioned response and reflexive-

response learning methods is not that humans have new responses to new stimuli, but that they offset the effects of social variables.

Skinner uses induction and a step-by-step approach to interpret an answer, using multiple experiments to strengthen the effect of the behavior. However, the reason for a behavior may be due to complex factors brought about by human subjective expectations and reinforced values. For example, children learn socially by observing the expressions on the faces around them. Bandura believes that in the process of growing up, children use observations to learn from people around them. The people are objects for observation and imitation and children learn by imitating what they observe. They may be rewarded for these imitations, which they then repeat because it's been positively reinforced.

After Bandura developed this set of social learning theories, he did not draw a gap with traditional learning theories but used this set of theories to form a communication bridge between behaviorism and cognitivism. This is because it focused on how psychological factors are involved in learning. Bandura (1977) therefore believed that his theory is an intermediary process and that humans are processors of information and not responders. Humans think about the relationship between their own behavioral responses and their behavioral consequences (Wicker 1969). This is because the discovery of mirror neurons in primates have promoted the development of social learning theories (Williams 2008; Cook et al. 2014). If

Fig. 5.2 The social theory of environmental learning (modified after Bandura 1977) (revised and illustrated by Wei-Ta Fang)

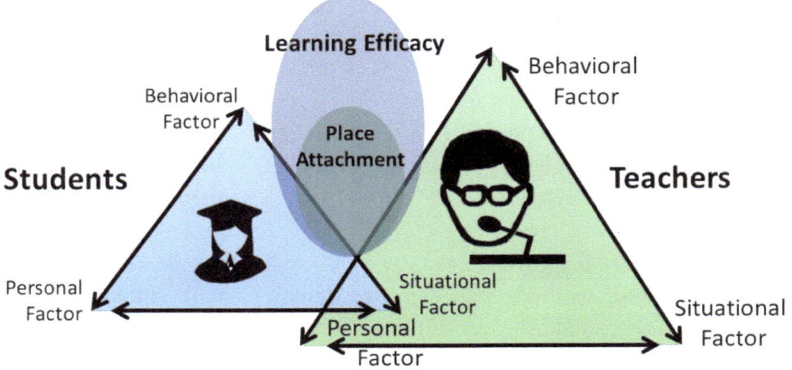

Fig. 5.3 We must draw our attention to social behaviors. Photo by Max Horng

these psychological factors interfere in the learning process, then the following intermediary processes will be produced:

5.1.3.1 Note

How much we pay attention to the actions of others. Individuals do not automatically observe the behavior of others and imitate them. We must draw our attention to social behaviors that require imitation. Therefore, we observe many behaviors every day, many of which we ignore. Therefore, attention is the first step in learning (Fig. 5.3).

5.1.3.2 Reservations

We remember the extent of this behavior. When we observe behavior, we produce visual and memory stimuli, but do not react immediately. We may notice this behavior, which is not performed immediately after memory retention and short/long-term memory formation.

5.1.3.3 Copy

We imitate after remembering the behavior shown by others. Although humans see many behaviors every day that we want to be able to imitate, this is not always feasible. We are limited by our bodies, our capabilities, and our technology. We can copy memories, but it takes time to repeat behavior.

5.1.3.4 Critical Evaluation

Social learning methods consider the thinking process, and thinking determines behavior, which requires patterns of behavior dominated

by thoughts and feelings at the time. Humans have considerable cognitive control over their own behavior. In particular, environmental constraints have a major impact on human behavior. Therefore, if environmental education is based on environmental information and teaching, the cultivation of environmentally friendly citizens can be limited. This is because the promotion of environmental education, cannot underestimate the complexity of human behavior. Pro-environmental behavior may be due to the natural biological factors and our interactions with the environment, resulting in a better explanation of better social behavior.

5.2 Personality Traits

Personality traits are the key to influencing human environmental behavior and one of the important elements in the decision-making process. In the past, there were few studies on environmental behaviors and personality traits (Kollmuss and Agyeman 2002; Dalvi-Esfahani et al. 2020). Scholars believed that personality traits and environmental behaviors were only related and not causal (Ajzen and Fishbein 1977; Fraj and Martinez 2006; Brick and Lewis 2016). In the study of personality traits, the generation of responsible environmental behavior is an important subject in environmental education (Harland et al. 2007; Dalvi-Esfahani et al. 2020). Although the causes of responsible environmental behaviors are complex (Hines et al. 1987), we hope that through different epistemological methods, we will gradually identify and understand the factors that influence environmental behavior, which will help to explain social behavior.

Tracing back the process of human behavior, personality traits are the key factors that affect human behavior (Ajzen 2005). According to the theory of social learning (Keen 2005), the emergence of responsible environmental behavior is not conceivable by a simple correlation path among a set of mindsets (Wals 2011). Therefore, we proceed from the past theory of responsible environmental behavior, explore the causal relationship among personality traits, and further explore the structural path from personality traits to responsible environmental behavior.

5.2.1 Personality Traits Responsibility

After constructing a research framework based on the theory of responsible environmental behavior, we examine the relationship between personality traits and environmental behavior. Among them, Pettus and Giles (1987) found that there is an inseparable positive correlation between good environmental attitudes and personality characteristics of self-control and determination. If we look for personality traits from a psychological perspective, then we can find that the Big Five personality traits have been widely used in psychological research (McCrae and Costa 1987). They are: openness to experience, conscientiousness, extraversion, agreeableness, and neuroticism. The characters have been used in Myers-Briggs type indicator and most other personality trat tests (Myers 1962).

5.2.1.1 Openness to Experience
Openness to experience including extensionality, shows that a person is curious, eager to learn, creative, and imaginative. They may possess active or physical skills such as independence, imaginative, responsive to art and beauty, attentive to their own feelings, willing to try new activities, intellectually curious, and advocate to freedom. However, they may not be restricted by tradition, like novelty, boldness, taste new, and intuitive thinking. They tend to be highly conservative, down-to-earth, observant, conventional, and a simple, but are considered to be lack creativity, think narrowly, lack curiosity, unwilling to take risks, non-artistic, and lead a routine life.

5.2.1.2 Conscientiousness
A Conscientious person has passion for work and a sense of responsibility in their work as a "tendency to respond in certain ways under certain circumstances" (Roberts et al. 2009,

2014). They are organized, efficient, systematic, practical, pragmatic, responsible, reliable, stable, self-disciplined, unyielding in matters, and self-reliant. When rigor is viewed from the opposite angle, it is called **indirectness** (Thurstone 1929).

5.2.1.3 Extraversion
Extraverts like group life, fun, talkative, active, and communicative, with spontaneity, dominance and vitality, explicit emotions, affectionate and warm characteristics, not lonely, outstanding performance in the group, vitality, and good at interpersonal relationships.

5.2.1.4 Agreeableness
Agreeable people are cooperative, trustworthy, and helpful. In addition, high affinity is often considered to be compassionate, soft-hearted, highly cooperative, tolerant, helpful, generous and trustworthy, warm, kind, selfless, broadminded, good-natured, and tolerant, forgiveness, flexibility, cheerfulness, modesty, courteous, comfortable life, straightforward and easy to be deceived. However, people with low agreeableness are considered cold, heartless, rude, selfish, mean, serious, low cooperation, high criticality, narrow-minded, harsh people, often vengeful, stubborn, cynical, thoughtful, and proud.

5.2.1.5 Neuroticism
The opposite angle of emotional stability is called neuroticism. Neuroticism is often considered anxious and self-doubtful; or tends to be irritable, stubborn, impatient, impulsive, emotional, often irritable, tense, suspicion, jealous, self-pity, and insecure, often making people worried and considered weak and subjective. On the other hand, people with high emotional stability are often considered calm and comfortable, often relaxed, and safe, patient, easily self-satisfied, impulsive, not jealous, strong and with objective.

5.2.2 The Green Personality Traits

The relationship between personality traits and environmental behavior is complicated and we discuss research performed by several scholars. For example, people with a high degree of openness to experience can appreciate nature with a wider perspective (Hirsh 2010). In environmental protection and green consumption behavior, personality traits are positively correlated with ecological consumption behavior (Fraj and Martinez 2006). In the eco-consumption market, extroversion, affinity, and rigor are the characteristics of consumers. In addition, agreeableness and openness to experience can highly predict the degree of environmental attention (Hirsh 2010). Conscientiousness is positively correlated to waste management behaviors such as recycling and waste reduction. Extraversion, agreeableness, conscientiousness, and neuroticism affect environmentally friendly tourism behaviors (Kvasova 2015). People with high emotional stability were significantly associated with environmental values and neuroticism was associated with saving electricity (Milfont and Sibley 2012).

People with "green personality" traits are generally open to experience, conscientious, and extraverted influenced on their environmental attitude to touch and experience in the real world (Figs. 5.4 and 5.5; Brick and Lewis 2014).

Fig. 5.4 Openness to experience will affect pro-environmental behavior. Photo by Max Horng

Fig. 5.5 People with high personality traits can affect attitudes and thus have a positive effect on the concept of control. Photo by Wei-Ta Fang

Openness to experience will affect pro-environmental behavior attitudes through environmental awareness (Poškus 2018); neuroticism will be affected by attitudes toward pro-environmental behavior through environmental awareness and expectations of environmental conditions in future (Chiang et al. 2019). We should employ this neurotic group to change their behavioral defaults and to empower their pro-environmental behavior (Poškus 2020). However, extroversion will affect environmentally-friendly attitudes through environmental awareness (Brick and Lewis 2016).

Among the five personality traits, they have an explicit nature, such as explained the connection between nature connectedness and personality (Lee et al. 2015). The relationship between nature and human's common personality correlates. Specifically, Lee et al. (2015) declared that the openness to experience and honesty–humility personality. Although some personalities can be hidden, whether it is emotional stability, extraversion, openness, affinity, and rigor, they

can be perceived in human daily life (Eaton and Funder 2001; Hermes and Riedl 2021). If these personality traits are related to attitudes, control concepts, and personal responsibility, then teachers can adjust the content of the material being taught by observing the personality traits of learners so that the content is relevant.

5.3   Social Norms

In Sect. 5.2 of this chapter, we focused on the personality traits of people that could affect environmental behavior. Social norms are important and need to be considered, especially at a time when global development is emphasizing sustainable development, it is important to develop environmentally friendly behaviors.

5.3.1   Specification

In Chap. 5, we pointed out that personal attitudes affect behavior, but in addition to personal factors, behavioral performance can be modified by social requirements (Fig. 5.6). In many cases, even if someone does not want to participate or comply, the need/desire to fit-in or peer pressure drives compliance (Heberlein 2012). Among the social norms that affect environmental behavior are broken down into subjective, injunctive, and descriptive norms.

5.3.1.1   Subjective Norms
Subjective norms can affect one of the variables of environmentally friendly behavior. The greater the pressure of normative support or opposition, the stronger the impact on behavioral intentions (Bock et al. 2005; Vermeir and Verbeke 2006). Subjective norms are related to the people around you.

5.3.1.2   Injunctive Norm
Injunctive norm can affect environmentally friendly behaviors when there are penalties and/or rewards for compliant/non-compliant behaviors. In an experiment of discarding advertising bills, the parking lot walls were affixed with orders and specifications to maintain cleanliness, and the

Fig. 5.6 Social norms are the
expectations that people
around us such as close
partners and friends impose
on us, which then modifies
our behavior. Photo by Max
Horng

parking lot was kept clean. In the parking lot
without ordering specifications, more than 30% of
the subjects would litter posting advertisements
(Keizer et al. 2008). Among the experiments
describing the specifications, the experiment
group is a description specification experiment
that maintains cleanliness, and the control group is
a description specification experiment that randomly discards advertisements. The former experiment can maintain the environment clean and
tidy in the parking lot, and the latter is a scattered
advertisement list (McKenzie-Mohr 2011).

5.3.1.3 Descriptive Norm

Descriptive norms are people's personal feelings.

5.3.1.4 Subjective Norms

Subjective norms are people's perception of a
particular behavior, which is influenced by the
judgment of important others. With this norm,
one personally feels people around them (Hernández et al. 2010; Thøgersen 2006).

Box 5.1: The Types of Norms

1. **Injunctive norms:** Injunctive norms
   is based on the principle of rewards
   and punishments for environmental
   protection, conduct recognized activities, and comply or violate relevant laws and regulations with

rewards or penalties (Heberlein 2012;
Hernández et al. 2010; McKenzie-
Mohr 2011; Thøgersen 2006)

2. **Descriptive norm:** Individuals observe
   whether important people around them
   will engage in a specific behavior
   (Goldstein et al. 2008; Heberlein 2012;
   Hernández et al. 2010; McKenzie-
   Mohr 2011; Thøgersen 2006)

3. **Personal norms:** Personal moral
   norms/personal ethics, personal self-
   expectation, a moral obligation feeling
   to perform certain actions, is regarded
   as a concept of self-worth (Bamberg
   and Möser 2007; De Groot and Steg
   2009; Heberlein 2012; Hernández et al.
   2010; McKenzie-Mohr 2011; Stern
   2000; Thøgersen 2006)

5.3.2 Direct and Indirect Paths for Social Norms to Predict Environmentally Friendly Behavior

Social norms affect environmentally friendly
behaviors, including direct and indirect effects.
Social norms can directly affect environmentally

friendly behaviors such as saving energy, maintaining environmental cleanliness, and protecting the natural environment.

Social norms can be predictive variables for environmentally friendly behavior (Mobley et al. 2010). In the discussion of the impact of social norms on environmental behavior, scholars believe that social norms produce a direct path that can directly predict behaviors, and some scholars believe that social norms produce an indirect path (Wiidegren 1998). The path relationship is through the influence of individual psychological variables, and psychological variables to influence behavior. However, depending on the type of environmental behavior, there are also gaps in the way social norms influence behavior (Hernández et al. 2010; Thøgersen 2006). In the experiment of saving water in hotels, the slogans of order specification and description specification were set up separately, requiring guests to use towels repeatedly to avoid unnecessary washing. The results of the experiment show that both injunctive norms and descriptive norms can affect environmental behavior (Keizer and Schultz 2018); however, conflict between the injunctive and descriptive norm produces weaker intentions (Smith et al. 2012). Descriptive codes are slightly more powerful than command codes (Goldstein et al. 2008). In addition, in household resource recycling behaviors, it was found that the descriptive norms are the same as the three variables of attitude, perceived behavior control, and subjective norms, and have predictive power for environmental behaviors (Wan et al. 2017). Descriptive norms can predict behavioral intentions in the use of public transportation's environmental behavior (Bissing-Olson et al. 2016).

This research incorporates descriptive norms into a psychological model to understand the impact of norms on behavioral intentions and environmentally friendly behaviors from environmental concern (Minton and Rose 1997). In the Park and Sohn (2012) study it was found that both injunctive and descriptive norms affect environmental behavior. According to a literature review, Thøgersen (2006) found a low correlation between injunctive norms and personal norms and subjective norms were highly correlated with personal norms. In his correlation analysis, subjective norms, description norms, and environmentally friendly behaviors were moderately correlated. The results of regression analysis showed the subjective and description norms all affected personal norms, and personal norms affected environmental behaviors. Descriptive norms affect not only the pathways that influence individual norms and then environmental behaviors, but also the direct pathways that directly affect environmental behaviors (Thøgersen 2006). Hernández et al. (2010) explored social norms, personal norms, and environmental behavior. According to the results of their path analysis, the order norms affected the subjective norms, the subjective norms influenced the individual norms, and the individual norms then influenced the path of the environmental behaviors, and the descriptive norms directly affect environmental behaviors. Command norms can affect subjective norms and personal norms in an indirect way, which then affects environmental behaviors. Subjective and descriptive norms can influence personal norms and environmental behaviors in a more direct way than command norms.

5.4 Development Pressure

Do you have or feel any environmental pressure from human development (Larkham 1990)? Why human development has been led a pressure? The concept of development pressure is clearly determined by Larkham during 1990s. Many developing countries have experienced unprecedented economic growth decoupled from resource conservation under complex environmental loadings. Human development could be represented as a total anthropogenic pressure (Poikane et al. 2017), and many scholars also developed their indices proposed in their papers, i.e., environmental pressure index (Gómez-Navarro et al. 2009); total anthropogenic pressure intensity index (Poikane et al. 2017), and human sustainable development index (in a positive way) (Bravo 2014) to measure an intensity index for urban development planning. For example, urban sprawl puts pressure on land and natural resources, resulting in undesirable outcomes with development pressure. However, pressure indicators could be more responsive and more sustainable policy relevant in many countries.

Why do we need to study human settlements? And how can we improve human settlements? Whey we have "development pressure" from human settlements? We need to know some issues why occur and how occur since human settlement with a poor quality of life (QoL) and low level of functioning (LoF) could be related to some pressures from developing issues dedicated to socio-demographic factor related to populations about their mental and physical health (Grassi et al. 2020).

If it is said that almost 11,000 years ago, Jericho, a Palestinian city in the West Bank in the Middle East, appeared to be the first settlements in human history. The first visitors tried to cool the hot weather, avoid floods, prevent deserts, and other natural disasters, and strive to change human behavior in space use. Scholars' research on settlements is a space concept that describes human dwellings and their surrounding aggregates (Larkham 1990). This humanistic space and environment are the research area of settlement environment. However, the settlements are dominated by village-type houses, and their formation and development have become the main component of the human landscape, such as the "webs of skyscrapers" in the surface block, see Fig. 5.7.

In the early days of human history, the Bible recorded that human had various dwellings 1656 years before the flood. The Bible recorded that after the flood, humans built the tower of Babel, which displeased God (Genesis 11: 3–4). An important filial piety of Buddhism, *Kṣitigarbha Bodhisattva Pūrvapraṇidhāna Sūtra*, Budda told to Kṣitigarbha Bodhisattva: "When born, but do good deeds, enrich the house, make the land happy, support the son and mother, have a great peace, and benefit."

Construction is an art to mankind, while architecture is a benefit (Taylor 2009; Labonnote et al. 2016). At a time when population is growing, water resources are becoming depleted, arable land, is disappearing, and ecological disasters are becoming more prevalent, the environmental sciences are emerging in the world could be considered an historic moment. We need to understand how humans multiply and our relationships with the environment (Fig. 5.8). Currently, the natural and anthropogenic environments are in conflict. Sustainability in urban and rural environments is a discipline that exposes the impact of human settlements on the natural environment (Yassi et al. 2001; Barton 2005).

Humans have overcome environmental pressures, showing human victory over the environment. However, because the subjects and objects of environmental pressures are unknown, environmental scientists and sociologists call it environmental pressure, while psychologists call it environmental stress due to development. The content is different, and the analysis is as follows:

5.4.1 Developing Pressure Defined by Environmental Scientists

Environmental pressure indicators reflect the impact of human activities on the environment, including environmental impacts caused by anthropogenic environmental pressure, which has caused substantial environmental problems. Environmental pressure includes exhaust gases, wastewater, and waste we generate. The system recovery is due to elasticity and resilience to restore its carrying capacity (Fig. 5.9). Humans and the pollution produced have caused environmental impacts and environmental protection problems that all depend on environmental scientists to propose solutions for environmental protection. But science alone cannot provide the solutions. We need to consider the social and economic aspects. Due to human resistance to environmental stress events and the pursuit of survival and happiness, humans will think about how to reduce stress and solve environmental problems. When stress occurs, there will be periodic reactions such as alertness, resistance, and fatigue (Taylor et al. 2006; Olson 2007). These stresses have physiological and

Fig. 5.7 Construction is an art to mankind; architecture is a benefit to mankind, the 101 tower and a spider are two of the symbolic conflict images in Taipei, Taiwan. Photo by Max Horng

Fig. 5.8 Sustainable urban and rural environment is a discipline that exposes the impact of human settlements on the natural environment (Women dressed Japanese kimono visited old buildings with full of curiosity, peace, and joy in Kyoto, Japan). Photo by Max Horng

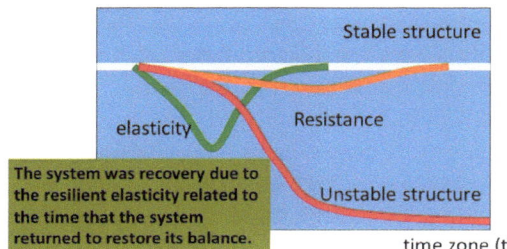

Fig. 5.9 Environmental pressure defined by environmental scientists could be applied to an entire ecosystem or an organism from stable structure toward unstable structure. Illustrated by Wei-Ta Fang

psychological effects. Next, we will take counter measures to the attitudes, ethics, environmental protection experience, and awareness of the consequences of environmental stressors.

5.4.2   Marine Impact and Stress Defined by Environmental Scientists

According to Fig. 5.9, we countered by nature, and caused depletion of marine resources. As marine resources dried up, fish stocks and coral reefs were declining (Figs. 5.10 and 5.11). The importance of environmental awareness to sustainability in fishing has been emphasized. The purpose of environmental awareness is to promote minimal impact on the environment and to advance the sustainability of the marine environment. Marine sustainability relies on developing positive environmental perspectives and fostering pro-environmental behaviors among fishing. To increase the awareness of biodiversity, we should provide training and education at legal fishing, which aim at informing fishers about improper behaviors that damage natural features they encounter. Take legal fishing as an example. Because of the emphasis on environmental resources, fishermen have adopted legal fishing methods, which have no environmental impact on traditional fishing. However, because of depletion of marine resources, fish stocks have been greatly reduced, and fishers cannot catch fish. Reduced income has created a social phenomenon of poverty from blue economy (Farmery et al. 2021).

We evaluate resources, strategies, and fishing effects if fishermen cope with stress from government. Inequality is likely to be an important factor in fisheries governance outcomes across a range of different societies (Fabinyi et al. 2015). We find that illegal fishing has created greater environmental problems, and the fishermen's economy has become richer, but the way they cope with the pressure has produced a wealthy situation, but the coral reef resources have been more exhausted.

Human beings use action to control or change psychological adaptation. One is to be comfortable with poverty, thinking that poverty is the result of morality and wisdom; the other is to take risks and take direct action to change the bad relationship between humans and the harsh environment around them. For example, at the risk of being arrested, trawling could be hidden into marine protected areas. Of course, fishermen may also adopt an adaptive way to adjust their emotions and cognition. For example, during the whale watching without illegal fishing (Chen and Chang 2017), they try to relax their emotions and watch the whale dolphin hunt their prey; or the whale watching dolphin who switches to ecotourism to protect coral reefs toward sustainable life (Fig. 5.11). Work to solve livelihood issues. However, if the income of marine ecotourism is not abundant, fishermen themselves cannot improve their meager income. If they think that switching from fishing to the tourism industry will not improve their final family income results, they will then have "I listen to experts and scholars' suggestions to change;" "Ecotourism, but it's useless." If the economy's income is extremely short, then it's cognitively felt that a single person's actions to protect marine resources will not affect the final depletion of marine resources, and it will become even more emotionally frustrated, "learning helplessness."

5.4.3   Climate Change Impact and Stress Defined by Environmental Scientists

Into the twenty-one century, human society has undergone tremendous changes. Due to the continued warming of the climate, seven severe typhoons hit Taiwan in 2001. By 2003, heat waves were raging in Europe, and the Greater Paris area warmed to 41.9 °C. In Europe, 70,000 people died of heat exhaustion (Robine et al. 2008). Hurricane Katrina invaded the United States in 2005, killing 1245 people and causing losses of up to US$ 125 billion. In 2008, severe snowstorms occurred in mainland China, and national traffic was paralyzed. In 2011, a severe tropical hurricane Yasi struck Australia. As the river skyrocketed, earth flows, sediments, and pesticides also polluted the ocean, the Great Barrier Reef, was affected. Heavy waves hit southern France again in 2019, reaching temperatures as high as 45.9 °C.

From the cataclysm of the natural environment, climate change, the environmental resources are becoming increasingly scarce. We may illustrate that the current climate change has caused a catastrophe in the fisheries and has created environmental pressure. This pressure

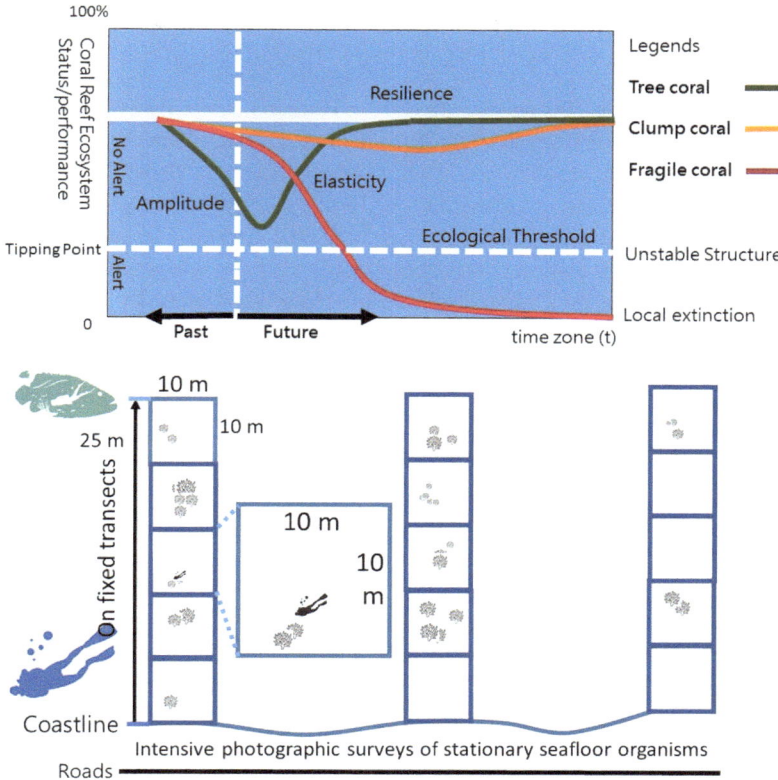

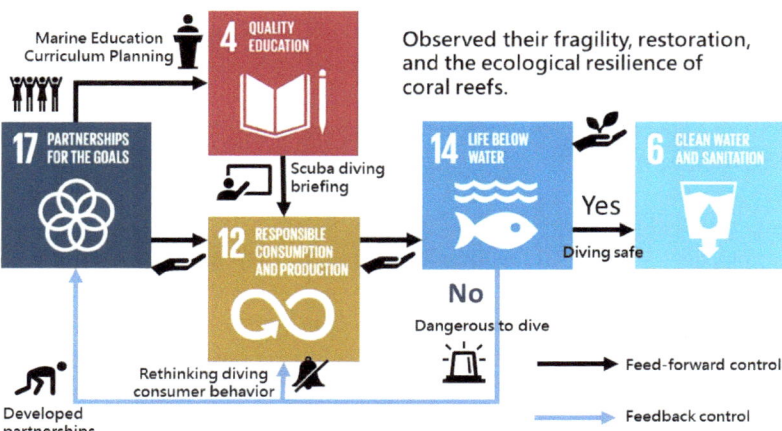

Fig. 5.10 Bottom trawls and fishing gears could be led harm to fisheries and to the marine environment by catching juvenile fish, damaging the seafloor, and leading to overfishing as well as coral reef damage. Bottom trawl nets can also harm sharks, and sea turtles we have detected in coral reefs. Illustrated by Wei-Ta Fang

Fig. 5.11 Environmental harm to deep-sea coral reefs on seamounts has been widely attributed to be lack of management from fishing and marine sports. Therefore, we need to train useful diving skills to learn more environmentally friendly behaviors for the seafloors. Marine sport training activities may also adopt an adaptive way to adjust their works beyond illegal fishing as a legal diving tour guided to instruct coral reef watch underwater. Illustrated by Wei-Ta Fang

has produced changes in the ecosystem. Obviously, marine species will be affected. However, to what extent climate change has caused environmental impacts; and which species are most vulnerable, scientists still have diverse opinions and still cannot give definitive answers. This illustrates that species have physiological limits, and they all migrate due to ocean warming.

Of course, we will look for answers from the fossil record, and most species have continued to exist on Earth during past climate changes. For the prediction of future impacts, scientists predict a wide range of species declines and regional extinctions, and species will also try to adapt or migrate. Therefore, through the climatic changes of the twentieth century, many species enter the twenty-first century in a state that reflects the changes of the environment, but will change the scope, limits, and type representation of survival. In addition, the reasons for the decline in species diversity are still largely unclear. However, recent climate-related reports that the species are gradually declining can still be used as a reference for recent research results, while helping guide current indicators for ecosystem management.

5.5 Healing Environment

The environment can cause human stress. Under the influence of the environment, we try to resist, slow down, or adjust the pressure brought by the harsh environment; however, in a comfortable situation, the environment can also heal the tired mind of human beings.

In the East, the healing environment is the place to create physical and mental healing. The goal of healing is mainly to reduce stress, which can reduce physical health, mood swings, and adjust logical thinking processes. For physical and mental symptoms of stress, the traditional eastern beliefs in the healing power of nature, such as Ayurveda in India, is an ancient healing tradition. In traditional India, it is recommended to spend each day in nature to let the senses experience the wonders of human existence. Nature can shift our attention beyond our own

sphere and allow the ego to be inextricably connected with the universe (Vaillant 1995; Gardner 2006) (Fig. 5.12).

Humans have always been interested in the healing power of nature. In the West, the most famous case is Henry David Thoreau (1817–1862), who spent two years of quiet time in Lake Walden, Concord, Massachusetts. When he wrote, it was also about life and nature that is the classic meditation process. In 1845, Thoreau felt the outdoor environment of the forest, allowing him to enjoy peace of mind and improve his health.

In addition, natural connection can enhance the spiritual level and strengthen the deep self-awareness. We can think about ourselves as being part of a larger universe than we can imagine. We can think about the warmth of entering my motherhood in my childhood and feeling the comfort of my mother. After you grow up, you often need to enter the natural environment to observe yourself and experience simple things to reconnect with natural motherhood. For example, we can walk barefoot in a mossy forest and feel "the sun spreads the net to catch the low temperature in the moss"; a natural connection of foam itching on the soles of the feet.

In recent years, eco-psychology has explored the relationship between humans and the natural world. Spending time in nature can reduce human stress and help relieve stress, thereby forming a healing environment to improve overall happiness. In healing environment, it can be basically divided into natural environment and man-made environment.

5.5.1 Natural Environment

When humans are exposed to nature, people often feel more generous, more connected to the community, and even more socially aware. Therefore, even just looking at photos of nature can enhance the sense of human-biological connection, thereby reminding humans of basic environmental values, such as generosity and care. A series of studies in the *Journal of Environmental Psychology* proved that people who are exposed to

Fig. 5.12 Assembly photos. In the East, the healing environment is the place to create physical and mental healing. You need to take a rest, and understand that you'll probably feel the best to your physical conditions eventually. Left (Wei-Ta Fang, Taipei, Taiwan, June 1, 2021): This photo explores the physiological correlates of a highly practiced Kundalini Yoga as a self-protected meditator (Arambula et al. 2001; Jung 2012) from thoracic and abdominal breathing way under the guidance of a competent teacher, Master Yen-Chen Cheng; Right (Wei-Ta Fang, Taipei, Taiwan, April 25th, 2021): You need to run by yourself, and running alone is still the best way to reduce your risk from Covid-19 during 2020 to 2023. Wei-Ta Fang is keeping getting in 30–60 min of moderate activity, with full daily nutrition from Angel Wings House directed by Max Horng, and he also got his booster shot as an additional dose of a vaccine for three shots to increase his immune system to keep viruses at bay. His philosophy is how to survive to live alone and not feel lonely or socially isolated? You need to learn self-isolation staying out of your room, that means your need to stay outdoors and completely avoiding contact with other people with your face mask during the COVID-19 Pandemic (Martinelli et al. 2021) (Self-photos by Wei-Ta Fang)

nature every day, although only 20 min, have a higher overall energy level and a better mood (Fig. 5.13). The reasons are as follows.

5.5.1.1 Vitamin D

It is a good way to spend sunny days outdoors (Beute and de Kort 2013). Let's put a smile on our faces. Sunshine provides us with vitamin D, calms emotions, regulates the nervous system, and improves seasonal affective disorder (SAD) in cold and gloomy winters. Because human beings surrounded by natural light have higher productivity and healthier lives. In addition, vitamin D can also promote the absorption of calcium in the body, and an appropriate amount of vitamin D can minimize the risk of hypertension, cancer, and other autoimmune diseases.

5.5.1.2 Phytoncide

What we smell in a forest, especially a conifer forest are phytoncides that have been volatilized. Phytoncides are antimicrobial compounds, called allelochemicals and these compounds can inhibit the growth of bacteria fungi, and other plants around the trees that produced the allelochemicals. In addition, smelling volatilized phytoncides can reduce blood pressure, stress, strengthen the function of parasympathetic nerves, improve sleep quality, and self-healing (Li et al. 2009; Putra et al. 2018). Walking through forests, enjoying the smell, and relaxing is a concept now called forest bathing. Regardless of the type of forest pinene is usually the most common and has bactericidal and exhilarating human health and wellness effects (Fig. 5.14).

Fig. 5.13 Sunshine provides us with vitamin D nourishment (Yehliu, New Taipei City, Taiwan)(Photo by Principal C.H. Chang of Yehliu Primary School)

Fig. 5.14 Phytoncide is highest in Taiwan fir, cedar, and camphor tree in the waterfall group (left: forest trails; right: waterfalls). Photos by Max Horng

5.5.1.3   Saliva Starch

When humans are stressed, they release hormones, including glucocorticoids and catecholamines. Non-invasive measurements of saliva, such as salivary alpha-amylase (sAA) and salivary cortisol (sC), can be used (Rohleder et al 2006; Nater et al. 2007; Bach et al. 2021).

To quantify the stress caused by the human, a social psychological stress response occurs as a biomarker of experienced fear (Buchanan et al. 2010). In a quiet and safe healing environment, psychological stress and saliva alpha-amylaseare are reduced (Minowa and Koitabashi 2012).

5.5.2 The Comfortable Environment

From the perspective of psychological cognition, we explored the environment based on the five senses, and then perform environmental cognition, which will be manifested in other aspects, from personality traits to external performance, and affected by social norms. Modeling external behavior should go under the influence of the environment and how do we resist the harsh environmental pressures? How can we achieve the effect of environmental healing in a comfortable environment?

The pleasant leisure environment can be indoor or outdoor environments. Human beings need to be free from daily life and enter a resting and recharging state where they can rest, meditate, and quietly experience their surroundings, sounds, and smells. A basic form of meditation is to bring self-attention to the present, rather than indulging in the past or worrying about the future. That is, stress and anxiety can be minimized.

To eliminate environmental stressors the healing environment needs to be connected to nature. And to enhance your sense of control and reduce the background noise of daily life, we need to find a quiet place to spend time to relax.

5.5.3 The Postnatal Environment

After human beings are born, they are torn from the mother's body and need to seek opportunities for social support (Martell 2001). At the moment of contact with the air, alienation of oneself begins. This kind of alienation produces self-perception from cognition, such as the development of touch, vision, and hearing; it begins to learn to classify (Given 2002).

The human baby then felt deprived of him, felt strange to the environment, and gradually alienated from his mother. As the mother and child are separated, the baby develops a sense of weakness, isolation, and maternal alienation. Due to the social environment, human babies need to learn peers, perform peer demonstrations, and learn about the alienation of the postnatal environment. Therefore, in the postpartum environment, it is necessary to provide a space for privacy for mothers and infants, provide an interactive context for mothers (Keefe 1988), and accompany the mother music developed for the healthcare environment to provide postpartum depressed mothers and anxious babies. The feelings of peace, hope, and connection, are providing opportunities for relaxation, peace, and relief.

5.5.4 Surfing in a Safe Virtual Environment After COVID-19 Pandemic?

Nowadays, smartphones are becoming increasingly popular that have brought many changes to our day-to-day "surfing lives," particularly with the ease of access to a vast variety of mobile applications for the purpose of internet browsing, gaming, social networking, communication after Covid-19 pandemic (Fang et al. 2021). This phenomenon has seen the number of smartphone users grow steadily from 2.5 billion in 2016 to 2.9 billion in 2018, and is expected to reach 3.8 billion by 2021 globally. If the smartphone technology has changed a human's growth and development, then there is evidence to support a close correlation between health, behavior and smartphone use (Fang et al. 2021). The adults and children who spent extended amount of time on smartphone in a day are more likely to exhibit a stronger negative pro-environmental behavior (Kesebir and Kesebir 2017). A. Several different aspects of smartphone usage have been detected, including: the effects of smartphone usage on children's academic performance, the relationship between smartphone usage and stress, and relationship between smartphone usage and social anxiety and loneliness (Gao et al. 2016). This is particularly important because our experiences gained through interactions with the environment during the early human development phase can have a considerable influence on a person's perception of the environment (Bandura 1986), not just a virtual environment (Fang et al. 2021).

5.6   Summary

Environmental education is the process of mastering the knowledge of environmental science and environmental psychology from the teaching experience, integrating different types of knowledge in different natural and social disciplines, and applying it to the teaching of "psychologize"—teaching objectives. Therefore, in this chapter, through environmental cognition, personality traits, social norms, environmental stress, and healing the environment, the experience under the framework of environmental psychology is embodied in understanding and realizing the characteristics of the natural environment and environmental cultivation. Environmental cultivation is a self-evident natural feeling (Fig. 5.15).

The Shakyamuni Buddha slaps flowers and His Holiness smiles. This is a kind of silent education, which completely transforms the human beings' understanding of humanity from an external situation. This requires not only the teachers' understanding of the diversity of environmental education, but also the value of the environment and the purpose of life. These two levels are important for teachers. In addition to the learner's understanding, the teachers also needed to inspire the learner's actions. At the same time, the human population has continued to expand. Although it has driven unprecedented economic growth, the earth has paid a price for human development. The rich and the poor are obviously uneven, and pollution is everywhere. To this day, environmental protection is not just a national and social issue, but an environmental psychological issue that has become part of an individual's values. The key question at the present time is whether human beings can develop the economy in a way that respects the ecological boundaries of the planet and supports the estimated 9.7 billion people in the middle of the twenty-one century. It has become the greatest collective psychological pressure on humankind to face sustainable development worldwide.

Fig. 5.15 The experience under the framework of environmental psychology is embodied in understanding of the natural environment. See Monica Kuo's work, she is the Department Head of Landscape Architecture, Chinese Culture University, Taiwan (Pratas Island, in the northern part of the South China Sea). Photo by Wei-Ta Fang

References

Ajzen I, Fishbein M (1977) Attitude-behavior relations: A theoretical analysis and review of empirical. Res Psychol Bull 84(5):888–918

Ajzen I (2005) Attitudes, personality and behaviour. McGraw-Hill, London

Arambula P, Peper E, Kawakami M, Gibney KH (2001) The physiological correlates of Kundalini Yoga meditation: a study of a yoga master. Appl Psychophysiol Biofeedback 26(2):147–153

Bach A, Ceron JJ, Maneja R, Llusià J, Penuelas J, Escribano D (2021) Evolution of human salivary stress markers during an eight-hour exposure to a Mediterranean Holm Oak Forest. A pilot study. Forests 12 (11):1600. https://doi.org/10.3390/f12111600

Barton H (2005) A health map for urban planners. Built Environ 31(4):339–355

Bamberg S, Möser G (2007) Twenty years after Hines, Hungerford, and Tomera: a new meta-analysis of psycho-social determinants of pro-environmental behaviour. J Environ Psychol 27(1):14–25

Bandura A (1977) Social learning theory. Prentice Hall, Hoboken

Bandura A (1986) Social foundations of thought and action: a social cognitive theory. Prentice Hall, Hoboken

Beute F, de Kort YAW (2013) Let the sun shine! Measuring explicit and implicit preference for environments differing in naturalness, weather type and brightness. J Environ Psychol 36:162–178

Bissing-Olson MJ, Fielding KS, Iyer A (2016) Experiences of pride, not guilt, predict pro-environmental behavior when pro-environmental descriptive norms are more positive. J Environ Psychol 45:145–153

Bock GW, Zmud RW, Kim YG, Lee JN (2005) Behavioral intention formation in knowledge sharing: examining the roles of extrinsic motivators, social-psychological forces, and organizational climate. MIS Q 29(1):87–111

Bravo G (2014) The human sustainable development index: new calculations and a first critical analysis. Ecol Ind 37:145–150

Brick C, Lewis GJ (2016) Unearthing the "Green" personality. Environ Behav 48(5):635–658

Buchanan T, Bibas D, Adolphs R (2010) Salivary α-amylase levels as a biomarker of experienced fear. Commun Integrative Biol 3(6):525–527

Chen CL, Chang YC (2017) A transition beyond traditional fisheries: Taiwan's experience with developing fishing tourism. Mar Policy 79:84–91

Chiang Y-T, Fang WT, Kaplan U, Ng E (2019) Locus of control: the mediation effect between emotional stability and pro-environmental behavior. Sustainability 11(3):820. https://doi.org/10.3390/su11030820

Cook R, Bird G, Catmur C, Press C, Heyes C (2014) Mirror neurons: from origin to function. Behav Brain Sci 37(2):177–192

Dalvi-Esfahani M, Alaedini Z, Nilashi M, Samad S, Asadi S, Mohammadi M (2020) Students' green information technology behavior: beliefs and personality traits. J Clean Prod 257. https://doi.org/10.1016/j.jclepro.2020.120406

De Groot JIM, Steg L (2009) Morality and prosocial behavior: the role of awareness, responsibility and norms in the norm activation model. J Soc Psychol 149:425–449

Eaton LG, Funder DC (2001) Emotional experience in daily life: valence, variability, and rate of change. Emotion 1(4):413–421

Fabinyi M, Foale S, Macintyre M (2015) Managing inequality or managing stocks? An ethnographic perspective on the governance of small-scale fisheries. Fish Fish 16(3):471–485

Fang W, Ng E, Liu S, Chiang Y, Chang M (2021) Determinants of pro-environmental behavior among excessive smartphone usage children and moderate smartphone usage children in Taiwan. PeerJ 9:e11635. https://doi.org/10.7717/peerj.11635

Farmery AK, Allison EH, Andrew NL, Troell M, Voyer M, Campbell B, Eriksson H, Fabinyi M, Song AM, Steenbergen D (2021) Blind spots in visions of a "blue economy" could undermine the ocean's contribution to eliminating hunger and malnutrition. One Earth 4(1):28–38

Fraj E, Martinez E (2006) Influence of personality on ecological consumer behaviour. J Consum Behav 5 (3):167–181

Gao Y, Li A, Zhu T, Liu X, Liu X (2016) How smartphone usage correlates with social anxiety and loneliness. PeerJ 4:e2197. https://doi.org/10.7717/peerj.2197

Gardner H (2006) Changing minds: the art and science of changing our own and other peoples minds. Harvard Business Review Press, Cambridge

Given BK (2002) Teaching to the brain's natural learning systems. ASCD, Alexandria

Goldstein NJ, Cialdini RB, Griskevicius V (2008) A room with a viewpoint: using social norms to motivate environmental conservation in hotels. J Consumer Res 35(3):472–482

Gómez-Navarro T, García-Melón M, Acuña-Dutra S, Díaz-Martín D (2009) An environmental pressure index proposal for urban development planning based on the analytic network process. Environ Impact Assess Rev 29(5):319–329

Grassi L, Caruso R, Da Ronch C, Härter M, Schulz H, Volkert J et al (2020) Quality of life, level of functioning, and its relationship with mental and physical disorders in the elderly: results from the MentDis_ICF65+ study. Health Qual Life Outcomes 18(1):1–12

Harland P, Staats H, Wilke HA (2007) Situational and personality factors as direct or personal norm mediated predictors of pro-environmental behavior: questions derived from norm-activation theory. Basic Appl Soc Psychol 29(4):323–334

Heberlein TA (2012) Navigating environmental attitudes. Oxford University Press

Hermes A, Riedl R (2021) Influence of personality traits on choice of retail purchasing channel: literature review and research agenda. J Theor Appl Electron Commer Res 16(7):3299–3320

Hernández B, Martín AM, Ruiz C, Hidalgo MdC (2010) The role of place identity and place attachment in breaking environmental protection laws. J Environ Psychol 30(3):281–288

Hines JM, Hungerford HR, Tomera AN (1987) Analysis and synthesis of research on responsible environmental behavior: a meta-analysis. J Environ Educ 18(2):1–8

Hirsh JB (2010) Personality and environmental concern. J Environ Psychol 30(2):245–248

Jung CG (2012) The psychology of kundalini yoga. In Shamdasani S (eds) The psychology of Kundalini Yoga. Princeton University Press, Princeton

Keefe MR (1988) The impact of infant rooming-in on maternal sleep at night. J Obstet Gynecol Neonatal Nurs 17(2):122–126

Keen M, Brown V, Dyball R (2005) Social learning in environmental management. Earthscan, London

Keizer K, Lindenberg S, Steg L (2008) The spreading of disorder. Science 322(5908):1681–1685

Keizer K, Schultz PW (2018) Social norms and pro-environmental behaviour. In Steg L, de Groot JIM (eds) Environmental psychology: an introduction, pp 179–188

Kesebir S, Kesebir P (2017) A growing disconnection from nature is evident in cultural products. Perspect Psychol Sci 12:258–269

Kollmuss A, Agyeman J (2002) Mind the gap: why do people act environmentally and what are the barriers to pro-environmental behavior? Environ Educ Res 8(3):239–260

Kvasova O (2015) The big five personality traits as antecedents of eco-friendly tourist behavior. Personality Individ Differ 83:111–116

Labonnote N, Rønnquist A, Manum B, Rüther P (2016) Additive construction: state-of-the-art, challenges and opportunities. Autom Constr 72:347–366

Larkham PJ (1990) The use and measurement of development pressure. Town Plann Rev 61(2):171–183

Lee K, Ashton MC, Choi J, Zachariassen K (2015) Connectedness to nature and to humanity: their association and personality correlates. Front Psychol 6. https://doi.org/10.3389/fpsyg.2015.01003

Li Q, Kobayashi M, Wakayama Y, Inagaki H, Katsumata M, Hirata Y, Hirata K, Shimizu T, Kawada T, Park BJ, Ohira T, Kagawa T, Miyazaki Y (2009) Effect of phytoncide from trees on human natural killer cell function. Int J Immunopathol Pharmacol 22(4):951–959

Martinelli L, Kopilaš V, Vidmar M, Heavin C, Machado H, Todorović Z et al (2021) Face masks during the COVID-19 pandemic: a simple protection tool with many meanings. Front Public Health 947. https://doi.org/10.3389/fpubh.2020.606635

Martell LK (2001) Heading toward the new normal: a contemporary postpartum experience. J Obstet Gynecol Neonatal Nurs 30(5):496–506

McCrae RR, Costa PT (1987) Validation of the five-factor model of personality across instruments and observers. J Pers Soc Psychol 52(1):81–90

McKenzie-Mohr D (2011) Fostering sustainable behavior: an introduction to community-based social marketing. New Society Publishers, Gabriola

Milfont TL, Sibley GG (2012) The big five personality traits and environmental engagement: associations at the individual and societal level. J Environ Psychol 32(2):187–195

Minowa C, Koitabashi K (2012) Salivary alpha-amylase activity—an indicator of relaxation response in perioperative patients. Open J Nursing 2(3). https://doi.org/10.4236/ojn.2012.23032

Minton AP, Rose RL (1997) The effects of environmental concern on environmentally friendly consumer behavior: an exploratory study. J Bus Res 40(1):37–48

Mobley C, Vagias WM, DeWard SL (2010) Exploring additional determinants of environmentally responsible behavior: the influence of environmental literature and environmental attitudes. Environ Behav 42(4):420–447

Myers IB (1962) The Myers-Briggs type indicator: manual. Consulting Psychologists Press, Palo Alto

Nater UM, Rohleder N, Schlotz W, Ehlert U, Kirschbaum C (2007) Determinants of the diurnal course of salivary alpha-amylase. Psychoneuroendocrinology 32(4):392–401

Olson K (2007) A new way of thinking about fatigue a reconceptualization. Oncol Nurs Forum 34(1):93–99

Park SY, Sohn SH (2012) Exploring the normative influences of social norms on individual environmental behavior. J Glob Scholars Market Sci 22(2):183–194

Pettus AM, Giles MB (1987) Personality characteristics and environmental attitudes. Popul Environ 9(3):127–137

Poikane S, Ritterbusch D, Argillier C, Białokoz W, Blabolil P, Breine J et al (2017) Response of fish communities to multiple pressures: development of a total anthropogenic pressure intensity index. Sci Total Environ 586:502–511

Poškus MS (2020) Normative influence of pro-environmental intentions in adolescents with different personality types. Curr Psychol 39(1):263–276

Poškus MS (2018) Personality and pro-environmental behaviour. J Epidemiol Community Health 72(11):969–970

Putra RRFA, Veridianti D, Nathalia E, Brilliant D, Rosellinny G, Suarez C, Sumarpo A (2018) Immunostimulant effect from phytoncide of forest bathing to prevent the development of cancer. Adv Sci Lett 24(9):6653–6659

Roberts BW, Jackson JJ, Fayard JV, Edmonds G, Meints J (2009) Conscientiousness. In: Leary MR, Hoyle RH (eds) Handbook of individual differences in social behavior. The Guilford Press, New York, pp 369–381

Roberts BW, Lejuez C, Krueger RF, Richards JM, Hill PL (2014) What is conscientiousness and how can it be assessed? Dev Psychol 50(5):1315–1330

Robine J-M, Cheung SLK, Le Roy S, Van Oyen H, Griffiths C, Michel J-P, Herrmann FR (2008) Solongo. CR Biol 331(2):171–178

Rohleder N, Wolf JM, Maldonado EF, Kirschbaum C (2006) The psychosocial stress-induced increase in salivary alpha-amylase is independent of saliva flow rate. Psychophysiology 43(6):645–652

Skinner BF (1967) In: Boring EG, Lindzey G (eds) A history of psychology in autobiography, vol 5. Appleton-Century-Crofts, New York, pp 385–413

Smith JR, Louis WR, Terry DJ, Greenaway KH, Clarke MR, Cheng X (2012) Congruent or conflicted? The impact of injunctive and descriptive norms on environmental intentions. J Environ Psychol 32 (4):353–361

Stern P (2000) Toward a coherent theory of environmentally significant behavior. J Soc Issues 56(3):407–424

Tagore R (1916) Stray birds. Macmillan, New York

Taylor A (2009) Linking architecture and education: sustainable design for learning environments. UNM Press, Albuquerque

Taylor AH, Dorn L (2006) Stress, fatigue, health, and risk of road traffic accidents among professional drivers: the contribution of physical inactivity. Annu Rev Public Health 27:371–391

Thøgersen J (2006) Norms for environmentally responsible behaviour: an extended taxonomy. J Environ Psychol 26(4):247–261

Thurstone LL (1929) Theory of attitude measurement. Psychol Rev 36(3):222–241

Tilbury D (1995) Environmental education for sustainability: defining the new focus of environmental education in the 1990s. Environ Educ Res 1(2):195–212

Vaillant GE (1995) The wisdom of the ego. Harvard University Press, Cambridge

Vela MR, Ortegon-Cortazar L (2019) Sensory motivations within children's concrete operations stage. Br Food J 121(4):910–925

Vermeir I, Verbeke W (2006) Sustainable food consumption: exploring the consumer "attitude–behavioral intention" gap. J Agric Environ Ethics 19(2):169–194

Wals AE (2011) Learning our way to sustainability. J Educ Sustain Dev 5(2):177–186

Wan C, Shen GQ, Choi S (2017) Experiential and instrumental attitudes: interaction effect of attitude and subjective norm on recycling intention. J Environ Psychol 50:69–79

Wicker AW (1969) Attitudes versus actions: the relationship of verbal and overt behavioral responses to attitude objects. J Soc Issues 25(4):41–78

Wiidegren Ö (1998) The new environmental paradigm and personal norms. Environ Behav 30(1):75–100

Williams JH (2008) Self–other relations in social development and autism: multiple roles for mirror neurons and other brain bases. Autism Res 1(2):73–90

Yassi A, Kjellström T, De Kok T, Guidotti TL (2001) Basic environmental health. Oxford University Press, Oxford

Ecocentrism goes beyond biocentrism with its fixation on organisms, for in the ecocentric view people are inseparable from the inorganic/organic nature that encapsulates them. They are particles and waves, body and spirit, in the context of Earth's ambient energy.

J. Stan Rowe, Ecocentrism,
1994.

Abstract

The framework of environmental ethics is built, challenging the way we view or interpret environmental education through the eyes of different stakeholders. In this chapter we consider aspects of land and ecological ethics as well as pedagogy as they relate to environmental ethics to form modelling. We classify that environmental ethics are "anthropocentrism," or the human-centered approach; "biocentrism," or the life-centered approach; and "ecocentrism," or the ecosystem-centered approach. Environmental paradigms are explored, which include the theories and practices regarding to environmental ethics, new environmental, ecological and behavioral paradigms, and paradigm shifts. Regarding to our choices from environmental values and concerns, we may use a model to detect our problem-solving approach to identify environmental problems we face and, find our practical needs and implement solutions toward sustainability.

6.1 What is Environmental Ethics?

Environmental ethics is an epistemological doctrine that is philosophically grounded that explores the relationship between humans and the environment. Many philosophical hypotheses relative to environmental ethics has established such as: *All things have intrinsic value* (Mazzucato 2020; Carney 2021). The social and natural sciences clearly have an influence on ethics (Bellah 1983; Schwartz 1987). Where is the ethics coming from, and when? Following birth of the life myth, we become detached from the warmth of motherhood, resulting in alienation and anxiety. We are born, grow, age, and before we die, we all look for ways to find connections between humans. Finding connections gives us status, identity, and value in the world that is closely related to Mother earth.

6.1.1 Beliefs of Land Ethics

Land ethics is a theory of environmental philosophy and considers how humans view and/or use the land in a moral sense (Callicott 1989, 2010). The term was coined by Aldo Leopold in *A Sand County Almanac* (1949) and in the middle twentieth century, it was considered a classic text of the environmental movement (Callicott 2005; Callicott et al. 2011). Leopold believed humans urgently needed a new ethic that dealt with the relationship between humans and land. He wrote (1949):

> The first ethics dealt with the relation between individuals; the Mosaic Decalogue is an example. Later accretions dealt with the relation between the individual and society. The Golden Rule tries to integrate the individual to society; democracy to integrate social organization to the individual.... There is as yet no ethic dealing with man's relation to land and to the animals and plants which grow upon it. The land-relation is still strictly economic, entailing privileges but not obligations....

This narrative provided an ecology-based land ethic that protected nature, developed the idea of a self-renewing ecosystem. and rejected the human-centered view of the environment. *A Sand County Almanac* is the first systematic introduction of an ecologically-centric method of environmental protection. While Leopold is credited with coining the term land ethics, numerous philosophical theories that explain how humans should treat the land followed (Callicott 1989). Economically-based utilitarianism, libertarianism, egalitarianism, and ecological land ethics were all considered (Callicott 1989; Noll 2017). Despite the plethora of definitions for the same concept, the UNEP in 1972 adopted Leopold's definition for designing the curriculum content for environmental education in different countries (Gruenewald 2004; Tete and Ariche 2021).

6.1.2 Beliefs of Anthropocentric Value System

The shift in emphasis from humans to nature is important. Theists believe that human beings exist on the earth and that they are superior to other forms of life and occupy a superior position. Therefore, all other forms of life are present to serve humanity's needs because human beings are superior and created in the image of God (Burdett 2015; Kilner 2015). But this doctrine is challenged by biblical commentators who believe that God wants human beings to be stewards or protectors of life on earth, thereby highlighting the plasticity of biblical interpretations.

Aristotle and Kant believed that only humans are moral creatures, because only humans have the ability to think rationally (Hurka 1996; Taylor 2010). This point of view is now referred to as anthropocentrism.

However, compared with the arbitrariness of western culture, oriental culture has its modesty. As Mencius (孟子)(Meng Tzu, 372– 289 BC), one of the Chinese Confucian philosophers, believed that humans and animals are very different from each other and only humans have a moral nature. Most people do not know the value of kindness. So, Mencius in the book of *Mencius* said (Lau 2004; Huang 2010): "Slight is the difference between man and beasts." Therefore, "They often abandon benevolence and morality, and only gentlemen know morality valuable." He tried to defend his claim of the innate goodness of human beings and claimed the human heart contains the sprouts of the four central Confucian virtues: benevolence (*ren*), righteousness (*yi*), propriety (*li*), and wisdom (*zhi*). Mencius believed that these sprouts needed to be nourished. The Taoist Zhuangzi (莊子)(Chuang Tzu, 369–286 BC) also often used animals to metaphorically view the world in his dreams. For example, he used the image of butterfly in his dream and when he woke (Möller 1999; Lee 2007), he thought about his lucid self and said: "Is myself coexisted in a butterfly's dream?" Zhuangzi argued that he'd rather to be a turtle playing in the mud. He talked to the vice-chancellor:

> I am told there is a sacred tortoise offered and canonized three thousand years ago, venerated by the prince, wrapped in silk, in a precious shrine on an altar in the temple. What do you think? Is it better to give up one's life, and leave a sacred shell, as an object of cult in a cloud of incense for three thousand years, or to live as a plain turtle dragging its tail in the mud?

"For the turtle," said the vice-chancellor,
"better to live and drag its tail in the mud!" "Go home!" said Zhuangzi.
"Leave me here to drag my tail in the mud."

Willing to be an official is equated with being a companion with a tiger, because being an official is a loss of human freedom. Literally it means staying close to the emperor and serving him is as risky as living with a tiger, for he may be killed by the emperor anytime. It carries a sense of worry and concern. Therefore, neither traditional Chinese Confucianism nor Taoism is the anthropocentric theorist in their human-centered theory. Even Zhuangzi's ideal is to learn swamp pheasants. "Pheasants in swamps have stepped ten steps, with one peck; one hundred steps, with one drink, and the muddy pheasants do not care about living in a fed cage associated with animal husbandry." This is a thought that pursues the freedom of humans and species (Wenzel 2003), rather than the only thought that respects humans. However, Zhuangzi is also similar in his anthropocentric theory on a dualism from this "radical critique of power and ultimate spiritual life" defined through human criteria from his theory (Kim 2009) since some other scholars regarding Zhuangzi as anti-anthropocentric thinker (Parkes 2013; D'Ambrosio 2022). This is quite controversial for an ancient Chinese scholar from his paradoxes of comments on radical critique of power and ultimate spiritual life in Western philosophy (Barrett 2011; Moeller 2015).

6.1.3 Beliefs of Biocentric Ethic Value System

In the past, this discourse on anthropocentrism had not been challenged until modern times. However, because of Darwin's theory of the evolution, the position of humans as a "superior" species has changed. Biocentrism is a moral point of view that extends the intrinsic value of life to all living things. The center of life is to explain how the earth works, especially what is related to biodiversity. Biocentrism encompasses all living things, extending the status of moral objects from humans to all living things in nature. As such biocentric ethics requires rethinking the relationships between humans and nature because nature no longer exists exclusively for the use or consumption by humans. Biocentrists believe that all species have intrinsic value and that humans are not morally or better than other species. The four pillars of biocentrism are:

(1) Humans and all other species are members of the earth;
(2) All species are part of an interdependent system;
(3) All creatures pursue their own advantages (good) in their own way; and
(4) Human beings are not better than other creatures.

Biocentrism does not imply an idea of equality between animals, as this phenomenon has not been observed in nature due to differences in their capacities (Singer 1997). Biocentrism is based on natural observations, not biased in favor of the human (Sterba 1998). Biocentrism should not treat humans as superior species (Sterba 1995). Proponents of biocentrism often promote biodiversity conservation, animal rights, and environmental protection. Biocentrism combines deep ecology with opposition to industrialism and capitalism (Johns 1992; Orton 1996; Barnhill and Gottlieb 2010; Farida et al. 2019) (Fig. 6.1).

6.1.4 Beliefs of Ecological Ethic Value System

Biocentrism contrasts strongly with anthropocentrism (Flores and Clark 2001). Anthropocentrism is centered on human values; however, biocentrism extends intrinsic value to the entire natural world (Bennett 2004). Because humans are one of many species in the world's ecosystems, any behaviors that negatively affect these ecosystems then negatively impact humans. Therefore, do we maintain a biocentric worldview or expand the moral category in the world? It depends how to extend all things to have

Fig. 6.1 Proponents of biocentrism often promote biodiversity conservation, animal rights, and environmental protection (Photo by Max Horng)

intrinsic value to strengthen the concept of ecological ethics (Sandler 2012).

The debate on environmental ethics with respect to *an Interconnected World* has become increasingly acute because of its interconnectedness and vulnerability to the ecosystem (Droz 2021). We previously intimated humans are part of nature and now we are intimating there is a *human ecosystem*. The *human ecosystem* should be regarded as *an organizing concept in ecosystem management* (Machlis et al. 1997).

What is Ecocentrism? Do we need to concern humans? Ecocentrism is the broadest term for worldviews that also recognize intrinsic value (Bennett 2004) in all lifeforms and ecosystems themselves, including their abiotic components (Washington et al. 2017). Proposed by 1990s (de Figueiredo et al. 2022), Rowe (1994b) declared ecocentrism puts a new interpretation on community from traditional ecological knowledge. The ideas of ecocentrism are focused on the entire biological community and committed to

maintaining the composition and ecological processes of the ecosystem (Shrivastava 2008; Fios 2019). Therefore, ecocentric approach to environmental ethics uses an *eco-holistic* perspective with the widest visions (Steverson 1991).

However, how is it different from biocentrism? Ecocentrism goes beyond biocentrism since ecocentrism having the widest vision. Biocentrism is implicitly establishing an equality among life-forms that favors or values all animals. Ecocentrism has been concerned about taking a broader view of our common home— planet Earth. Why eco-centrism is the key pathway to sustainability (Washington et al. 2017)? In a sense, Washington et al. (2017) declared eco-centrism has been with humanity since it underpins what can be called the *'old' sustainability*. Why we need to examine Leopold's principle of eco-centrism? Is this an 'old' sustainability to be detected from conservation biology? To answer this question, we may read one of the papers: *Making the law more eco-centric: Responding to Leopold and conservation biology* (Kuhlmann 1996).

Leopold (1949), recognized that all species, including humans, were the products of long-term evolutionary processes that interrelated in their life processes. His views on land ethics and environmental management are the key elements of ecological ethics. Rolston (1975) considered the responsibilities of the biota in their ecosystems, illustrating the philosophy of nature, and suggested nature needed to be protected according to ethical decisions and processes.

Ecocentrism is not an argument that all living things are of equal value (Washington et al. 2017). It does not deny the existence of countless important core issues, such as Nature Needs Half movement (Kopnina et al. 2018). Unlike many species, human beings are a resilient species in a rigid situation of under a climate-mediated mechanical change (Madin et al. 2008). However, human beings need to learn how to survive from their social networks and their living environment. Therefore, an ecocentric epistemology for ecosocialism can be reproduced social relations, sustaining habitat for sustainability (Salleh 2022). We may consider that Disinger (1990) described environmental world views as placed on an "ecocentric-anthropocentric continuum." While the dominant social paradigm follows the anthropocentric view. In addition, ecocentric practices also offer an alternative episteme for building a life-affirming civilization from resilience ethics (Bravo-Osorio 2022). This is one of the sound-science roots to support of a growing number of conservationists for ecocentric-based approaches addressing human concerns and directing human action regarding to the environment by the concepts of social-ecological resilience (Piccolo et al. 2018).

6.1.5   From Deep Ecology to Animal Rights

Deep ecology is opposed to the worldviews that emerged in the eighteenth century and proponents believe that the world is not a freely exploitable resource for humans (Gladwin et al. 1995). Therefore, the ethics of deep ecology holds that the survival of any ecosystem depends on this struggle for their lives for its overall well-being (Næss 1973; Bradford 1989). Deep ecology states (Næss 1973, 1985a, b, 1986, 1987, 1989):

(1)   The life of human beings or other living things on the earth itself have "value." This life value is not determined by the contribution of the non-human world to the human world;

(2)   Life forms have value in themselves; moreover, the richness and diversity of life forms contribute to the "realization" of these life values in themselves (Næss 1986, 2011);

(3)   Human beings have no right to reduce richness and diversity, except for the essential basic needs for sustaining life;

(4)   The prosperity of human life and culture is compatible with the small human p opulation. To maintain the abundance of other organisms, a small population needs to be maintained;

(5)   At present, human beings' excessive interference with other living things is rapidly deteriorating;

(6)   Humans must change policies that affect basic economic, technological, and ideological structures. As a result, the situation will be very different from what it is now;

(7)   Based on the natural value of life, the change in ideology is mainly due to the appreciation of "life quality" (Næss 1986, 2011), not the pursuit of a higher standard of living. We will be profoundly aware that there is a difference between "big ness" and "greatness." (McElroy 2002; Næss 2011); and

(8)   Anyone who agrees with the above viewpoints has the obligation to participate directly or indirectly in the necessary reforms (Næss 1986, 2011).

Deep ecologists have written an ambitious statement to change the current political and economic system (Devall 1980; Næss 1986; McLaughlin 1993; Pepper 2002; Zimmerman 2020). Næss (1984, 1986) emphasized the intrinsic value based on the relations to individual living being with its sense in a holistic system (Katz 1987). He believed that the connection of ecological phenomena affects the whole body in a Gaia sense (Næss 1995). Therefore, he believed that human beings should adjust their attitudes towards nature and use ecological worldviews for macro-control, otherwise the global environment will suffer.

Bill Devall (1938–2009) and George Sessions (1938–2016) cite *New Physics* in their book entitled *Deep Ecology,* said *the ultimate norms of deep ecology suggest a view of the nature of reality* in 1985 (Devall and Sessions 1985), described the new physics as the *view of reality* (Sessions 1987) as smashing Cartesian (René Descartes, 1596–1650) and Newton's (Sir Isaac Newton, 1643–1727) cosmic vision.

Devall and Sessions (1984) agreed to deny the empty image of nature as created by the human, and the *New physics* created by them denies that nature is a simple linear causal machine. Devall and Sessions (1985) argued that nature was in a state of constant change and rejected the notion that the observer was independent of the environment. They referred to the new physics presented in *The Tao of Physics* and the impact of new physics on the interconnectedness of metaphysics and ecology (Capra 1975). According to Capra, this should make deep ecology the framework of future human society. Devall and Sessions (1985) talked about ecological science itself and emphasized the links between ecosystems where is thus closely related to a rigorous determinism (Capra and Luisi 2014). They point out that in addition to scientific viewpoints, ecologists and natural historians have developed a profound ecological consciousness, including political and spiritual consciousness (Devall and Sessions 1985). They criticize anthropocentrism because ecocentrism is a discourse beyond human perspective.

Capra believed that this kind of complex network organization model will lead to a novel and systemic way of thinking. The ecosystem will be a form of autopoiesis and the structure and function of all ecosystems are complementary so they are indispensable (Fig. 6.2).

Ecosystems are unbalanced dynamic structures, but at the same time they can maintain dynamic stable structures. At a time when the ecosystem is constantly seeking to improve itself, continuously absorbing energy and matter from the environment, and releasing "entropy" to the environment. The ecosystem can even adopt a model of environmental destruction that exchanges "entropy" with the external environment to maintain its own stable. This presumes that the biotic elements of an ecosystem have the ability to adapt (Mersereau 2016; Feliciotti et al. 2018). Finally, the ecosystem uses social networks for system information exchange and repairs between systems (Capra and Luisi 2014).

Deep ecology affects the *Animal Liberation* movement (Flükiger 2009). Experts on animal liberation, such as Tom Regan (1938–2017) and Peter Singer (1946–), put forward the theory of animal protection. Regan inferred from the theory of benefit that human beings are not morally unique and equal judgments based on theory (Singer 1975). Regan wrote *The Case for Animal Rights*, which argues that humans cannot use rationalism as the principle of supremacy and only grant rights to those who have reason. In fact, these rights should be given to infants, vegetative, and non-human. These rights are intrinsic values, and humans should put the case of animals in moral considerations (Regan 1983).

Peter Singer's 1975 book, *Animal Liberation*, severely criticized anthropocentrism, and Singer disagreed with deep ecology's belief in the "inner value" of nature (Vilkka 2021). Singer took a more practical stand, called "effective altruism," meaning that protecting animals can bring greater benefits from utilitarian basis (Regan 1980).

Deep ecology and animal rights are in relation to environmental education (Kopnina and Gjerris 2015). However, animal welfare (AW), animal rights (AR), and deep ecology (DE) have often been absent within environmental education and education for sustainable development (Kopnina and Cherniak 2015). Therefore, we may try to realize the concept from "biocentric equality," according to Devall and Sessions, entails that *all organisms and entities in the ecosphere, as parts of the interrelated whole, are equal in intrinsic*

Fig. 6.2 The ecosystem will be a form of autopoiesis, and the structure and function of all ecosystems are complementary, so they are indispensable. The ecosystem uses social networks for information exchange with species like humans and animals. (*Bambusicola sonorivox*, as the common name Taiwan bamboo partridge, is a subspecies bird endemic to Taiwan (Hong et al. 2014) occurs at Qixingshan Trail, Yangmingshan National Park, Taipei, Taiwan) (Photo by Max Horng)

worth (Devall and Sessions 1985) and the *nature of reality*, ultimately, "intimately connected" with the environment (Borgmann 1995). Environmental education, therefore, is intimately related and their connection with nature should be intimately involved in the learning process from animal welfare (AW), animal rights (AR), and deep ecology (DE).

6.2 Environmental Paradigm

We talked about the philosophical basis of environmental ethics in the aforementioned section, mainly to provide philosophical concepts for environmental education. In addition, this section discloses a new environmental paradigm in modelling, utilizing as a measuring scale of the measurement for an evaluation approach on the environmental education research.

Do we need environmental paradigm? How do we need, and why? What is the relationship from environmental ethics to environmental paradigm? It should be noted, however, that a few studies have examined the relationship between environmental ethics and environmental paradigm (Dunlap and Liere 1984), but it does not seem obvious that we, humans, may not clearly understand our relationships with the biosphere where we depend on, and we also do not find out the fate and all well-beings on earth in the future.

Are we smart? May be not. Or we are only just a bug living on earth?

From the previous discussion, environmental ethics is a basic model of human morality; it belongs to a kind of "self-respect" and "external respect" for things that are deep inside and exposed.

In this section we will talk about "paradigm." What is paradigm? The term of paradigm in *the American Heritage Dictionary of the English Language* is defined as *A set of assumptions, concepts, values, and practices that constitutes a way of viewing reality for the community that shares them, especially in an intellectual discipline*. A paradigm for model could be fitting with applications to a real world. Therefore, in social science we do not replace necessarily an old one; various paradigms could be existing side-by-side (Kornai 2002). We try to extend the models for our multidimensional paradigm in many refining works on our studies. We tried to study conceptual model for environmental ethics to be constructed in several years.

However, what is a "model"?

Examples of a **model** include all concepts, assumptions, values, approaches, and benchmarks used to test truth in human activities. The word paradigm is derived from the Greek *paradeigma* and has the meaning of pattern, model, or plan, and refers to all applicable experimental situations or procedures. Therefore, a **paradigm** could be a significant scientific view of how to look at the world, which should be recognized by a community that provides a model. Plato (429–347 BC) coined the word paradigm, hoping to use it in the idea of its ideas or forms to resolve the way in which disputes over truth are discussed. The German philosopher Georg Lichtenberg (1742–1799) believes that the "paradigm" is an exemplary achievement. We can use this achievement as a model and use an analogous process to answer questions. Later, Ludwig Wittgenstein (1889–1951) talked about "paradigms" in the concept of language games, hoping to follow the process of analogy, let the questions be answered, and seek the truth in this world. This truth-based paradigm has allowed Riley E. Dunlap, a professor of sociology at Oklahoma State University, to study the nature and origin of environmental problems for 40 years.

6.2.1 Traditional Beliefs and Values

Dunlap and Liere (1984) emphasized the links between environmental issues, public opinion, and environmental decision-making. When he developed the New Environmental Paradigm Scale, he called the opposing paradigm the Dominant Social Paradigm.

What is **Dominant Social Paradigm?**

Dominant Social Paradigm (Pirages and Ehrlich 1974) was coined as one of the world views in human society, representing that *humans are superior to other all other species, the Earth provides unlimited resources for humans, and that progress is an inherent part of human history*. This term was developed by Pirages and Ehrlich (1974) and has been elaborated further by Dunlap and Liere (1984). In their studies, Dominant Social Paradigm of western industrial society could be containing political, economic, and technological institutions from a capital domain. It is these institutions that determine both the quality of life and environmental constructs within any society (Kilbourne 2006).

Dunlap examined the associations between traditional American beliefs and values (e.g., individualism, laissez-faire, and progressivism), environmental attitudes and behaviors (Dunlap 2022). Concerned about the beliefs and values of the Dominant Social Paradigm and his concern for environmental quality, Dunlap and his colleagues developed core elements for measuring environmental models and worldviews, and has applied his work in many countries (Dunlap et al. 1983; Dunlap and Liere 1984; Dunlap 2022).

6.2.2 New Environmental Paradigm

Dunlap's idea of a New Ecological Paradigm was developed in the 1980s, after he developed a New Environmental Paradigm Scale during 1976 and later published in 1978 (Dunlap and Van Liere 1978); and later published the New Ecological Paradigm Scale in 2000 (Dunlap et al. 2000; Dunlap 2008). His work is currently focused on an analysis of public opinion on climate change, the polarization of climate science and policy, and the analysis of negative sources and nature of climate change (Dunlap and McCright 2015).

We tried to introduce the **New Environmental Paradigm** (NEP) scale. The earliest model of environmental norms was proposed by environmental sociologist Riley Dunlap and his colleague (Dunlap and Van Liere 1978), and it's currently the most widely used environmental attitude assessment tool (Lalonde and Jackson 2002). Lalonde and Jackson (2002) argued that the NEP scale is limited *with respect both to the anachronistic wording of items and its inability to capture people's increasingly thorough understanding of the nature, severity, and scope of environmental problems* over the last four decade from now. The new environmental paradigm model is centered on human development and emphasizes the interaction between humans and nature. After presenting the first version of the scale, Dunlap merged it into a streamlined version of a set of six items by modifying the vocabulary, and a simplified version was used by John Pierce, who shared the information.

New Environmental Paradigm is related to the responses to individual environmental attitude questions (Pienaar et al. 2013). Environmental attitudes in our studies are measured by one-order constituents such as caring or not caring in a moral root. Later, environmental attitudes adopted a multi-component concept and were adopted in many studies (Cherdymova et al. 2018; Sorokoumova et al. 2021). Therefore, we may propose this New Environmental Paradigm model to be refined through history lesson that has some relevance on their multi-component concept to the topic from an idea of the paradigm shift in the theories of behaviors.

6.3 Paradigm of the Theory of Behaviors

We learned from Riley Dunlap's concepts of "The creation of paradigm" and "searching to the truth" that we need to rely on environmental psychologists to adopt human attitudes and

methods that are different from normal science to conduct experiments on human behavior. British scholar Edmund Burke (1729–1797) said (Burke 1790):

> The world would then have the means of knowing how many they are; who they are; and of what value their opinions may be, from their personal abilities, from their knowledge, their experience, or their lead and authority in this state.

Because, traditionally, we have thought that as long as humans have knowledge, their attitudes and values, will change their behaviors (Fig. 6.3). However, this argument is not absolute (Dunlap 1975). To explore the relationship between knowledge, attitude, and behavior is to find out whether the relationship has been wrong. That is, verifying that having environmental knowledge does not necessarily affect environmental attitudes, and that having attitudes does not necessarily affect pro-environmental behaviors. The relationship among them is very complicated.

The impact of environmental knowledge and environmental attitudes on people's indirect actions may be greater than that of people's direct pro-environmental behaviors (Kollmuss and Agyeman 2002). Economic factors, social norms, emotions, and internal logic have a great impact on people's decision on pro-environmental behavior. We conduct a review of human environmental behavior, including good behavior and bad behavior. We answer the question: "Why do we do what we should do?".

First, the moment a behavior occurs is a neurobiological explanation. That is, what kind of vision, sound, or scent, when a behavior occurs, causes the nervous system to produce this behavior? Then, what hormones respond to the stimulation of the nervous system in human individuals? In these sensory worlds of neurobiology and environmental endocrinology, we can try to explain what thoughts, attitudes, and behaviors will take place in the next moment (Sapolsky 2017) (see Fig. 6.4).

Of course, all behaviors can be traced back to the effects of structural changes in the nervous system, including adolescence, childhood, fetal

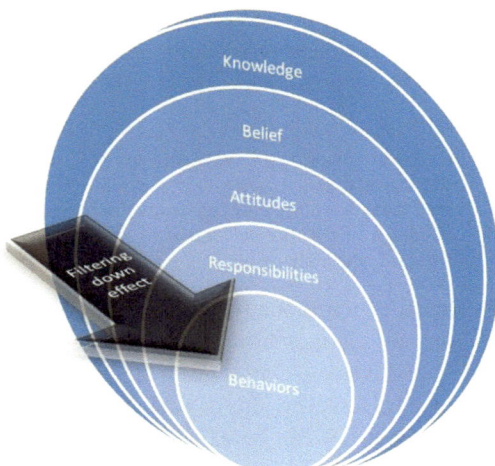

Fig. 6.3 Environmental efficacy by measured the function of behavioral change could be detected a lower level until the effect has reached the bottom from the educational and/or learning market in a civic society and/or at schools. We argued that in this complex area by presenting the conventional models of the linkages between knowledge and behaviors (Kerkhoff and Lebel 2006). The filtering down effect warned that the transfer function of the output, i.e., pro-environmental behaviors, less the scaling gain from the input of environmental knowledge investment enforced in a civic society and/or at schools. Marcinkowski and Reid (2019) argued that many attitude-behavior (A-B) relationships with substantial evidence were determined to be regarding as a relatively moderate strength (Illustrated by Wei-Ta Fang)

life, and genetic makeup. Finally, we should extend the perspective of environmental protection to social and cultural factors. Because, how does environmental protection culture shape personal environmental perceptions, and what ecological factors have formed this kind of environmental protection culture? From the perspective of environmental protection, pro-environmental behavior is one of the dazzling human behavior sciences. These issues involve the biophilia hypothesis, social norms, moral obligations, altruism, free will, and human values (Dunlap et al. 1983). All the achievements of environmental protection are human performances. We emphasize that practice itself is a symbol of an unknown hero because environmental protection is a nameless and lonely job. The following is explanation of the paradigm of

behavioral theories, including theoretical models such as the Theory of Planned Behavior (TPB).

6.3.1 Theory of Planned Behavior (TPB)

The Theory of Planned Behavior (TPB) is a behavioral decision model that was used to predict and understand human behavior (Ajzen 1985, 1991). The model is mainly composed of environmental attitudes, subjective norms, perceived behavior control, behavioral intentions, and behaviors theory (Fig. 6.5). It specifies the nature of the relationship between belief and attitude. According to the model, human evaluation or attitude to behavior depends on their belief in behavior, where belief is defined as the subjective probability that the behavior produces some result. Specifically, the evaluation of each outcome helps shape the behavior. In other words, a positive environmental attitude strengthens the pro-environmental intention.

(1) Subjective norm: An individual's perception of a particular behavior is influenced by the judgment of important others (such as parents, spouses, friends, teachers).

(2) Perceived behavioral control: The degree to which an individual perceives the difficulty of performing a particular behavior. Here we assume that perceived behavioral control is determined by the total set of accessible control beliefs.

In assessing important factors such as normative beliefs, social norms, attitudes, and perceived behavioral control, we may complete the development of the scale under social and cultural causes. While we clarify the causal relationship between important factors, we will understand the importance of social influence.

The concept of social impact is assessed through the social norms and beliefs. Humans' detailed thinking on subjective norms is based on whether their friends, family members, and society expect them to perform specific behaviors. Social influence is measured by assessing various social groups. For example, we provide in the

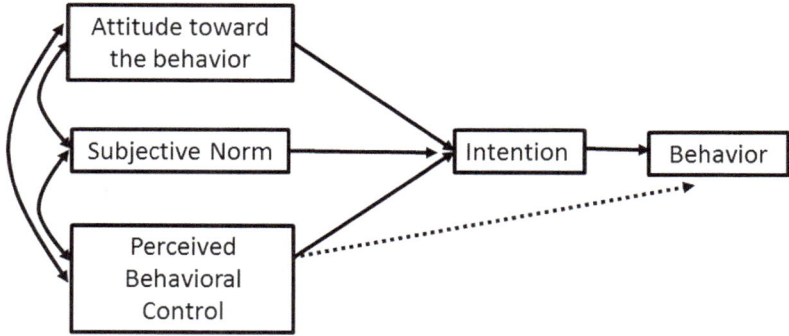

Fig. 6.4 In these sensory worlds of neurobiology and environmental endocrinology, we can try to explain what gestures and postures will take place in moments beyond planned behaviors (Merlion, as an official mascot of Singapore, named and designed by Fraser Brunner) (Elegant look by model by Chiao-Yen Chang; Photo by Max Horng)

Fig. 6.5 The theory of planned behavior (Modified after Ajzen, 1991; Illustrated by Wei-Ta Fang)

Attitude toward the behavior → Subjective Norm → Perceived Behavioral Control → Intention → Behavior

case of smoking (Ajzen and Manstead 2007). Subjective norms from peer groups, including ideas such as: "Most of my friends smoke," or "I feel ashamed to smoke in front of a group of non-smoking friends," and subjective norms of the family, such as: "The idea that family members smoke and it seems natural to start smoking; or "My parents were really mad at me when I started smoking"; and subjective norms from society or culture, including things like:

"Everyone is against smoking," and ideas like "we just assume everyone is a non-smoker."

Although most models are conceptualized in the individual's cognitive space, planned behavior theory is based on collectivistic culture-related variables to consider social influences, such as social norms and normative beliefs. Whereas, individual behaviors (including health-related decisions such as diet, condom use, smoking cessation, and alcohol consumption), may be built on social networks and organizational knowledge (for example, peer groups, family members, school faculties, and the workplace colleagues). Social influence has a great influence on the Theory of Planned Behavior. Therefore, in the social norms that affect environmental behavior, in addition to subjective norms, describing norms may also be one of the important variables.

At present, the Theory of Planned Behavior has been applied in research fields related to environmental protection and public health (Fang et al. 2017; Liu et al. 2018; Woo et al. 2022) as well as the similar research modelling studies, such as Fang et al. (2021a, b). The research found that the most important psychological variables that affect behavioral intentions are different in various groups and regions. Respondents have different conditions, such as those with a high degree of environmental care. Perceived behavioral control is an important variable, while those with a low degree of attitude are important variables that affect environmental behavioral intentions. In addition, different regions and interviewees have different conditions and the important intermediary variables that directly affect behavior are also different (Bamberg 2003). For example, when buying environmentally friendly products. At the national level, attitudes are the most important variable in Spain (Nyrud et al. 2008). Take the example of switching to public transportation without a car. In Frankfurt, Germany, perceived behavioral control is the most obvious, and in Bochum, Germany, attitude is the most important variable (Bamberg et al. 2007).

The Theory of Planned Behavior holds that subjective norms can directly influence behavioral intentions (Ajzen 1991), but does not discuss whether descriptive norms affect behavioral intentions. In terms of environmentally friendly behavior, in recent years, researchers have tended to include description norms and subjective norms (such as the expectations and support of important people around them) as social norms. Social norms affect individual psychological variables, such as social norms affecting attitudes, and attitudes in turn affect environmentally friendly behavioral intentions. A little different from the Theory of Planned Behavior, social norms influence environmentally friendly behaviors in an indirect way (Thøgersen 2006; Bamberg and Möser 2007; Hernández et al. 2010; McKenzie-Mohr 2011).

6.3.2 The Motivation-Opportunity-Abilities Model

The Theory of Planned Behavior emphasizes environmental attitudes, subjective norms, and perceived behavioral control. Another type of integration model is the "Motivation-Opportunity-Abilities" (MOA) model proposed by Ölander and Thøgersen (1995). The important structural feature of the MOA model is to integrate motivation, habits, and background factors into a single model of pro-environmental behavior. Because environmental protection behaviors are mainly habitual behaviors, they are not necessarily conscious behaviors based on conscious decisions.

Ölander and Thøgersen (1995) point out that the improvement of behavioral ability can be predicted by combining the concept of capability to strengthen conditions and transforming behavior into a model through opportunity (Fig. 6.6). In the model, in addition to the behavioral environmental attitude, subjective norms, and perceived behavioral control are the contents of the original model of planned behavior theory, the MOA model adds the following:

(1) **Motivation**: As each person's value system is different, personal needs and desires may

Fig. 6.6 Motivation-opportunity-ability theory (Modified after Ölander and Thøgersen 1995; Thøgersen 2009; Illustrated by Wei-Ta Fang)

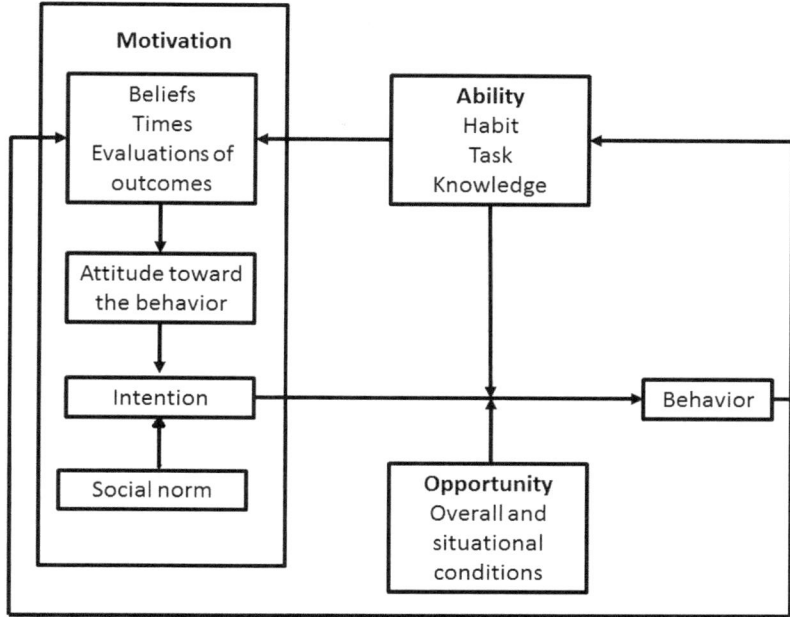

affect their behavior in some way. The so-called motivation is the motivation of behavior. Motivation is a prerequisite for generating incentives and rewards through behavior types and behavior outcomes that are beneficial to the environment. Because of praise or other encouragement, human beings can encourage pro-environmental behaviors based on rewards. For example, motivational rewards can be as simple as volunteers' efforts to promote environmental education and gain recognition from the general public.

(2) **Opportunity**: Opportunity is a limitation of the availability of time and resources. The opportunity composition of the MOA model belongs to the "objective prerequisites for environmental behavior." This model also has some similarities with the concept of perception in planned behavior theory. Often, we look for opportunities to accomplish a task that will benefit us or others.

(3) **Ability**: Ability is a strength of a person's cognitive, emotional, technical, or social resources that can be used to perform specific actions. The concept of competence

regarding to ability should include knowledge, habits, and tasks. Among them, habit is an independent behavior, and it is also one of the main items that determine the intention of the environment.

6.3.3 The Value-Belief-Norm Theory

The Value-Belief-Norm theory (VBN) is the development of decision theory by Stern et al. (1999, 2000) improved communication between stakeholder groups by establishing consensus on important behaviors affecting the environment (Stern et al. 1999; Stern 2000). The main structure is through the individual variables linked by the causal chain, he developed the VBN theory (Fig. 6.7), which is connected by the causal chain of five variables: values, ecological world view, awareness of consequences, ascription of responsibility, pro-environmental personal norms, and pro-environmental behaviors. Each chain directly affects the next variable, and each variable may also indirectly affect the next variable. Values affect beliefs, beliefs affect personal norms, and personal norms affect pro-

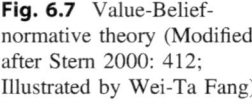

Fig. 6.7 Value-Belief-normative theory (Modified after Stern 2000: 412; Illustrated by Wei-Ta Fang)

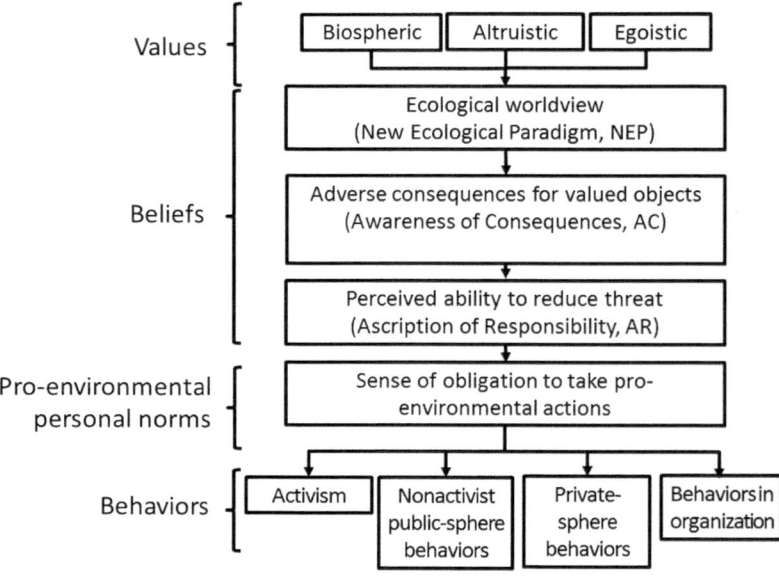

environmental behaviors. Values are divided into ecological values, altruistic values, and biosphere values; beliefs are derived from the ecological world view, human's awareness of the consequences of the adverse environment, and the ascription of responsibilities, so that people believe that their actions can slow the negative factors of the environment; the previous factors affect personal norms. Personal norms are the only variables that affect environmental behavior. Environmental behaviors include activism, non-activist public-sphere behaviors in the public domain, behavior in the private sphere, and behavior within the organization, as described below (Fig. 6.7):

(1) **Ecological Worldview**: This is a world view of sustainable development. Its purpose is not to maintain the status quo, but to strengthen the health, adaptability, and evolution potential of a fully integrated global social ecosystem. The ecological worldview is a kind of self-regeneration, thus creating conditions for the prosperity and rich future of the ecological environment, including the integrity of the ecological environment, social relations, and the

transformative nature of the economy. These models can strengthen the ecological environment of regeneration and sustainability.

(2) **Awareness of Consequences (AC)**: awareness of the impact of environmental issues (Hansla et al. 2008; Fang et al. 2019).

(3) **Ascription of Responsibility (AR)**: The attribution of responsibility is the reason for the occurrence of environmental problems, summarize their causes, and bear the negative facts that need to be assumed, attributed, dealt with, or controlled by the environment. This is the environmental importance influence factors in behavior (Hines et al. 1986/1987; Kaiser et al. 1999; Fang et al. 2019; Chao et al. 2021).

(4) **Pro-environmental Personal Norms**: Personal norms are often discussed with morality (De Groot and Steg 2009; Fang et al. 2019; 2021a), and are also regarded as a concept of self-value extension. Personal norms are simply the recognition of obligations and morals, and are considered to be a self-disciplined consciousness that may be related to the generation of environmental behavior.

(5) **Activism**: committed environmental activities and actively participate in environmental organizations.

(6) **Non-activist Public-sphere Behaviors**: Support or accept public policies is like the willingness to pay higher environmental protection tax. Non-aggressive behavior in the public domain affects public policy and may have a significant impact on the environment, as it can immediately change the behavior of many people or organizations,

(7) **Private-sphere Behaviors**: The purchase, use, and disposal of personal and household products that have an impact on the environment will have direct environmental consequences, but the effects will be small.

(8) **Behaviors in Organizational**: Individuals may significantly influence the goodness of the environment by, for example, affecting the behavior of their affiliated organizations. For example, developers may use or ignore environmental standards in their development decision-making process, and may do so because of right or wrong things. Make decisions to reduce or increase pollution from commercial buildings. Organizational behavior is the largest direct source of many environmental problems.

The Value-Belief-Normative theory uses intent-oriented definitions that focus on human beliefs and motivations in order to understand and change target behaviors. Value-Belief-Normative theory provides a description of the reasons for the general tendency to environmental behavior. Environmental behavior depends on a wide range of contingencies; therefore, Stern (2000) argued that the general theory of environmentalism may not be very useful for changing specific behaviors. Because different kinds of environmental behaviors have different reasons, and their causal factors for causality may be very different between behaviors and individuals, each target behavior should be theoretically separated. If the above causality affects each other, attitude reasons have the greatest predictive value for individual behaviors from different backgrounds. However, for more difficult environmental protection behaviors, environmental factors and personal capabilities may cause more variation. Although VBN theory is concerned with explaining the reasons for environmental behaviors, VBN theory cannot explain all behaviors. He also suggests that future research can identify important behaviors and discuss the factors that affect them (Stern 2000).

6.3.4 Two-Phase Decision-Making Model

Hirose (1994) considered the process of forming behaviors and proposed that pro-environmental behavior can be explained by a two-phase decision-making model (Fig. 6.8). The first phase involves the formation of environmentally friendly attitudes and the second involves various behavioral assessments to determine environmental behavioral intention that will directly or indirectly influence the pro-environmental behaviors. An environmentally friendly attitude refers to "the intent to solve an environmental problem or make a contribution" that supports ecofriendly behavior that is accompanied by a degree of respect for the environment and express concern for ecological issues. It involves three factors:

(1) **Perceived Seriousness:** This represents the perception of the consequences of environmental problems (Chao et al. 2021). The perceived seriousness emphasizes the perception of environmental risk, the severity of environmental pollution, the likelihood of occurrence, and the perceptions and expectations of the likelihood of occurrence of the environmental problem and the severity of the problem. However, individuals may feel that their power is insignificant despite the impact they have on larger-scale problems.

(2) **Ascription of Responsibility:** Ascription of responsibility refers to the recognition of the cause of responsibility (Fang et al. 2019, 2021b; Chao et al. 2021), that is, the perception of responsibility. Specifically, who or what causes environmental pollution and damage.

Fig. 6.8 The two-phase
decision-making model
(Modified after Hirose 1994;
Illustrated by Wei-Ta Fang)

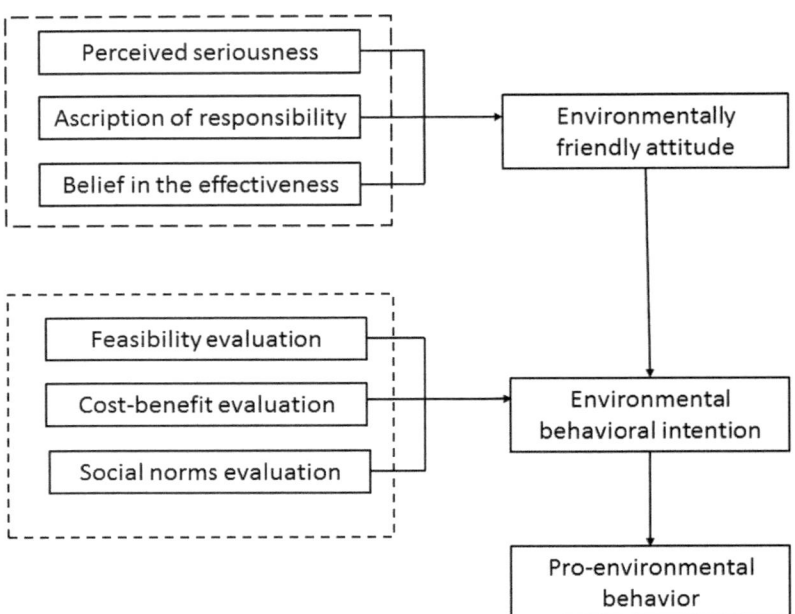

Although it is easy to attribute the cause of complex environmental issues to natural phenomena, residents often place the blame on themselves and therefore different actions can be taken to solve environmental problems depending on the responsibility involved.

(3) **Belief in the Effectiveness:** This is the recognition of validity of a counter measurement to solve the environmental problem. For instance, a sense of effectiveness can arise if one considers the environmental problem to be solvable by an individual and/or collective efforts of other people. In contrast, if one feels that there will be limited or no effect on addressing the environmental problem regardless of the commitments and efforts put in, then a sense of effectiveness will not substantiate.

(4) **Feasibility Evaluation:** The non-economic factors that are considered when determining if it is practicable to adopt a pro-environmental action. It also helps to assess whether individuals can engage in pro-environmental behaviors when opportunities arise externally and internally.

(5) **Cost–Benefit Evaluation:** This type of evaluation assesses the benefits of adopting

pro-environmental actions and the costs involved. The main evaluation criteria for comparing the two are the personal benefit and cost evaluation, such as convenience and comfort. If the reduction in personal benefits and the increase in costs of taking pro-environmental actions are significant, then no action is taken and vice versa.

(6) **Social Norm Evaluation:** The assessment of whether an individual's behavior conforms to the norms and expectations of an organization or society. In the theory of planned behavior, social norm evaluation corresponds to the subjective norm, so the two-phase decision-making model derived from the theory of planned behavior also uses the subjective norm as an assessment item.

(7) **Environmental Behavioral Intention:** This refers to the extent to which individuals are willing to consider taking appropriate actions to protect the environment, and this is directly linked to the formation of the target "pro-environmental behavior."

Chao et al. (2021) revised the application of the two-phase decision-making model to include the variables of social needs to explain citizen science engagement behaviors. The three influencing variables of social needs were social networks, learning and growth, and belonging and contribution. The results indicated that both the development of an environmentally friendly attitude in the first phase and the series of behavioral assessments generated in the second phase were influenced by the social needs. Therefore, a two-phase decision-making model was developed to incorporate the variables of social needs was proposed (Fig. 6.9). There was evidence from Chao et al. (2021) to indicate the occurrence and effects of social networks and needs in the two-phase decision-making model. Thus, the two-phase decision-making model that incorporated the key variables (i.e., social networks, learning and growth, and belonging and contribution) of social needs had provided a more comprehensive understanding about the citizen science participation behaviors.

6.4 Paradigm Shift

Environmental protection has been underway for over 60 years and although we've made great strides on environmental issues, environmental pollution, reduced biodiversity, and global warming are based on symbols of this era. Some people insist on an early worldview and refuse to deal with the reality that our environment is changing. However, whether this an issue of young people that are environmentally conscious of a younger generation versus that of an older generation needs to be further assessed.

6.4.1 Dominant Social Paradigm

Dominant Social Paradigms advocate economic growth, but its popularity in the policy world is relatively short lived (Fang 2020: 12). In 1940, Western governments used gross domestic production to measure economic growth and to support employment goals. In 1950, economic growth became the focus of government policy. This "growth" is currently a goal supported by the Organization for Economic Cooperation and Development. The Dominant Social Paradigm (DSP), however, is too optimistic about social development regarding to concern for environmental quality (Dunlap and Liere 1984). In order to solve environmental pollution, a mainstream person proposes to improve the efficiency of resource utilization through technological improvement and sustainable communication

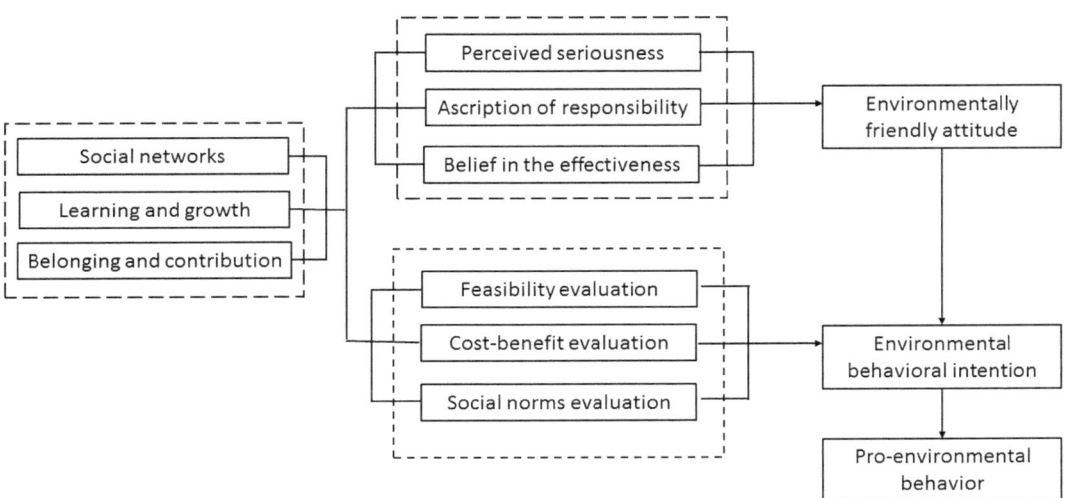

Fig. 6.9 Extended two-phase decision-making model with social needs (Modified after Chao et al. 2021; Illustrated by Wei-Ta Fang)

(Kilbourne 2004), but the consumption of unit resources will produce more products. Therefore, most people believe that the mass production of goods will reduce energy consumption and achieve energy saving and reduction of unit goods. DSP defines the basic belief structures and practices of marketplace actors and is manifested in existing exchange structures (Gollnhofer and Schouten 2017). Now, of course, it is also possible to reduce the waste discharge per unit resource and increase the recycling rate, which can also have a slowing effect. The paradigm of mainstream society emphasizes the following characteristics:

(1) Human beings are different from the creatures they control.
(2) Human beings are the masters of their own destiny; they can choose their goals and learn to achieve them.
(3) The world is vast and offers unlimited opportunities for humankind.
(4) Human history is progressive, and every problem can be solved, so progresses endless.

Due to the instruction of DSP, technologies developed in many fields have harmed the environment. For example, the large-scale promotion of fuel-efficient cars has actually increased the total mileage of human beings, but has caused the total amount of fuel consumption to rise, resulting in more carbon emissions. However, more and more people are beginning to realize that economic growth cannot solve all problems in society. The idea of the DSP formed the Jevons paradox (Ruzzenenti et al. 2019). This is because the mainstream person's dependence on science and technology has led to the misconception that science can solve all problems. However, the new ecological paradigm is to solve environmental problems and consider what actions to take effectively. However, it still has its limitations, and we must continue to transfer paradigms.

6.4.2 New Ecological Paradigm

When people raise their living standards, population growth will slow down and fertility will decline (Day and Dowrick 2004). The current global economic challenge is how to use the earth's resources economically. With rates of fertility declining in every region of the world (Connelly 2003), it is now possible to begin to see the end of the *limit population growth by order*. For example, adopting the one-child policy in mainland China from historical judge (Feng et al. 2013), does not play a sustainable role from social consequences (Cai and Feng 2021). The increasing proportion of elderly in China is producing social pressures (Zhang and Goza 2006). Who will care for the elderly in China?

Of course, the idea of the Garden of Eden is a myth (Delumeau 2000); mankind will never return to the original state of nature. When we are getting old, we may not remember same enchanting natural world where it was, but located this time on our side of death, is described as the kingdom of heaven, does urge the age in our turns while we should be getting old.

We need to explain the issues and take action to protect high-quality air, water, soil, sunlight, and biodiversity from all generations. We believed that the elderly people will manifest a higher level of endorsement of the New Ecological Paradigm (NEP) (Costache and Sencovici 2019). However, the NEP scale could be limited with respect to the anachronistic wording of items (Lalonde and Jackson 2002). Some questions of the New Ecological Paradigm (NEP) could be starting to get into some very complicated, ethical issues that readers will believe we may support or refute. Some wording of items could be hard to capture people's increasingly thorough understanding of the nature, severity, and scope of environmental problems (Lalonde and Jackson 2002). Hawcroft and Milfont (2010: 143) have documented this kind of abuse among previous studies using the NEP scale (Dunlap et al. 2000; Cruz and Manata 2020). We may also claim to explain this relatively new focus with a meaningful construct toward **sustainability paradigm**, including increasingly more pervasive and global environmental issues, changing societal expectations, and educational reform (Hart 2013).

6.4.3 Sustainability Paradigm

Fifty years after the birth of neoliberal economic policies, the debate over how to properly address global environmental issues continues. It is worth noting that, as of now, the proponents of the Dominant Social Paradigm (DSP) and the New Ecological Paradigm (NEP) have each held their own words. Our goal is to guide on dialogue and action on environmental issues. We try to achieve sustainable development through environmental education, communication, and advocacy as a **Sustainability Paradigm** (see Fig. 6.10).

6.5 Summary

Global climate change (GCC) represents a world-historical opportunity for the emergence of a common global society (Broadbent et al. 2016), with failure to do so likely to bring intensifying calamities for all economic developed and/or emerging economics. This is the time represented a global field to discuss our modelling for some choices in certain ways of media discourse (Broadbent et al. 2016). To prevent a global extinction crisis and achieve a sustainable society requires rethinking our social values. Environmental education can help learners understand the connections of living environments during a pandemic, become creative problem solvers and active environmental citizens associated with climate governance (Chen and Lee 2020) to participate in shaping a common future (Fig. 6.11). Therefore, experiential learning and critical pedagogy will provide learners with opportunities for transformative and sustainable development. While we create a worldwide community of critical thinking, we may try to remember how relative to other forms of teaching, generated by critical educational research. We will engage in critical pedagogy in diverse and creative ways and in different settings (Kincheloe 2008). Environmental education is a modern education paradigm that inspires civic responsibility, constructs a positive social status, and promotes a healthy lifestyle. Therefore, we are not convincing all environmental ethics from rigid lessons, but we tried to apply modelling from ethical theories to behave humans' capacities to follow pedagogy's notion of praxis—informed action from practical knowledge. This required to be gained through learning anything adopting by day-to-day hands-on experiences from personal theories. We may encourage that you may learn skills of "knowing-how" in all empirical condition from your effective motivation. In this praxis-based context, we gain the ability to change ourselves relative to other forms of teaching and learning.

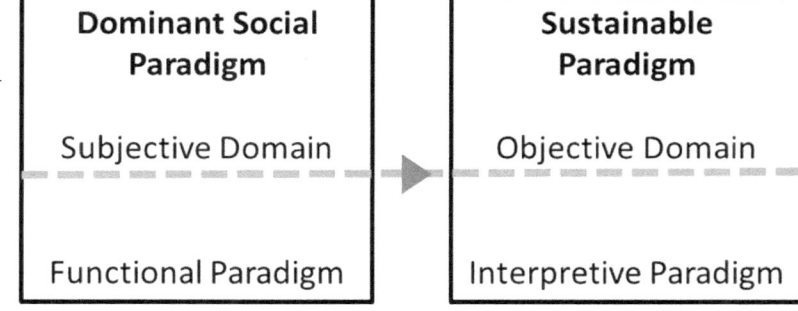

Fig. 6.10 A paradigm returning from functional paradigm toward interpretive paradigm (Illustrated by Wei-Ta Fang)

Fig. 6.11 Environmental education can help learners understand the connections of living environments (Qixingtan Beach, Hualien, Taiwan, 2019) (Photo by Dennis Woo)

References

Ajzen I (1985) From intentions to actions: a theory of planned behavior. In: Kuhl J, Beckmann J (eds) Action control: from cognition to behavior. Springer, Berlin Heidelberg, pp 11–39

Ajzen I (1991) The theory of planned behavior. Organ Behav Hum Decis Proc 50(2):179–211

Ajzen I, Manstead AS (2007) Changing health-related behaviours: an approach based on the theory of planned behaviour. In: The scope of social psychology. Psychology Press, pp 55–76

Bamberg S (2003) How does environmental concern influence specific environmentally related behaviors? A new answer to an old question. J Environ Psychol 23(1):21–32

Bamberg S, Möser G (2007) Twenty years after Hines, Hungerford, and Tomera: a new meta-analysis of psycho-social determinants of pro-environmental behaviour. J Environ Psychol 27(1):14–25

Bamberg S, Hunecke M, Blöbaum A (2007) Social context, personal norms and the use of public transportation: Two field studies. J Environ Psychol 27(3):190–203

Barnhill DL, Gottlieb RS (eds) (2010) Deep ecology and world religions: new essays on sacred ground. SUNY Press, New York

Barrett NF (2011) Wuwei and flow: comparative reflections on spirituality, transcendence, and skill in the Zhuangzi. Philosophy East and West 61(4):679–706

Bellah RN (1983) The ethical aims of social inquiry. In: Haan N, Bellah RN, Rabinow P, Sullivan WM (eds) Social science as moral inquiry. Columbia University Press, New York, pp 360–382

Bennett M (2004) Different shades of green. Coll Lit 31(3):207–212

Borgmann A (1995) The nature of reality and the reality of nature. In: Soulé ME, Lease G (eds) Reinventing nature. Island, Washington DC

Bradford G (1989) How Deep is Deep Ecology. Times Change, Ojai

Bravo-Osorio (2022) Towards an ethic of ecological resilience. Ethics, Policy Environ https://doi.org/10.1080/21550085.2022.2054648

Broadbent J, Sonnett J, Botetzagias I, Carson M, Carvalho A, Chien Y et al (2016) Conflicting climate change frames in a global field of media discourse. Socius 2:2378023116670660

Burdett MS (2015) The image of God and human uniqueness: challenges from the biological and information sciences. Exposit times 127(1):3–10

Burke E (1790) Reflections on the revolution in France and on the proceedings in certain societies in London relative to that event in a letter intended to have been sent to a gentleman in Paris. https://socialsciences.mcmaster.ca/econ/ugcm/3ll3/burke/revfrance.pdf

Cai Y, Feng W (2021) The social and sociological consequences of China's one-child policy. Ann Rev Sociol 47:587–606

Callicott JB (1989) In defense of the land ethic: essays in environmental philosophy. Suny Press, New York

Callicott JB, Parker J, Batson J, Bell N, Brown K, Moss S (2011) The other in a sand county almanac: Aldo Leopold's animals and his wild-animal ethic. Environ Ethics 33(2):115–146

Callicott JB (2005) Turning the whole soul: the educational dialectic of a sand county almanac. Glob Relig Cult Ecol Worldv 9(3):365–384

Callicott JB (2010) The conceptual foundations of the land ethic. Technol and Values: Essent Read 438–453

Capra F, Luisi PL (2014) The systems view of life: a unifying vision. Cambridge University Press, Cambridge

Capra F (1975) The Tao of Phys. Shambhala, Boulder

Carney M (2021) Value(s): building a better world for all. Public Affairs, New York

Chao S-H, Jiang J, Wei K-C, Ng E, Hsu C-H, Chiang Y-T, Fang W-T (2021) Understanding pro-environmental behavior of citizen science: an exploratory study of the bird survey in Taoyuan's farm ponds project. Sustainability 13(9):5126. https://doi.org/10.3390/su13095126

Chen RS, Lee HC (2020) Assessing climate governance of Tainan City through stakeholder networks and text mining. In: Chou K-T, Hasegawa K, Ku D, Kao S-F (eds) Climate change governance in Asia. Routledge, Oxfordshire, pp 256–281

Cherdymova EI, Ukolova LI, Gribkova OV, Kabkova EP, Tararina LI, Kurbanov RA, Belyalova AM, Kudrinskaya IV (2018) Projective techniques for student environmental attitudes study. Ekoloji 27(106):541–546

Connelly M (2003) Population control is history: New perspectives on the international campaign to limit population growth. Comp Stud Soc Hist 45(1):122–147

Costache A, Sencovici M (2019) Age, gender and endorsement of the New Ecological Paradigm. Int Multi Sci GeoConf: SGEM 19(5.1):11–22

Cruz SM, Manata B (2020) Measurement of environmental concern: a review and analysis. Front Psychol 11:363. https://doi.org/10.3389/fpsyg.2020.00363

D'Ambrosio PJ (2022) Non-humans in the Zhuangzi: animalism and anti-anthropocentrism. Asian Philos 32(1):1–18

Day C, Dowrick S (2004) Ageing economics: human capital, productivity and fertility. Agenda: J Pol Anal Reform 11(1):3–20

Delumeau J (2000) History of paradise: the Garden of Eden in myth and tradition. University of Illinois Press

Devall B (1980) The deep ecology movement. Nat Resour J 20:299

Devall B, Sessions G (1984) The development of nature resources and the integrity of nature. Environ Eth 6(4):293–322

Devall B, Sessions G (1985) Deep ecology: living as if nature mattered. Gibbs Smith

Disinger JF (1990) Environmental education for sustainable development? J Environ Educ 21:3–6

Droz L (2021) The concept of milieu in environmental ethics: individual responsibility within an interconnected world. Routledge, Oxfordshire

Dunlap RE (1975) The impact of political orientation on environmental attitude and action. Environ Behav 7(4):428–454

Dunlap RE (2008) The new environmental paradigm scale: from marginality to worldwide use. J Environ Educ 40(1):3–18

Dunlap RE (2022) Understanding opposition to the environmental movement: the importance of dominant American values. Edward Elgar Publishing, In Handbook of Anti-Environmentalism

Dunlap RE, Van Liere KD (1978) The new environmental paradigm. J Environ Educ 9(4):10–19

Dunlap RE, Van Liere KD (1984) Commitment to the dominant social paradigm and concern for environmental quality. Soc Sci Q 65:1013–1028

Dunlap RE, Grieneeks JK, Rokeach M (1983) Human values and pro-environmental behavior. In: Conn WD (ed) Energy and material resources: attitudes, values, and public policy. Boulder, pp 145–168

Dunlap RE, Van Liere KD, Mertig AG, Jones RE (2000) Measuring endorsement of the new ecological paradigm: a revised NEP Scale. J Soc Issues 56(3):425–442

Dunlap RE, McCright AM (2015). Challenging climate change. In RE Dunlap and RJ Brulle (eds) Climate change and society: sociological perspectives. Oxford, Cambridge

de Figueiredo MD, Marquesan FFS (2022) Back to the future: ecocentrism, organization studies, and the Anthropocene. Scand J Manag 38(2):101197

Fang W-T, Ng E, Wang C-M, Hsu M-L (2017) Normative beliefs, attitudes, and social norms: People reduce waste as an index of social relationships when spending leisure time. Sustainability 9(10):1696

Fang W-T, Chiang Y-T, Ng E, Lo J-C (2019) Using the norm activation model to predict the pro-environmental behaviors of public servants at the central and local governments in Taiwan. Sustainability 11(13):3712

Fang W-T, Huang M-H, Cheng B-Y, Chiu R-J, Chiang Y-T, Hsu C-W, Ng E (2021a) Applying a comprehensive action determination model to examine the recycling behavior of Taipei city residents. Sustainability 13(2):490

Fang W-T, Ng E, Liu S-M, Chiang Y-T, Chang M-C (2021b) Determinants of pro-environmental behavior among excessive smartphone usage children and moderate smartphone usage children in Taiwan. PeerJ 9:e11635. https://doi.org/10.7717/peerj.11635

Fang W-T (2020) Envisioning environmental literacy. Springer Singapore, p 12

Farida I, Permadi Y, Adelia T, Liviani N (2019) Considering all (Non) living things: a biocentric orientation in Blair Richmond's the lithia trilogy. Lingua Cultura 13(2):87–92

Feliciotti A, Romice O, Porta S (2018) From system ecology to urban morphology: towards a theory of urban form resilience. In: International Forum on Urbanism. 2018-12-10–2018-12-12, UIC School of Architecture, Chicago

Feng W, Cai Y, Gu B (2013) Population, policy, and politics: how will history judge China's one-child policy? Popul Dev Rev 38:115–129

Fios F (2019). Building awareness of eco-centrism to protect the environment. J Phys Conf Ser 1402 (2):022095 (IOP Publishing, Bristol)

Flores A, Clark TW (2001) Finding common ground in biological conservation: beyond the anthropocentric vs. biocentric controversy. Yale Sch for Environ Stud Bull Ser 105:241–252

Flükiger JM (2009) The radical animal liberation movement: some reflections on its future. J Stud Radicalism 2(2):111–132

De Groot JIM, Steg L (2009) Morality and prosocial behavior: the role of awareness, responsibility and norms in the norm activation model. J Soc Psychol 149:425–449

Gladwin TN, Kennelly JJ, Krause TS (1995) Shifting paradigms for sustainable development: implications for management theory and research. Acad Manag Rev 20(4):874–907

Gollnhofer JF, Schouten JW (2017) Complementing the dominant social paradigm with sustainability. J Macromark 37(2):143–152

Gruenewald DA (2004) A foucauldian analysis of environmental education: toward the socioecological challenge of the earth charter. Curric Inq 34(1):71–107

Hansla A, Gamble A, Juliusson A, Gärling T (2008) The relationships between awareness of consequences, environmental concern, and value orientations. J Environ Psychol 28(1):1–9

Hart P (2013) Environmental education. In: Handbook of research on science education. Routledge, pp 703–740

Hawcroft LJ, Milfont TL (2010) The use (and abuse) of the new environmental paradigm scale over the last 30 years: a meta-analysis. J Environ Psychol 30:143–158

Hernández B, Martín AM, Ruiz C, Hidalgo MdC (2010) The role of place identity and place attachment in breaking environmental protection laws. J Environ Psychol 30(3):281–288

Hines JM, Hungerford HR, Tomera AN (1986/87) Analysis and synthesis of research on responsible environmental behavior: a meta-analysis. J Environ Educ 18(2):1–8

Hirose Y (1994) Determinants of environmental conscious behavior. Jap J Soc Psychol 10(1):44–55

Huang Y (2010) Confucius and Mencius on the motivation to be moral. Philos East and West 65–87

Hung C-M, Hung H-Y, Yeh C-F, Fu Y-Q, Chen D, Lei F et al (2014) Species delimitation in the Chinese bamboo partridge Bambusicola thoracica (Phasianidae; Aves). Zoolog Scr 43(6):562–575

Hurka T (1996) Perfectionism. Oxford University Press, Oxford

Johns D (1992) The practical relevance of deep ecology. Wild Earth 2:62–68

Kaiser FG, Wolfing S, Fuhrer U (1999) Environmental attitude and ecological behaviour. J Environ Psychol 19:1–19

Katz E (1987) Searching for intrinsic value: pragmatism and despair in environmental ethics. Environ Ethics 9 (3):231–241

Kerkhoff LV, Lebel L (2006) Linking knowledge and action for sustainable development. Annu Rev Environ Resour 31:445–477

Kilbourne WE (2004) Sustainable communication and the dominant social paradigm: can they be integrated? Mark Theory 4(3):187–208

Kilbourne WE (2006) The role of the dominant social paradigm in the quality of life/environmental interface. Appl Res Qual Life 1(1):39–61

Kilner JF (2015) Dignity and destiny: Humanity in the image of God. Eerdmans Publishing, Grand Rapids, Wm. B

Kim T (2009) Reading Zhuangzi eco-philosophically. J Daoist Stud 2(2):1–31

Kincheloe JL (2008) Critical pedagogy primer, vol 1. Peter Lang

Kollmuss A, Agyeman J (2002) Mind the gap: why do people act environmentally and what are the barriers to pro-environmental behavior? Environ Educ Res 8 (3):239–260

Kopnina H, Cherniak B (2015) Cultivating a value for non-human interests through the convergence of animal welfare, animal rights, and deep ecology in environmental education. Educat Sci 5(4):363–379

Kopnina H, Gjerris M (2015) Are some animals more equal than others? Animal rights and deep ecology in environmental education. Can J Environ Educ 20:108–122

Kopnina H, Washington H, Gray J, Taylor B (2018) The 'future of conservation' debate: defending ecocentrism and the nature needs half movement. Biol Cons 217:140–148

Kornai J (2002) The system paradigm. Voprosy Economiki 4. https://ideas.repec.org/a/nos/voprec/2002-4-1.html

Kuhlmann W (1996) Making the law more ecocentric: responding to leopold and conservation biology. Duke Envtl l Pol'y f 7:133

Lalonde R, Jackson EL (2002) The new environmental paradigm scale: has it outlived its usefulness? J Environ Educ 33(4):28–36

Lau DC (2004) Mencius. Penguin UK, London

Lee JH (2007) What is it like to be a butterfly? A philosophical interpretation of Zhuangzi's butterfly dream. Asian Phil 17(2):185–202

Leopold A (1949) A sand county almanac. Oxford University Press, Oxford

Liu S, Chiang Y-T, Tseng C-C, Ng E, Yeh G-L, Fang W-T (2018) The theory of planned behavior to predict protective behavioral intentions against PM2. 5 in parents of young children from urban and rural Beijing, China. Int J Environ Res Publ Health 15 (10):2215

Machlis GE, Force JE, Burch WR Jr (1997) The human ecosystem part I: the human ecosystem as an organizing concept in ecosystem management. Soc Nat Resour 10(4):347–367

Madin JS, O'Donnell MJ, Connolly SR (2008) Climate-mediated mechanical changes to post-disturbance coral assemblages. Biol Let 4(5):490–493

Marcinkowski T, Reid A (2019) Reviews of research on the attitude–behavior relationship and their implications for future environmental education research. Environ Educ Res 25(4):459–471

Mazzucato M (2020) The value of everything. Public Affairs, New York

McElroy M (2002) Deep knowledge management and sustainability. Centre for Sustainable Innovation. Retrieved on April 30, 2022. Center for Sustainable Innovation (www.sustainableinnovation.org)

McKenzie-Mohr D (2011) Fostering Sustainable Behavior: an introduction to community-based social marketing. New Society Publishers

McLaughlin A (1993) Regarding nature: industrialism and deep ecology. Suny Press, New York

Mersereau MR (2016) The abiotic Internet: internet mediation through organizational practice at Na-Me-Res. University of Toronto, Toronto

Moeller HG (2015) Paradoxes of health and power in the Zhuangzi. New Vis Zhuangzi 70–81

Möller HG (1999). Zhuangzi's "Dream of the Butterfly": a daoist interpretation. Phil East and West 439–450

Næss A (1973) The shallow and the deep, long-range ecology movement: a summary. Inquiry 16(1–4):95–100

Næss A (1984) A defence of the deep ecology movement. Environ Ethics 6(3):265–270

Næss A (1986) The deep ecological movement: some philosophical aspect. Philos Inq 8:10–31

Næss A (1987) Self-realization: an ecological approach to being in the world. Trumpeter 4(3):35–42

Næss A (1989) Ecology, community, and lifestyle. Cambridge University Press, Cambridge

Næss A (1985a) Identification as a source of deep ecological attitudes. In: Deep ecology. In: Tobias M (ed) Avant books, San Diego, pp 256–270

Næss A (1985b) Ecosophy T: deep versus shallow ecology. In: Pojman (ed) Environ Ethics, pp 151–153

Næss A (1995) Self-realization. An ecological approach to being in the world. In: Sessions G (ed) Deep ecology for the twenty-first century Shambhala, Boston, pp 225–239

Næss A (2011) The deep ecological movement: Some philosophical aspects. In: Bhaskar R, Næss P, KG Høyer (eds) Ecophilosophy in a world of crisis; critical realism and the Nordic Contributions,. Routledge, Oxfordshire, pp 96–110

Noll S (2017) Climate induced migration: a pragmatic strategy for wildlife conservation on farmland. Pragmatism Today 8(2):24–40

Nyrud AQ, Roos A, Sande JB (2008) Residential bioenergy heating: a study of consumer perceptions of improved woodstoves. Energy Policy 36(8):3169–3176

Ölander F, Thøgersen J (1995) Understanding of consumer behavior as a prerequisite for environmental protection. J Consum Policy 18(4):345–385

Orton D (1996) Left biocentrism. Green Web 49:1–10

Parkes G (2013) Zhuangzi and Nietzsche on the human and nature. Environ Philos 10(1):1–24

Pepper D (2002) Eco-socialism: from deep ecology to social justice. Routledge, Oxfordshire

Piccolo J, Washington H, Kopnina H, Taylor B (2018) Why conservation scientists should re-embrace their ecocentric roots. Conserv Biol 32(4):959–961

Pienaar EF, Lew DK, Wallmo K (2013) Are environmental attitudes influenced by survey context? An investigation of the context dependency of the New Ecological Paradigm (NEP) Scale. Soc Sci Res 42(6):1542–1554

Pirages DC, Ehrlich PR (1974) Ark II: social response to environmental imperatives. Freeman, San Francisco

Regan T (1980) Utilitarianism, vegetarianism, and animal rights. Philos Public Aff 9(4):305–324

Regan T (1983) The case for animal rights. University of California Press

Rolston H III (1975) Is there an ecological ethic? Ethics 85(2):93–109

Rowe JS (1994b) Ecocentrism and traditional ecological knowledge. http://www.ecospherics.net/pages/Ro993tek_1.html

Ruzzenenti F, Font Vivanco D, Galvin R, Sorrell S, Wagner A, Walnum HJ (2019) The rebound effect and the Jevons' paradox: beyond the conventional wisdom. Front Energy Res 90https://doi.org/10.3389/fenrg.2019.00090

Salleh A (2022) An ecocentric epistemology for ecosocialism. the routledge handbook on ecosocialism. Routledge, Oxfordshire, pp 57–66

Sandler R (2012) Intrinsic value. Ecology, and conservation. Nat Educ Knowl 3(10):4

Sapolsky RM (2017) Behave: the biology of humans at our best and worst. Penguin

Schwartz B (1987) The battle for human nature: science. Morality and Modern Life. WW Norton & Company, New York

Sessions G (1987) The deep ecology movement: a review. Environ Rev 11(2):105–125

Shrivastava P (2008) Corporate citizenship and the environment. Handb Res Glob Corp Citizensh, 166–184

Singer P (1975) Animal Liberation. HarperCollins

Singer P (1997) Neither human nor natural: ethics and feral animals. Reprod Fertil Dev 9(1):157–162

Sorokoumova EA, Cherdymova EI (2021) Developing structural components of ecological consciousness to promote civic identity formation. Psychol Sci Educ 26(1):102–112

Sterba JP (1995) From biocentric individualism to biocentric pluralism. Environ Ethics 17(2):191–207

Sterba JP (1998) A biocentrist strikes back. Environ Ethics 20(4):361–376

Stern P (2000) Toward a coherent theory of environmentally significant behavior. J Soc Issues 56(3):407–424

Stern PC, Dietz T, Abel TD, Guagnano GA, Kalof L (1999) A value-belief-norm theory of support for social movements: the case of environmentalism. Hum Ecol Rev 6(2):81–97

Steverson BK (1991) A critique of ecocentric environmental ethics, Doctoral dissertation, Tulane University

Taylor C (2010) Aristotle. In: Skorupski J (ed) The routledge companion to ethics. Routledge, Oxfordshire, pp 67–77

Tete F, Ariche CK (2021) Virtue ethics as philosophical foundation for environmental education. GNOSI: An Interdisc J Human Theo Praxis 4(1(May)):83–90

Thøgersen J (2006) Norms for environmentally responsible behaviour: an extended taxonomy. J Environ Psychol 26(4):247–261

Thøgersen J (2009) Promoting public transport as a subscription service: effects of a free month travel card. Transp Policy 16(6):335–343

Vilkka L (2021) The intrinsic value of nature. Brill, Leiden

Washington W, Taylor B, Kopnina HN, Cryer P, Piccolo JJ (2017) Why ecocentrism is the key pathway to sustainability. Ecolog Citizen 1(1):35–41

Washington H, Chapron G, Kopnina H, Curry P, Gray J, Piccolo JJ (2018) Foregrounding ecojustice in conservation. Biol Cons 228:367–374

Wenzel CH (2003) Ethics and zhuangzi: awareness, freedom, and autonomy. J Chin Philos 30(1):115–126

Woo S-K, LePage B, Chiang Y-T, Fang W-T (2022). Predicting the protective behavioral intentions for parents with young children that possess different levels of education in Hong Kong using the theory of planned behavior for air polluted with PM2. 5. BMC Public Health 22(1):1–11

Zhang Y, Goza FW (2006) Who will care for the elderly in China? A review of the problems caused by China's one-child policy and their potential solutions. J Aging Stud 20(2):151–164

Zimmerman ME (2020) Contesting earth's future. University of California Press, Oakland, Contesting Earth's Future

Part III

Living Media Lab: Searching a Means to an End

Environmental Learning and Communication

7

The field of environmental communication consists of the following seven research and practical areas: environmental discourse, environmental news media, public participation in environmental decision-making, social marketing, and advocacy activities, environmental cooperation and conflict resolution, risk communication, and nature in pop culture and green marketing expression.

J. Robert Cox, Environmental Communication and the Public Sphere, 2010.

Abstract

Environmental learning is an act of communication. Whether it is self-directed learning, learning through teachers or professors, or learning through an online platform, all need a learning medium and content. Therefore, environmental learning and communication in this chapter refer to how individuals, institutions, social groups, and cultural communities produce, share, accept, understand, and properly use the environmental information, and then utilize the relationship between human society and the environment through using environmental communication. In the interaction of the social network of human society, from interpersonal communication to virtual communities, modern humans need to participate in environmental decision-making to understand the problems that occur in the world's environment through environmental media reports. Therefore, this chapter could be focused on "learning as process" and, see how to learn from theorized fields of studies. We may encourage that you may learn from spoken, written, audio-visual, image, and information exchanges through carriers such as learning fields, learning plans, learning mode, information transmission, and communication media. It is hoped that environmental learning and communication, through creation, adopt diverse communication methods and platforms to establish the correct environmental information pipeline.

7.1 What Should We Learn?

Sorohan (1993) declared "We do; therefore, we learn." What we learn is a key point, and could be detected from scientific experiments, handworks, and any practices specializing in *learning by doing* for environmental studies. Efficacy studies, therefore, can be informed and scaled-up to benefit broader populations if we want to learn (Christakis 2009). Why we, or our students, need to learn? This is the topic issue we should ask ourselves before you read this chapter since we are requested today to face many critical situations vis-à-vis global hazards (Bisoffi et al. 2021). In such a learning situation, motivation from students is never a problem to examine disasters they could probably face (Schank 1995); we learn because something must be solved from our needs in our daily lives and/or the nearby future. All hard lessons should be learned during the current and future crisis. In our opinions, one should like to live in this critical world doing hard work for her/his stronger motivation. We need to face in the foreseeable future, therefore, only a short-sightedness should not prevail over a long-term perspective if you do not want to understand from an emergency case without any desirable intention to learn.

"Learning to survive," learning from lessons requires to address important needs for our students and/or our colleagues (Grant 2002; Keenan 2020), whereas learning the school courses and/or homework does not act on the basis of that learning; it should be innovated related to any programs helping students how to survive. Keenan (2020) argued the wicked problem education for the Anthropocene age:

> Wicked problems transcend national, cultural and disciplinary boundaries. Eco-survival, international migration, destabilized global markets, shifts in the balance of strategic power, population pressures, cultural imperialism, post-secular quests for meaning-in-life, ambivalence of bio-scientific progress, to name a selection, are global.

That is, very few conditionings for students were more likely to survive provided that learning occurred since students are never interested on the courses, or they needed to study their "take home message" without learning to their future survival skills. We may argue that students could be attributed for a lack of interest, but rather to its perceived in lower rewarded and admirable traditional school homework. We know little about why students do not say what they learned directly from school when they are heading back home. It's very ironic to say, class students can be motivated by the desire for the rewards of their personal interests from their experiences and capabilities (Turner and Paris 1995; Finn 2010). All courses should be regarding to social competence and academic survival skills in the real world (Foulks and Morrow 1989). We would strongly argue that prospective teachers need to have their skills with capacity building, to communicate to their students. We believe (Delors 2013), *is why learning to be, learning to live together, and learning about the paths followed by humanity and sustainability.* We all need to learn how to survive; so we have to learn to live sustainably (Combes 2005).

"Learning as process and outcome" is a theme in the field of environmental education (Rickinson 2006). Environmental learning could be led as an interdisciplinary approach through interactive communication (Rickinson et al. 2009). The students could be emphasized their studies on making communication accessible and useable through their environmental courses, from authoritative to dialogic pedagogies. We need to consider students' intentions and their active participation in the negotiation of both the content and structure of learning discourse (Aguiar et al. 2010). Since the discourse requires the students to be more actively involved, we need to take an active part in class and display how to answer questions. This process could become part of the stream of discourse (Boyd 2012).

Communication is a learning process for students. The process that we need mutually inclusive support, for example, a desire for mutual learning, and respect teachers and students (Senecah 2007). Environmental communication is one of the notable amplifications to deliver environmental messages (Jurin et al. 2010). Therefore, teachers should try to develop an interpretive program (Hansen 2011). We can conduct interpretive talks and show videos. The content could be prepared a set of interpretive storylines. Based on a clear stage or method, the content of a textbook is written to facilitate environmental communication. Therefore, we need to study the nature of environmental education and its connotation based on environmental psychology, environmental education, and environmental communication: "Why do we learn environmental education?" "How do we learn environmental education?" and "Where do we learn environmental education?".

7.1.1 Learning is a Binary Opposition Process

We know that learning is improved when people see a similar process of mediating between binary terms (Egan 1993). Binary opposition is a learning occurs through everyday action and experience (Shekarey 2006). Learning should be simple: "simple to understand the process" of teaching from instructors, and learning are among the important goals of sustainability (Martins et al. 2006). Therefore, the teachers served as research methods instructors, developing and reviewing curriculum (Lanahan and Yeager 2008), should design the pedagogic strategy for making these binary oppositions as an integral part of learning how to utilize sources outside beyond classrooms. For example, teachers may employ a case learning from the binary oppositions and thus help the students how to build a model from this learning process.

First, when we want to teach our students by *Learning is a Binary Opposition Process*. Then, should we face the focal-point issues whether environmental education be used place-based education, online teaching, or virtual reality

teaching? All teaching skills have set off a wave of discussion in the environmental education community in the twenty-first century. In our course, do we need to get used to thinking in terms of "the nature of sympathy life," from external cases of binary opposition for references (Scheler and McAleer 2017), such as online teaching vs. face-to-face meeting? In short, should students learn to think differently in order to better utilize their learning tools from a thinking step of binary opposition? This is the first step to enter our topic of *Learning is a Binary Opposition Process*.

We may provide a case study for our students. Do we need to reconsider our world as a place of values to be built up? Can we love and hate blindly, or with evidence to love or hate (Scheler and McAleer 2017)?

These two, differently-held theories and the biophilia hypothesis encourage localized biological education (Olivos-Jara 2020). If we are afraid of the biophobia hypothesis, we also need to discuss which field can we use to teach? Let students love life, and then enter the outdoor field of environmental education to discuss "biophilia." The term "biophilia" means "love for life or living systems." This term was first proposed by the well-known social psychologist named Erich Fromm (1900–1980) and was used to describe a psychological trend in which human beings are drawn to all "close to alive" lives. An Ecologist, Edward Osborne Wilson (1929–2021), argues that humans seek

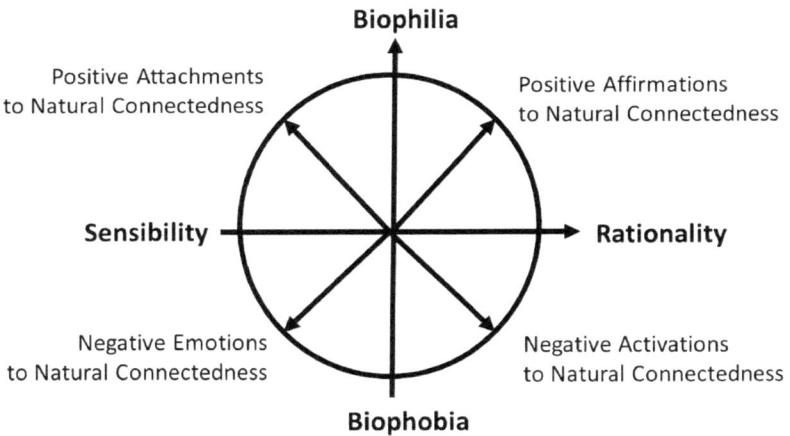

Fig. 7.1 "Biophobia" is the aversion and fear of nature and biophilia is opposite. Illustrated by Wei-Ta Fang

to connect with life subconsciously. He put forward the hypothesis "the innate tendency [in human beings] to focus on life and lifelike process" as the deep connection between human beings and other life forms and the whole nature is rooted in our inner biology (Wilson 1984). "Biophilia" is different from "biophobia," which is what humans have in the environment (see Fig. 7.1).

Humans have had a longing for life since ancient times because death is an end. Therefore, environmental education is the pursuit of a kind of life education that is necessary to sublimate from life and learn through recognition. From the observation of the world, you must learn from all things, including the Book of Changes, *I Ching*. "It's still," "it's easy to get through."

After watching life normally, watching the sunset, watching the tide of the four seas; when our heartbeat is synchronized with the pulse of the earth before there is no induction and realization, it is, of course, solid and silent, but once we feel it like a big bell ringing up through the earth, the heavens, and the earth suddenly open up, to penetrate the principles of the things in the world.

Box 7.1: Case Analysis
Biophilia hypothesis vs. marine notes related to biophobia hypothesis.

In August 2002, I was lying on the pier deck of a small island on the Caribbean Sea, listening to the sound of the sea waves, the white sandy beach revealed the silver light of stars, the blue sea water full of indigo, and the silver light of the stars was also reflected at night. Occasionally, skyline lightning suddenly shocked the sea level, lighting up the clouds like the bride in June, bright and invisible, but it was just a glimpse of glory. Did not hear the rumbling of lightning, but only the breaking wave quietly rubbed his feet. The tide here is only tens of centimeters. The sound of the tide in the early morning, beating against the deck, is like a kitten chirping, sweet and gentle. Like the stars covered

with oracles in the sky, it is as light as a shaker swaying under the wooden house.

The sky is full of stars, and the artificial satellites with dazzling yellow light confuse my understanding of the stars in the starry sky. The milky road in the sky spreads from the northeast to the southwest. Then, the meteor cross-country, shooting hundreds of millions of light-years of stars, and fell into the atmosphere, too late to make a wish. Heading north, The Big Dipper rotates in the northern sky. I like to look north, imagine the currents of the south, blow the north, and drift into the Gulf of Mexico. Then they parted ways, the ocean currents went downstream to Florida, meanwhile, the reverse currents went up to Mexico, the sea breeze blew the ears, and the distant coconut island was married under the light night sea breeze, tightly covering the sticky lingering and tenderness of summer. The sound of the seaside in the night always stirs the physical genes under the blood, as if the bloody light and shadow wiped out by the last night and the sunset, and then cast into another abyss in the distance.

I want to keep my thoughts still, but the heat of the day is still lingering in my brain, which stirs my centuries of thoughts, and then my emotions fly to the hundred-year taboo of the Caribbean Gulf of Mexico, the hurricane of the 1900s. It is said that a hundred years ago, torrential rain washed up a small isolated city in Texas, the big waves rolled up, and then the house was empty. That lone city never recovered, only staying on the small campus near the bay. It was a wild hurricane on the sea. After rising far away, it re-awakened the deep hatred on the seafloor. The stormy sea was staged at the same time 40 years ago. The wind and rain stormed, and all the island houses were involved in the sea waves. During the process, all the corals on the seafloor flipped. Tossed to the island

and accumulated into fossil bones. The little girl on that island clung to the towering coconut palms on the island and waited for the dawn and calm after the storm, only 40 years later, with silver hair to express the fear of that year. Because of the multi-story high waves, the sleeping evil spirits deep in the ocean floor have been awakened, and the corals trapped in the ocean floor have been panic-stricken. After exchanging six houses, big waves sat down on the coral seabed with satisfaction and carried out large-scale sea-land exchange and remediation operations. Tonight, I can't imagine that the sleeping ocean would be so irritated. I didn't imagine that the periodic waves would be so appalling. Then, I started to be intimidated by the ocean. Although I did not like to wear a rescue suit during the day, I had already boldly dropped into the deep sea in a roll-back posture. After a circle, I raised my head and stepped on the waves. A circle of currents rides in the deep-sea water (Fig. 7.2a, b). I don't know how long I can float and/or dive, so I dived into the depths of the coral and shook hands with the

starfish. During the day, the waves shone deep into the bottom of the sea. The manatee and turtle grass wandered and wandered. I swam over the fan-shaped corals and greedily wanted to pick one. It is a small piece but the corals of bright red blood and the air become pale and bloodless skulls, and there is a smell in the air. I know that greed only happens on land, and greed should not be brought to the bottom of the sea. I hope forgiveness My recklessness allowed the broken corals to return to the sea to heal, so I returned the pale guilt after the bloody greed to the island and did not dare to bring it to the mainland, as always.

(Wei-Ta Fang, Belize, 2002, unpublished notes)

7.1.2 Teaching is the Half of Learning

Education is to promote the way of learning by teaching; this also discussed the final way of ultimate eternity in our daily lives. This will be served some kinds of personal reflection and the capability to observe quietly. Confucius (Master

a

b

Fig. 7.2 a. Awaken from seafloors by scientists; you will see sleeping microbes that awaken up being buried for 100 million years in the seafloor. Photo by Chun-Yen

Lin). **b.** Observing coral reefs can provide crucial data that help scientists and managers to conduct coastal and marine management. Photo by Chun-Yen Lin

Kǒng; or commonly 孔子; Kǒngzǐ; c. 551–c. 479 BC) said in *the Way of Ultimate:*

The way of ultimate wisdom is the comprehension of absolute integrity, genial development of the common people, and endless pursuit of the perfection of humanities. Acknowledgment of such an ultimate terminus of trinity provides a focus; having a focus enables calmness; a calm demeanor brings about tranquility; a tranquil mind allows for clear deliberation; clear deliberation leads to the attainment of wisdom. There exist the fundamental and incidental, everything has a beginning and an end, knowing what comes first and after takes one closer to know the way.

Wisdom only when "focus, calmness, calm demeanor, tranquility, clear deliberation" are taken into consideration can he gain. We may remember three approaches, like "example education, precept education, and tacit education," are indispensable in the ancient oriental rules, and the three are indispensable for environmental education from the process of "focus, calmness,

calm demeanor, tranquility, and clear deliberation." The *Book of Rites Record (Liji)* 《禮記:學記》 said, "Take care of the questioner like a bell." Also, said by the chapter *Record on the subject of education* in *Liji*:

However fine the viands are if one does not eat, he does not know their taste; however perfect the course maybe, if one does not learn it, be does not know its goodness. Therefore, when he learns, one knows his deficiencies; when he teaches, he knows the difficulties of learning. After he knows his deficiencies, one can turn round and examine himself; after he knows the difficulties, he can stimulate himself to the effort. Hence it is said, 'Teaching and learning help each other;' as it is said in the *Charge to Yueh* 《兌命》, 'Teaching is the half of learning.'

Therefore, the teacher treats the student like a "hanging bell." The teacher itself is full and subtle (Fig. 7.3). If the student does not ask, then it depends on the student's natural experience. If the student does not have any questions, then there is no induction; once the student sees all

Fig. 7.3 The teacher itself is full and subtle, like a father and his son (left: Marinus Otte; right: Velsen Otte, 2013). Photo by Wei-Ta Fang

Fig. 7.4 The student sees all things in nature, the induction of the mind and the teacher's detailed explanation for all natural facts. Yi-Ju Yang (right), National Dong Hwa University, recorded geomorphic facts with students at Sulfur Vally, Yangmingshan National Park, Taiwan. Photo by Wei-Ta Fang

things in nature, the induction of the mind and the teacher's detailed explanation like a bell rings and will sound like a big bell ringing—to make students suddenly cheerful. This is one of the crucial pedagogic approaches of oriental education (Fig. 7.4).

Moreover, a few western scholars are emphasizing this kind of spiritual induction for science education. All inductions do not necessarily lie in the psychological interaction between teachers and students, or even the interaction between animal teaching and the psychological level of students. The book entitled *Children and Nature: Psychological, Sociocultural and Evolutionary Investigations* by Kahn and Kellert (2004), the importance of using animals for education is emphasized, especially for animals that are comfortable with the presence/interaction with humans so that children can develop

parenting relationships is explained. By observing and interacting with animals and plants in nature, we can develop the connection to nature/nature connectedness (Cheng and Monroe 2012).

It is important to appreciate and accept nature as important components of life and to strengthen nature relatedness, connectivity, and emotional affinity between humans and nature. In terms of cognitive composition, it strengthens the core of our natural connection and integrates humanity with nature. In terms of emotional composition, it strengthens personal care for nature (Uhlmann et al. 2018; Oh et al. 2020). Behaviorally, it strengthens personal commitment to protect the natural environment (Perkins 2010). In addition, animal appreciation can help children with autism disorders and strengthen the closed support system in the heart, which is a necessary condition to help humans maintain a healthy relationship with nature through life education.

Thus, the design of environmental education should be through the field design. In the field, having a living life provides environmental education materials. In 1998, the Environmental Protection Administration of the Executive Yuan (Taiwan EPA) invited the Ministry of Education, the Committee of Agriculture, and the Construction and Planning Agency (Taiwan CPA) to promote the ecological campus designed by one of the authors, Wei-Ta Fang. He believes that the concept of an ecological campus could be incorporated into course content using the materials in school gardens, such as aquatic plant areas, nectar plant areas, natural trails, nursery areas, and organic compost area. According to the concept of school environmental education, the eco-campus could use the materials "from cradle to grave in gardens. This would allow students see the circularity of water, good, and energy cycles. By growing vegetation and incorporating organic approaches to natural resource use and recycling, simple to complex food/ecological webs are created. Moreover, impacts to these ecosystems such as a lack of water, heat, and exposure to pathogens can illustrate the sensitivity of what we think are robust systems. This project was conducted through careful observation and children's notes to strengthen the positive interactions between students and the campus' natural resources to cultivate environmental awareness, such as acuity, care, respect, and a sense of attachment to environments that we often take for granted. The first ecological campus was built on the gardens at Yonghe Primary School, New Taipei City, in Taiwan: "Water area ecological area, bird watching area, butterfly-attracting area, woodland area, grassland area". Since 2003, the Ministry of Education has promoted "The campuses of Water and Greenery", and under the circumstances of social changes it's continued to promote/support school environmental education in the "Green School Project." It can be seen that the purpose of promoting an ecological garden-style project on green the campus is to provide

teachers and students a place to teach and learn about nature education, ecological conservation, and to carry out the following school environmental education elements.

7.2 How Should I Learn?

7.2.1 School Environmental Education Should be a Practical-Learning

Environmental knowledge should be gained by implementing theory in real-life activities. In the field of school environmental education, teachers should teach or encourage students to feel the connection between themselves and nature, study environmental education on campus, and arrange formal teaching outside the school to provide students with opportunities to care and protect the living environment.

Therefore, the teaching content of school environmental education is not the installation of knowledge. Since ancient times, Eastern-style teachers have emphasized traditional teaching activities, using mechanical calculations and exercises to force students to learn conceptual knowledge of environmental protection; however, this is futile to be embedded mainly into various science subjects to implement a practical program to solve issues (Chapman and Sharma 2001; Parker et al. 2018; Tariq et al. 2020). In the United States, teachers generally focus on promoting the creative ability of students in the environment and the ability to inquire about environmental knowledge (Sternberg and Williams 1996; Sternberg and Davidson 2005). However, this may be universal though. This may be available equitably to applied to all educators who love universality about truth. Sternberg and Williams (1996) believed that "… Although few, if any, our students will become great artists." The educator is based on maternal love based on the object-side of education is not the student, but first and foremost the thing that is studied in the classroom. Environmental educators love to teach in the particular while opening them to the universal—to learn from the universality in ecosystems. Students in Australian Universities tried to find toward open-ended, real life, and purposive inquiries from problem-based learning (PBL) (Thomas 2009). The diversity of environmental education learning curriculum activities not only helps students reflect and operate within the environment, but it helps provide a better understanding of the development of environmental protection processes and facilitate citizen discussion and participation. In the school environmental education, the basic concepts of the environmental education curriculum are met. You can teach in the environment, put students in the natural environment and observe environmental problems in person. Therefore, in addition to teaching environmental knowledge for students to observe and record these observations on their own, teachers should guide students to conduct field environmental analyses and comparisons. That is, encourage them to ask questions and build on concepts that are difficult to understand or test. To solve environmental issues, students need to participate in activities that are the subject of teaching and guiding students to think more critically, and to evaluate the data as part of the teaching process (Fig. 7.5).

Box 7.2: Case analysis

Nymphar shimadae Hayata cultivated on campus in Taiwan (Figs. 7.6, 7.7, 7.8 and 7.9).

In Taiwan, it is not difficult to cultivate *Nuphar shimadae* Hayata. *Nuphar shimadae* has a temperature requirement between 10 °C and 32 °C. However, if grown/maintained for a long time the growth rate water flow through a pond will slow.

Nymphae is an aquatic plant so where it needs to be well light. It's best to plant *Nymphae* in a pool that can be directly exposed to sunlight and surrounded by woods. Only the light formed by the natural sunlight is enough to make the leaves of *Nymphae* grow aquatic leaves and flowers (Fig. 7.6). The demand for water quality and fertilizer is not as high as that of water lilies and Guanyin lotuses are commonly seen in the market. If it is cultivated in a pond, the fertilizer in general agricultural soil is sufficient. If it is cultivated in an aquarium, add a proper amount of comprehensive liquid fertilizer every week, and use root fertilizer every month, such as the general care method of waterweed, you can grow a beautiful nuptial grass. Then, we talked about the choice of symbiotic

Fig. 7.5 A good teacher should be teaching to guide students to think, judge, and evaluate during the teaching process like Ben LePage's approaches at National Taiwan Normal University. Photo by Wei-Ta Fang

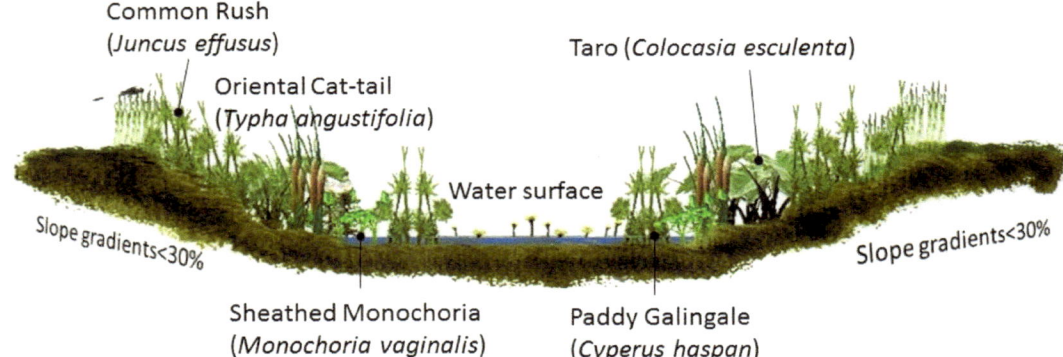

Fig. 7.6 Most aquatic plants inhabit the shallow water of the riparian areas. Therefore, aquatic plants could be growing along a shallow water's edge on campus. Illustrated by Wei-Ta Fang

Fig. 7.7 The yellow water-lily (*Nuphar shimadae* Hayata), with large, lily-pad leaves, is a endemic plant of still or slow-moving water and grows at ponds on campus. Illustrated by Wei-Ta Fang

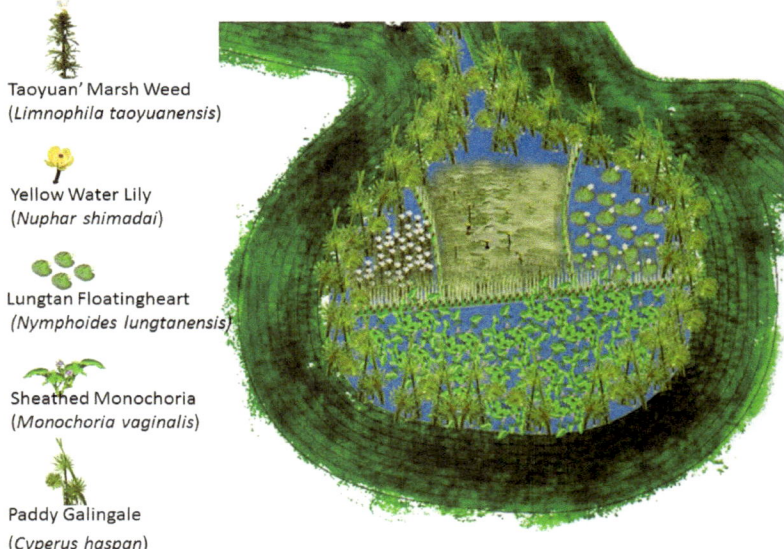

organisms. Because the young buds of *Nymphae* are particularly tender, they are a favorite food for herbivorous organisms. For example, grass carp and other organisms that eat the buds are not suitable for cultivation. Tilapia digs the bottom sand to build a nest during breeding. A large amount of overturning will cause the stems of *Nymphae* to be lifted. Therefore, it is necessary to carefully choose the organisms that symbiotic with *Nymphae*.

We are based on the four aspects of the "Three-year Implementation Plan for Strengthening School Environmental Education" issued by the Taiwan EPA of the Executive Yuan, including promoting school environmental management, implementing environmental teaching, promoting campus environmental protection, and providing environmental education facilities. The aforementioned content includes the planning and management of the learning fields, and in the aspect of teaching materials, promoting life-like environmental protection activities. In addition, according to the five aspects of the "Self-

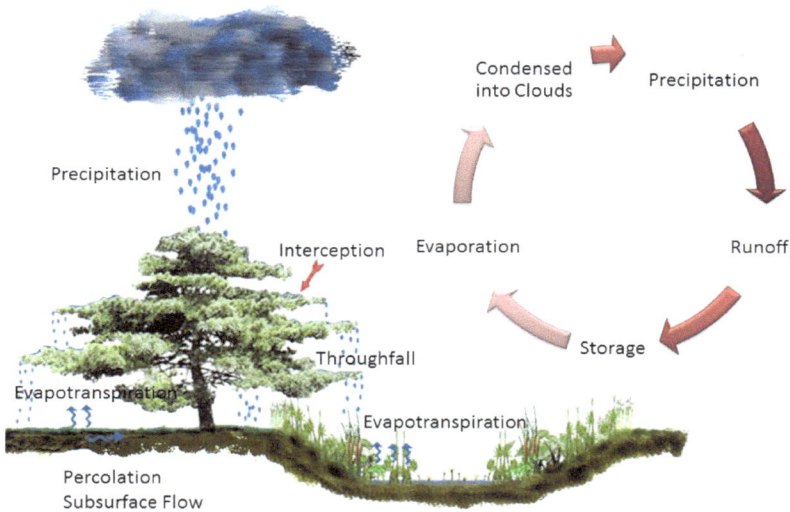

Fig. 7.8 Planting yellow water-lily is not just teaching horticulture and/or permaculture, and we may lead students to study hydrological cycling. The open water on campus may be recycling to irrigate fields, recharge aquifers, and school's ponds for potable and recharge purposes in aqua-life cycle. Illustrated by Wei-Ta Fang

Fig. 7.9 *Nuphar shimadae* Hayata. Photo by Wei-Ta Fang

Checklist Green School Plans" including the aspects of life, campus, buildings and facilities, teaching and outreach activities, administrative management, and the characteristics of school environmental education, all need to be properly implemented teaching at an appropriate level. These teaching activities need to pay attention to the following:

7.2.1.1 Environmental Educators Need to Understand Their Audience

We can use simple and old examples, because of the change of concept, the graph of the new example is converted for a simple and understanding explanation.

7.2.1.2 Environmental Education Should Try to Minimize Scientific Jargon

For example, we often say "paradigm shift" in academics, but in practice, this is a proprietary scientific term that few people understand (Fisk 2019).

7.2.1.3 Teaching Using Various Methods

- Use Microsoft presentation software (PowerPoint).
- Use videos for playback.
- Use lively hands-on activities.
- Perform nature experiences off-campus.
- Use citizen science surveys (Hsu et al. 2018; Chao et al. 2021).
- Use social media for teaching.

7.2.1.4 Teaching Demonstration of Artificial Wetland on Campus and Outdoor Wetland0.0.2

- Assist students to understand the role of wetlands in the carbon cycle and their role in mitigating climate change.
- Emphasize the concept of water storage = sponge.

- To teach the function of carbon sink in the coastal wetlands and/or reference wetlands of the inland (Otte et al. 2021), please do not use blue carbon, because this is scientific jargon.
- Emphasize that this is a wildlife habitat (Fig. 7.10). Red-crowned waterfowl, mallards, and other waterfowl have been observed in the urbanized areas (Fang et al. 2020).
- Amphibians were observed.
- Observe the diversity of vegetation.
- Check the drainage holes of the artificial wetland.

7.2.1.5 Teachers Need to Understand the Content of Teaching Content

To teach students according to curriculum standards, teachers need to understand subject matter deeply and flexibly so they can help students create useful knowledge to understand.

7.2.1.6 School Environmental Education Work is Dedicated to Environmental Education

The roles and functions of environmental educators are like liaisons between school teachers and policymakers. School environmental education personnel must be interested in environmental affairs. In addition, they need to have the ability to communicate in the community, resist stress, and be environmentally literate (Fig. 7.11). Through the cultivation and certification of the school's environmental education program by the Environmental Protection Agency of the Executive Yuan (Taiwan EPA), the liaison work of the resources, courses, and personnel of the school's environmental education activities is integrated. In the environmental education teacher team, a team of environmental educators who support and cooperate is established. All school teachers in the team can obtain the qualification of environmental personnel through the certification of the Taiwan EPA.

For school teachers, textbooks in environmental education-related fields are the most important teaching resource, and the most

Fig. 7.10 Emphasize that this is a wildlife habitat. Red-crowned waterfowl, mallards, and other waterfowl have been observed on pond areas. Photo by Max Horng

obvious constraint they face most often. For students and parents, textbooks are the most important medium for understanding the content of school curricula (Westbury 1990). Therefore, according to the main characteristics of school curriculum content, teachers need to strengthen their vocational skills, conduct student interaction, and communicate with parents, including the following three aspects:

- The teacher's **understanding of the subject content** of the environmental-related field is especially the scope and topics to that teacher who often teaches the subject for promoting environmental education (Liu et al. 2015).
- The teacher's **grasp and use of the content** representation of the above-mentioned environmental-related fields, such as what form (analog, example, metaphor, illustration, and demonstration, etc.) is used to express the content of the subject are effective, most

convincing, and most likely for students to understand.

- The teachers **understanding of environmental content learning** and learners, such as students' existing concepts, concepts before learning specific content, feeling easy or difficult concerning a certain content, whether it is easy to understand or misunderstand, and know what factors affect the student's learning progress, so use the contact book for parent communication.

7.2.2 Social Environmental Education Should be an Adaptive Learning

If we say that school environmental education is formal environmental education, then social environmental education is non-formal environmental education. By definition, the term social

Fig. 7.11 Textbooks in environmental education-related fields are the most important teaching resource, and the most obvious constraint they face most often (Affiliated Experimental School of University of Taipei, Taipei, Taiwan). Self-Photo by Wei-Ta Fang

environment refers to the education environment that is promoted through social interactions between social groups, teachers, families, and government agencies outside the school environment. An environmentally-friendly social environment helps foster positive peer relationships, generates good interactions between adults and children between generations, and provides adults with opportunities to support the achievement of their social goals.

We may define that adaptive learning should be sought to address differences in ability by targeting teaching practices that address the unique needs of an individual for citizens. Environmental education is fostering environmentally conscious citizens. In ancient Greece Aristotle started to promote social environmental education. From Jean-Jacques Rousseau to John Dewey, the progressive schools' movement started in the classroom and promoted nature studies and conservation education, and outdoor education-style social education. Fundamentally, social environmental education is a cross-disciplinary education. It draws on social and scientific studies and uses environmental protection as a model to further develop critical thinking and creativity. Creative thinking and integrated thinking can solve real environmental problems.

Therefore, in the process of cultivating environmentally conscious citizens, environmental education trains environmental citizens to compete in the global economy, has environmental protection skills, knowledge, and inclinations, can make wise choices, and exercise the rights and responsibilities of global citizens. To achieve this goal, social-environmental education must include the following:

7.2.2.1 Strengthen Affection Ability

Strengthen environmental sensitivity and environmental appreciation ability.

7.2.2.2 Strengthen Ecological Knowledge

Strengthen our understanding of major ecological concepts, including caring for individuals, species, ethnic groups, communities, ecosystems, and biogeochemical cycles.

7.2.2.3 Improve Socio-political Understanding

Strengthen our understanding how human cultural activities affect the environment, including an understanding of geography, history, and environmental aesthetics. In addition, regardless of the local, regional, and global environment, a sound global citizen should understand the interdependent relationship of economy, society, politics, and the ecological environment.

7.2.2.4 Strengthen the Basic Knowledge of Problem Occurrence

Strengthen social education needs to understand the knowledge of environmental issues.

7.2.2.5 Strengthen Skill Development for Environmental Protection and Analysis

Encourage the public to use information from primary and secondary sources for analysis, and strengthen the ability to synthesize and evaluate information on environmental issues.

7.2.2.6 Strengthening Personal Responsibility

Enable the public to understand the role of individuals and groups can strengthen the broad impact on society.

7.2.2.7 Strengthen Citizenship Skills and Strategies

Actively participate in social and environmental education activities. They are like participating in conferences, seminars, and work organized by international organizations, government agencies, or civil society organizations. Workshops and seminars can include events such as World Wetland Day on February 2nd, World Earth Day on April 22nd, and World Environment Day on June 5th as well as other commemorative activities. The following are the resource channels for social environmental education:

- **Learning using mass media**: TV, radio, newspapers, magazines, social media, and digital information (information and communications technology, ICT tools) (Wals et al. 2014; Chao et al. 2020; Stagg et al. 2022).
- **Learning with social and educational institutions**: Visiting museums, zoos, expositions, bird gardens, botanical gardens, planetariums, science education centers, local cultural centers, cultural parks/houses, nature education centers, tourist attractions, scenic areas, conservation area, aquarium, and national park (Fig. 7.12).

7.3 Environmental Learning Center Should be Suitable for the Best Quality Education

We understand that six basic elements influence the quality of education as a high-quality learning environment, such as: (1) The teacher and teaching methods; (2) Educational content; (3) Learning environment; (4) Place management; (5) Preconditions for pupils; and (6) Funding and organization. In the curriculum of a school's environmental education program including the non-formal curriculum associated with social environmental education, a type of environmental learning center forms where teachers can lead students to the outdoors to develop a student's environmental literacy. The environmental learning center is a place where environmental educators, students can congregate. It's also a symbol of the effectiveness of environmental education in a region. In the United States, a nature center usually displays small living animals, such as insects,

Fig. 7.12 Officers and visitors also can learn with social and educational institutions at the Endemic Species Research Institute, Council of Agriculture, Jiji, Nantou, Taiwan. Photo by Wei-Ta Fang

Fig. 7.13 Arba'at Hassan was here at SIUC Touch of Nature, Environmental Learning Center. Photo by Arba'at Hassan

Fig. 7.14 Nature and ecological conservation at Garden of Gods Nature Park, Harrisburg. Photo by Arba'at Hassan

reptiles, rodents, or fish. Therefore, the nature center also has a museum exhibition and has the function of displaying natural history. However, a domestic Environmental Education Center may have a wider range of exhibitions. In addition, the difference between an Environmental Education Center and a Nature Center is that museum exhibitions and educational course activities commonly require appointments (Figs. 7.13 and 7.14). However, many international nature areas and university campuses can provide self-directed

learning without a prior appointment (Figs. 7.15 and 7.16).

7.3.1 Features of Environmental Learning Center Implementation

An Environmental Learning Center teaches people to experience nature and establish a

Fig. 7.15 The purpose of the Environmental Education Center is to provide social environmental education in addition to dynamic display, static display, and ecological conservation, like from "trash pick-up" activities to "no littering" activities in Switzerland (The lake of the Leisee (2232 m) is Zermatt's beach and a popular outing for school education. With children's playground, barbecue areas, and picnic spots– and a glorious view of the Matterhorn, Switzerland, 2009). Photo by Wei-Ta Fang

relationship with nature and the environment (Figs. 7.17, 7.18 and 7.19). The function of research and learning of environmental education and school environmental education can be:

7.3.1.1   Have Hardware and Software Facilities and Personnel

Own land, building reductions and simplified hardware facilities, perfect activity plans; and paid full-time professional staff, as well as unpaid volunteer support.

7.3.1.2   Independently Operating Legal Organization Entity

It is a legally independent entity operated by a team with a clear vision of ecological conservation,

education, and rehabilitation. Although some centers allow free admission, small donations are encouraged to help offset expenses.

7.3.1.3   Professional Employees with Salary

To strengthen the breadth and depth of tourism, combined with the training of local commentators, create local employment opportunities through the principle of professional salary cooperation. Because these professional local employees not only have a good understanding of the local human environment, geographical landscape, and the climatic and safety conditions that must be paid attention to when interpreting services.

Fig. 7.16 The Environmental Education Center also provides school environmental education in outdoor learning venues or training bases and can be used for energy conservation curriculum planning and training activities. However, the Environmental Learning Center is not just a park, zoo, or museum, nor is it an ecological protection area, sometimes the university campus is a good example (Harvard University, Cambridge, Massachusetts, USA, 2015). Photo by Wei-Ta Fang

Fig. 7.17 The Environmental Learning Center could be designed for outdoor activities for protecting sea creatures. In addition, based on maintaining the local environmental quality, the local guide also protects the ecological environment with professional knowledge and the facilities of the environmental learning center from unknown damage (Zanpa Misaki Park, Okinawa, Japan). Photo by Wei-Ta Fang

Fig. 7.18 Environmental education requires experiential learning and needs to consider the place of education, curriculum planning, business plan, business strategy, and education principles (Audubon Society of Rhode Island, Rhode Island, USA). Photo by Wei-Ta Fang

Fig. 7.19 The Environmental Learning Center provides popular social education and professional training for the public to care for the natural environment (Audubon Society of Rhode Island, Rhode Island, USA). Photo by Wei-Ta Fang

7.3.2 Participatory Learning at the Environmental Learning Center

Environmental education is a process that advances with time. Through dialogue with educators, we identified five factors that are typical characteristics of symbolic and substantive participatory learning in the twenty-first century (Loh 2010).

7.3.2.1 Authenticity

Understanding the learner's identity, interests, and relevance of the curriculum can develop professional knowledge.

7.3.2.2 Creativity

Environmental education is a creative space for developing professional knowledge (Fig. 7.20). Therefore, participatory learning needs to improve learners' motivation, creativity, and lively participation through environmentally meaningful games and experimental activities.

7.3.2.3 Co-configured Expertise

Based on the professional knowledge jointly configured by teachers and learners, the teachers and learners collectively integrate the skills and knowledge of environmental protection and share the tasks of teaching and learning. You should learn to let go.

Fig. 7.20 The teachers, may be with their family members collectively integrate the skills and knowledge of environmental protection and share the tasks of teaching and learning (Left, Arba'at Hassan with his family members and friends at Touch of Nature, Southern Illinois University at Carbondale (SIUC), USA. Photo by Arba'at Hassan

7.3.2.4   Motivation and Engagement

Educators teach learners to use a variety of media, tools, and practical methods to use lively and interesting teaching methods to motivate participation and expand opportunities for creating and solving problems, such as through the carbon footprint calculations to discuss how to reduce greenhouse gas emissions.

7.3.2.5   Learning Ecosystem

Learning the ecological environment based on an integrated learning system can promote and encourage environmental connections between individuals, families, schools, communities, and the world. The structure of the environmental learning center is to show the interconnection between these five different aspects, which can be interdependent and complementary at the same time.

7.4   Learning Plans Toward Sustainability

Environmental learning centers are not the same as nature conservation areas because the purpose of setting up an environmental learning center is to emphasize the function of education in addition to conservation. At the same time, they shoulder social responsibility, educate local school students, parents and children, and the public, and take care of the local environment. The environmental learning center is an important field for the development of environmental education and the provision of resource services.

7.4.1   Program of Environmental Learning Center

7.4.1.1   Formal Education

The school's class cooperates with the Nature Center to run environmental protection or natural ecology summer camp. Leading or co-sponsored off-school teaching or summer activity courses are part of the school's formal education curriculum.

7.4.1.2   Non-formal Education

Parents participate in outdoor or indoor activities organized by the government, colleges, parenting classes, homeschool classes. These activities include science camps, field visits, ecotourism, food education, visits to other educational facilities or outdoor recreational activities. These are short-term courses without credits or non-school activities (Figs. 7.21, 7.22, 7.23, 7.24, 7.25 and 7.26).

The use of community-developed natural centers for the participation of surrounding residents and school teachers and students can enhance local environmental awareness and strengthen local emotional connections. The center of nature must act as a bridge between

Fig. 7.21 Outdoor camping. Photo by Arba'at Hassan

Fig. 7.22 Mountain climbing. Photo by Arba'at Hassan

Fig. 7.23 Boat recreation at SIUC Campus Lake. Photo by Arba'at Hassan

Fig. 7.24 Trail hiking. Photo by Arba'at Hassan

parent-child and nature. Using the theory of experiential learning, Kolb (1981, 2005) summarized empirical learning. The empirical learning mode reinforces the learning of practical experience, allowing learners to learn about environmental phenomena, discover natural mysteries, and gain spiritual growth and inspiration from different cultures (Joy and Kolb 2009). By directly comprehending and mastering skills through experience or by indirectly understanding the experience represented by symbols, you can transform thinking into learning theory and the learning style inventory (Kolb 1981, 2005). Finally, based on internal reflection, one can strengthen external action capabilities and understand the importance of environmental protection.

7.4.2 Course Content of the Environmental Learning Center

For the Nature Center to be able to implement curriculum programs with environmental education content and local environmental content, it needs to use different teaching strategies based on different age groups, objects, and make the overall teaching program run more smoothly to achieve the teaching goals. Therefore, the

Fig. 7.25. a, b Assembly photos, graduate student gatherings studying tribal knowledge from Tayal's elders (left and right) organized by Wei-Ta Fang at National Taiwan Normal University's formal educational activities at Xiakelo Historic Trail, Hsinchu, Taiwan. Photos by Max Horng

a

b

Fig. 7.26 Wei-Ta Fang guided students heading to riparian areas, Sakeyajin Creek to study tribal hot-spring knowledge from the Tayal tribe in Hsinchu, Taiwan. Photo by Max Horng

contents of the environmental education curriculum plan should include the following:

7.4.2.1 Environmental Learning Goals

The purpose of environmental education is for to learners acquire knowledge and action ability to improve the environment. Therefore, the environmental education teaching goals should include awareness, knowledge, affective level (environmental ethics), civic activity skills, and citizens' experience in active participation as the main axis of development

- **Awareness** is the ability to feel and respond to the environment. For example: "I feel the current situation of global warming."
- **Knowledge is** to assist learners with the ability to understand the operation of nature, learn the interaction between humans and the environment, and understand the basic operation of the natural environment. For example: "the cause of global warming and the problem of carbon emissions."

- **Environmental ethics** will assist learners to develop values and ethics for the environment. For example: "If I know that carbon emissions are too high, can I stop driving, sell my car, and take public transportation in Taipei?" Said one of the authors Wei-Ta Fang. The answer is yes.
- **Citizen action skills are** required to assist learners conduct surveys, prevent environmental pollution, and seek action to resolve environmental issues. For example: "Do I have a way to calculate my carbon footprint?"
- **Citizen action experience** will assist learners to improve environmental awareness, knowledge, attitudes, and action skills, and invest in environmental protection, prevent excessive carbon footprint, and explain their experience in solving environmental issues and problems. For example: "After I sold my car if I knew how to calculate carbon emissions, I started to take public transportation, and I would calculate how much carbon footprint I could reduce during the year, to reduce collective

pollution caused by air pollution from non-point sources," said by Wei-Ta Fang. In addition, we (the authors) preached a simple life of energy-saving and carbon reduction, and lived a daily life without a TV, private car, and locomotive in the family, and exempted the karmic forces from the use of cars and locomotives to generate carbon emissions from transportation), This is a kind of environmental protection action generated by good intentions.

Box 7.3: Learning and celebration at Dr. Hungerford's residence.

Harold R. Hungerford (1928–) taught at the Department of Curriculum and Instruction at Southern Illinois University, Carbondale (SIUC). He was in science and education topics, such as Teacher Education, Teaching Methods, Educational Technology, Higher Education, Language Education (Fig. 7.27).

Fig. 7.27 Assembly photos; defense party activities. Photos by Arba'at Hassan

7.4.2.2 Environmental Education Issues into School-Based Curriculum

Environmental education is integrated learning. Therefore, the curriculum cooperates with the surrounding environment, activities, and atmosphere to allow students achieve their educational goals. A school-based curriculum helps students achieve their learning and educational goals. Measures that schools can take include readjusting learning goals, changing the organization of teaching content, conducting optional studies, and conducting teaching assessment strategies. Therefore, the school-based curriculum is the result of a balance between the curriculum issued by the Ministry of Education and the autonomy of schools and teachers.

7.4.2.3 Interpretation of Environmental Education

Interpretation is the process of conveying information. It meets people's needs and curiosity through the transmission of information. In the process of arranging the tour experience, it uses different media, including speeches, tours, displays, practices, and pictures to develop a new understanding of environmental issues. The narration function includes information, communication, guidance, service, education, propaganda, art, entertainment, inspiration, and business promotion roles.

7.4.2.4 Purpose of the Narration Plan

The purpose of the narration program is to assist learners to understand the learning activities and use teaching resources wisely. It can be integrated with environmental resource, land use, and related plans to create a low-intensity environmental impact design, create an environmental learning center that is integrated with local ecology, and uses suitable commentary media to integrate various environmental and ecological characteristics.

7.4.2.5 Interpretation Object

Interpretation services are mainly to provide learners with relevant information and help learners to choose a variety of experience activities to obtain a high-quality learning experience.

7.4.2.6 Interpretation Media

The explanation methods can be distinguished through different types of media (Table 7.1).

- Self-guided narration (non-personal narration): Through self-guided narration facilities, the narration is performed through placards, displays, publications, audiovisual media, and self-guided trails.
- Guided commentary (personal commentary): Consultant services, live tours, special lectures, and performances (Fig. 7.28). Such communication is two-way, and the content and methods can be adjusted in the interpretation process at the appropriate time, and timely responses can be made.

7.4.3 Educational Interpretation Plan

Because the learning center has various zones, activities and environmental conditions are different, so a variety of interpretation services are provided in different activities as shown in Tables 7.2 and 7.3.

7.4.3.1 Ecotourism Commentary

- **Animal and plant knowledge:** For example, explain the ecology of frogs in Taomi Village, Puli, Nantou, Taiwan;
- **Aesthetic knowledge:** For example, explain the primitive features of a tribal community, such as slate houses, wooden structures, south island trunk-type buildings.
- **History, geography, literature, architecture, religion, folklore, and other knowledge**. For example, when conducting an

Table 7.1 Interpretation media of an environmental education center

Non-personal commentary	Commentary card	Use display boards with text or pictures to explain the main themes. There are two types of commentary cards: management and commentary	Landmark commentary Warning signs, direction signs
	Display	Use static modes such as text, tables, models, pictures, or dynamic combination modes such as audio and video, sound and light, models, and performances to attract learners through visual and auditory communication	Use of supplementary media, such as: signage, commentary cards, commentary leaflets, or a combination of audiovisual equipment to increase the commentary effect
	Publication	In addition to general guide maps or texts introducing ecology, it mainly introduces learners to the center's location and traffic conditions, humanities or natural environment, center service facilities or activities, etc., or displays the history and environmental evolution of other ecological studies and other information	Advisory publications, explanatory publications
	Audiovisual media	Audiovisual media is an environmental learning center that is a facility for learners to explain or provide other services. It uses dynamic and static or audiovisual media to attract relevant learners	Computer briefing system Indoor display facilities
Personnel commentary	Consultation service	Eco-narrators explain at specific and prominent locations, such as visitor centers and commentary stations	
	Site personnel	Eco-narrator guides tourists to follow the designed route	
	Keynote speech	Give lectures, seminars, seminars, etc. in specific locations as an ecological commentator or hire an expert	
	Live performance	Performed by specific personnel such as board games, magic, dance, performances, games, treasure hunts, questionnaires, role-playing, land games, orientation games, handicrafts, DIY and other activities	

ecological tour in Taiwan's aboriginal tribal areas with ecological characteristics, you can explain the local history and culture, such as in Taiwan: Hsinchu's Qalang Smangus of the Atayal history (Fang et al. 2016), Miaoli's Nanzhuang of the Saisiyat history, Hualien's Fata'an of the Amis history in wetland culture, the slate house construction of the Paiwan's history. In Sabah, Malaysia, you can find and see different aboriginal tribes of Malays, Bajau, Dusun Lotud, Kadazandusun, Cocos, and Murut live in different regions (Fig. 7.29).

They can be found in Semporna, Tuaran, Penampang, Sandakan, and Keningau. Sabah has

more than 33 ethnic tribes, each with a different culture and language.

7.4.3.2 Commentary During Travel and Study

Understanding the scenery and points of interest between locations being visited.

7.4.3.3 Opportunity Commentary on Accommodation Resources

- **Meal commentary**: For environmentally friendly local specialties, food education can be conducted (Fig. 7.30). For example, you

Fig. 7.28 Assembly photos. Consultant services, live tours, and live performances with commentator commentary guided explanations are conducted in the form of personnel explanations from self-guided narration (non-personal narration) to guided commentary (personal commentary) at outdoor activities (top: Yoro Tribe, Jianshi, Hsinchu, Taiwan; bottom: Sanxia, New Taipei City, Taiwan). Photo by Max Horng

Table 7.2 Interpretation plan

Description	Environmental learning center	Commentary card	Commentary publications	Display	Audiovisual media
Symbol legend					

Assembled and Illustrated by Wei-Ta Fang

Table 7.3 The displays of explanation facilities

Partition name	Interpretation facilities	
Waters and waterfront-related activities (View by boat)		Sightseeing and route explanation Water sports precautions
Cultural experience-related activities (Agricultural trip, Aboriginal culture, Temple trip)		Tour guide and introduction of agricultural product characteristics Organic tea planting and harvesting process Explain the production process of unique tea species Explanation of Aboriginal history and life Learn about Aboriginal farming instructions and Aboriginal traditional etiquette The characteristics of local temples and related customs
Natural experience related activity zone (Hiking tour, ecological trip)		Tourist attractions and route commentary Introduce the species and habits of special flora and fauna Promote the concept of environmental conservation
General event division (Aerial gondola, camping experience, bicycle view)		Camping notes and location explanations of related facilities Sightseeing and route explanation

Assembled and Illustrated by Wei-Ta Fang

can explain the location of the environment in which it is produced and design the menu with features such as health, freshness, hygiene, and price.

- **Accommodation explanation**: Emphasizes the characteristics of eco-friendly hotel accommodations, including location, scale, facilities, room size, specifications, and types, all of which meet the accommodation requirements set by the relevant regulations of the surrounding environment of the ecotourism location.

7.4.4 Teaching Activity Strategy

The choice of an explanation strategy must be based on the content and goals of the course. It was found that the narrative questioning method only improved environmental knowledge; however, the role-playing method (Gao et al. 2009) improved environmental knowledge, sensitivity, attitudes, internal control views, and environmental actions (Worth and Book 2014; Gordon and Thomas 2018).

Malay	Bajau	Dusun Lotud
Kadazandusun	Cocos	Murut

Fig. 7.29 Assembly photos. Here are few photos of different ethnics wearing their colorful dress, found and live in Sabah. They are Malays, Bajau, Dusun Lotud, Kadazandusun, Cocos, and Murut, captured during different occasion of cultural exhibitions. Photos by Arba'at Hassan

Fig. 7.30 Food education can be conducted with the indigenous tribes. Fishing and prey are mainly the sources of food, such as wild boar, deer, and goat wrapped by bamboo leaves at Tayal culture. Photo by Max Horng

Our strategy for integrating teaching activities is suggested to be applied flexibly to the commentary activities:

7.4.4.1 Music, Dance or Drama

The teaching content is presented in the form of music, dance or theater performance.

7.4.4.2 Role-Playing

Use role-playing to stimulate learners' understanding of environmental issues and teaching content.

7.4.4.3 Lecture or Movie Viewing

Hire a professional to give a direct lecture on a specific topic or teach through film watching.

7.4.4.4 Appreciation or Writing of Poetry

Use the appreciation of new poems and the writing of various poem styles to let learners show their feelings about environmental issues and problems.

7.4.4.5 Teaching of Cartoons and Pictures

Encourage learners to learn about the environment through interesting patterns, and use the pictures to discuss issues.

7.4.4.6 Guided Meditation

Allow learners to calm down and feel and think about the relationships and status of the environment through meditation.

7.4.4.7 Games

In addition to helping learners have novel feelings about the course activities, they can also learn about knowledge, attitudes or skills on specific environmental issues.

7.4.4.8 Values and Attitudes

Through diagrams, reading and listening, etc., to help learners recognize their values, and help them think and build positive environmental values and attitudes.

7.5 Learning

Environmental learning involves the social, physical, psychological, and cultural factors experienced by the learner, which will then affect the student's ability to learn. The assessment of environmental learning requires the determination of knowledge, attitudes, and practices, whether sustainable development is integrated into the overall teacher's readiness to assess integrated education, and whether it has been incorporated into the teaching and learning stages (Norizan 2010:41).

7.5.1 Triadic Reciprocal Determinism Explores the Dependence of Teachers and Students

Environmental sociologist Albert Bandura (1925–2021) proposed the theory of social cognition in the 1960s, which produced a connection between behaviorism and cognition. Bandura proposed the theory of social cognition in 1986, stating that "all behaviors are based on the psychological needs of satisfying feelings, emotions, and desires." Social cognitive theory is based on the idea that humans learn by observing others. This can only happen if the individual recognizes behavior and environmental factors that are conducive to learning. Human pre-set behaviors allow them to attach to anything that exists at a critical period of social development. After this attachment period, as children grow up, they will learn to imitate teachers, peers, and siblings to solve problems. These cognitive activities are carried out not only through thinking but also through social and emotional connections. Therefore, Bandura's social learning theory (Bandura 1977) has expanded into a comprehensive theory of human motivation and action, by analyzing cognitive, vicarious, self-regulatory, and self-reflective processes in psychosocial function (Bandura 1986).

Bandura first proposed the argument of reciprocal determinism on the social basis of

Fig. 7.31 Triadic Reciprocal Determinism. (Bandura 1986). Revised and illustrated by Wei-Ta Fang

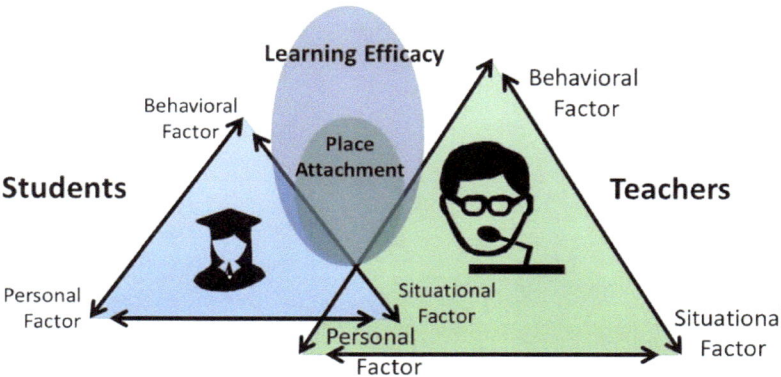

thought and action (Bandura 1986). Figure 7.31 we call Triadic Reciprocal Determinism (TRD). In short, ternary reciprocity determinism can be explained as people think, believe, and feel, which then affects their behavior. Conversely, the natural and external effects of human behavior determine their thinking patterns and emotional responses.

Therefore, social-emotional learning increases reduces the slow response of learning. In social cognitive learning theoretical models, behavioral factors affect environmental factors, and environmental factors affect behavioral factors. Environmental factors influence personal factors (cognitive, emotional, and other biological habits). Personal factors are also affected by behavior. Therefore, all factors are interconnected and interact to enable learning in the environment. Humans can use plans and elaborate the learning environment needed to trigger a response through behavioral changes and personal factors, and then strengthen the resilience of their learning (Fig. 7.32). In situations where peers learn together, the process by which teachers evoke personal and behavioral responses through teaching is called situational inducement. If the learning situation is tense and the learner is slow, then they will stop learning in the environment. Remember that the three aspects of social cognitive theory must work together to create a favorable learning environment. In addition, if context induction induces a high-stress response, then the attachment relationship between teachers and learners is broken, the

teacher-student relationship is affected, and the learning context begins to deteriorate.

7.5.2 Strengthen Voluntary Environmental Learning Behaviors

The goals of social cognitive and attachment theory are for teachers to accompany students, let students know where teachers are, and what positive feedback they provide (McLeod 2009). Students should think that teachers can be trusted not because the teacher has the authority to reward and punish, but because the teacher's relationship has received the following three major feedback phenomena. Behaviorist advocate Edward Lee Thorndike (1874–1949) summarizes **three** major laws of "trial and error":

7.5.2.1 Law of Effect

In the process of learner trial and error, if other conditions are equal, and after a specific response in the learning situation, a satisfactory result can be obtained, the connection relationship will be strengthened; if unhappy results are obtained As time goes by, its connection is weakened.

7.5.2.2 Practice Law

In the process of trial and error learning, the connection of any stimulus and response, once practiced, the strength of the connection will gradually increase; if not, the strength of the connection will gradually decrease.

Fig. 7.32 Attachment relationship between teachers and learners is closed to be affected (Shalun, Danshui, Taipei, Taiwan, 2021). Photo by Dennis Woo

7.5.2.3 The Law of Preparation

In the trial and error learning process, when the connection between stimulus and response has a state of preparation beforehand, if the learning is realized, you will be satisfied, otherwise you will be troubled; otherwise, when this connection is not When you are ready to implement, you will feel annoyed if you do. Therefore, the "try-and-error" learning model is still very useful for human learning.

It is recommended that teachers can give some easier homework to strengthen positive promotion in the promotion of attachment statement, affective statement, and arousal statement (Fig. 7.33). As a result, students have more confidence to continue their studies.

7.5.3 Learning Through Different Types of Learning Methods

In deep environmental education and learning, you can use multiple methods such as decision analysis, analysis of dilemma, value clarification, problem solving, false test, role-playing, game simulation, outdoor teaching, and group discussion (Zipko 1980; Errington 1991; Wheeler 2006; Raymond 2010; Fang et al. 2017).

Eco-environmental education is a "conceptual experience" that allows learners to study and understand the relationship between learning resources and the environment. The process of learning is through the skills learned in school,

Fig. 7.33 The object of manipulation restriction is the behavior that the individual has already voluntarily performed. Illustrated by Wei-Ta Fang

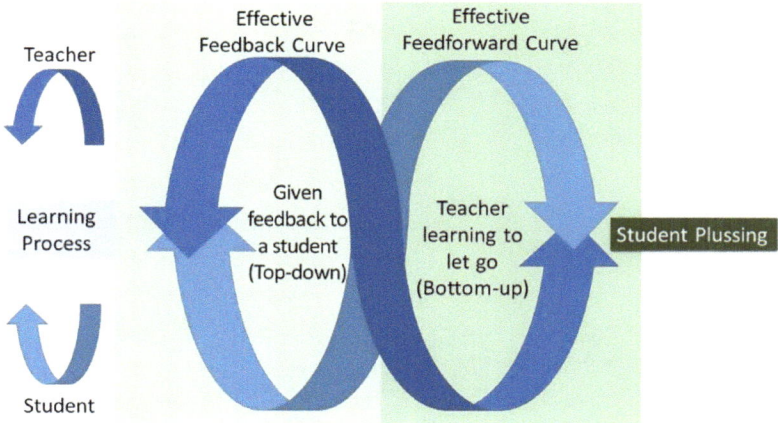

linking to the predetermined learning goals and the real needs of environmental protection, the use of focused thinking methods, coordination and problem-solving skills, concern for social impact, environmental pollution, ecological destruction, and other diverse issues. In the process of developing interpreters, it is necessary to set up internship courses for field activities. Based on experience, understanding, and appreciation of nature as the learning focus, through an in-depth understanding of outdoor ecology and cultural assets, students are trained to understand the ways of acting as ecological commentators. The integration and guidance of environmental education activities, the development of interpretive education in environmental learning centers.

In the course of the narrator's development, narration training is conducted through McGuire's Information Processing Theory (McGuire 1968). McGuire believes that there are three stages of information reception: "attention → understanding → acceptance." These three stages developed into six stages in the 1960s, and behavior changed due to "attention → understanding → acceptance."

McGuire later proposed 12 steps in 1989 to explain the factors that influence behavior, including:

(1) Contact information;
(2) Pay attention to the information;
(3) Like or become interested in the information;
(4) Understanding the message (what can be learned?);
(5) Skill acquisition (learn how to operate);
(6) Convincing message (change of attitude);
(7) Storage/consent of memory contents;
(8) Information search and retrieval;
(9) Making decisions based on retrieval;
(10) The behavior is consistent with decision-making;
(11) Strengthen ideal behavior;
(12) Strengthening after behavior.

The 12-point arguments advocated by McGuire focused on the psychological and behavioral aspects of "message → behavior," but the preaching is too strong and cannot explain non-informative education methods (McGuire 1968). In the information processing to learn in environmental education (McGuire 2006; 2015), "cognition, affection," in addition, to explaining the sources of rational "cognition and skills" experience from behavioral change (Fig. 7.34), the emotional experience is lacking insensibility.

The learning process of environmental education should cover three aspects: rationality, emotion, and ultimate care. They evaluated the university's environmental general courses to place too much emphasis on the teaching content of the cognitive domain, but the teaching goals of the affective domain and the action domain did not receive enough attention, because most of the courses did not improve students' affective

Fig. 7.34 In environmental education, "cognition/affection" could be explaining the sources of rational "cognition and skills" experience by student-centered learning activities. Photo by Max Horng

environmental literacy and environmental action (Fang et al. 2018; Liang et al. 2018).

Emotional experience is the psychological and physical state that the learner traces back after studying in the nature center. The process of tourism activities includes "Expectation-Outbound-Travel-Return-Recall" and other parts. In the "emotional experience," learners must learn to experience the environment with their hearts. This feeling is not something that can be experienced by looking at flowers. When the environment permits, the learner must carry travel notes and even make records with images, and use audiovisual equipment, such as telescopes, cameras, voice recorders, and video cameras, to leave landscapes and people in nature.

When the learner records all the situations as much as possible, they record the fact sheet with their own interpretation of their observation experience and realizes the experience of invisible mountains and forests in a mood of caring for the environment. In tourism learning, Gossling

(2006) used a simple coordinate to show the difference between ecological learning and mass tourism. With an "unpleasant" feeling, he showed that the experience of mass tourism is too superficial. In addition, hurried behavior cannot produce the joyful feeling of deep experience (Gossling 2006:93). Gosling believed that only by a deep and slow appreciation of nature's pulse and breathing can one enjoy the pleasure of reading the landscape pleasure. "Reading ecology" is like flipping through a good book, you can get the deepest baptism and feast in the soul.

7.6 Information Transmission

Looking at the evolution of social science from the perspective of environmental education information transmission, if the amount of information generated by an environmental social event is determined by the degree of social impact it can bring. In other words, the lower the frequency of social and environmental events the larger the impact of information will occur.

Take personal reception as an example. The value of information depends on the degree of accident caused by the impact of information. If the degree of accident is higher, it means that the individual finds that it is inconsistent with the original block impression. The change will be greater.

Box 7.5: Case Analysis: Strengthen from unknown information we need to discover

Let's take Taoyuan farm ponds, Taiwan, as an example (Fig. 7.35). Compared to the past 100 years, the temperature in Taipei has increased by 2 °C and Taichung has increased by 2.3 °C.

The temperature in the Taoyuan farm ponds area has not increased. The cooling effect of farm ponds is equivalent to sun-moon lake. Using data from the Central Weather Bureau, the temperature in Taoyuan has not changed much, but the fluctuations are severe. Based on the 100-year temperature record from Taichung, we can predict that the temperature in Taichung will rise from 23.9 °C in the next

century, at least to 26.2 °C in a hundred years. Because we know that we can almost survive by rising 1, 2, and 3 °C, there are only a few more typhoons and heavy rain. When it rises to 6 °C, the whole world will be in trouble.

"So, why does Taoyuan have so many ponds?"

About 20,000 years ago, an earthquake subsided in the Taipei Basin. The ancient Danshui River led the entire river from the ancient Shimen River to the Taipei Basin. The river in the Taoyuan platform became a beheaded river, so there was no source of irrigation water. Without irrigation water, development is relatively slow. From the past, Zimuliu (知母六), an aboriginal official of the Qing Dynasty (清朝), led the Hans to dig the first pond, Longtan Big Pond 274 years ago from 1748 to 2022. Later, the Taoyuan terraces have gradually continued. There are thousands of ponds (see Fig. 7.36). In the past, the ponds were accounted for about 11.8% of the total area of the Taoyuan Tableland. Now the land

Fig. 7.35 Taoyuan's Farm Ponds. Please refer to: Hsu et al. (2019). Habitat selection of wintering birds in farm ponds in Taoyuan, Taiwan, Animals 9, 113; doi:10.3390/ani9030113. Illustrated by Bai-You Cheng

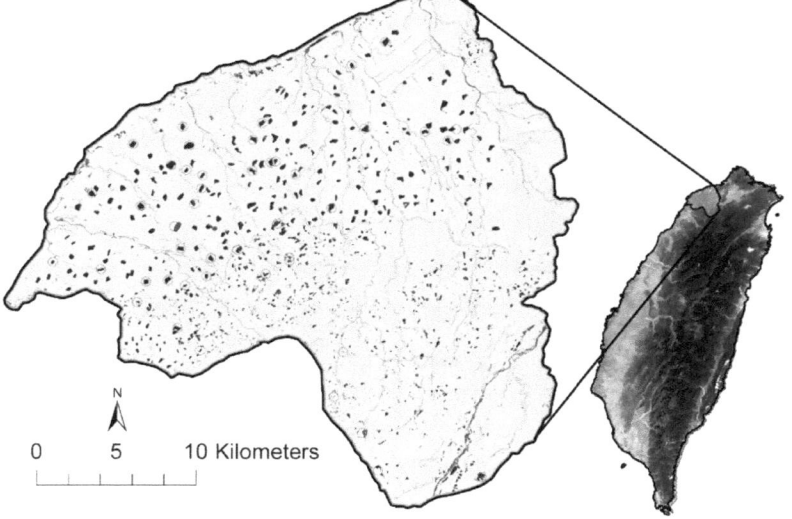

0   5   10 Kilometers

Fig. 7.36 Taoyuan has many ponds from field studies at GIEE, National Taiwan Normal University (from right to left: Shen-Yu Wu, an aquatic naturalist; Ling-Chu Wu (she; her), an EE officer, Taiwan EPA; Oi Lye Wong (she; her), an eco-traveler of sailing expeditions, Malaysia). Photo by Wei-Ta Fang

Fig. 7.37 The pond can provide leisure, fishing, and biodiversity. Taipei Farm Frog (*Hylarana taipehensis*) is a rare and endemic species in Taiwan. They were originally common frogs in the flatlands of Taiwan, but due to the recent abuse of pesticides in Taiwan, their numbers have been drastically reduced in Taoyuan farm ponds. Photo by Wei-Ta Fang

area is only accounts for 3.8%, and almost 90% of the pond area has disappeared.

"So, how do we do rehabilitation?"

In 2003, we hoped to restore *Nuphar shimadae*, so we also dug some ponds for re-cultivation. We found that the pond have some collective memories (Chao et al. 2021). The pond can provide leisure, fishing, biodiversity, and another function preserving Hakka culture (Fig. 7.37). Hakka culture is farming on sunny days and reading on rainy days, so the Hakkas are very hardworking people. The Hakka nationality also has the culture of building a Pavilion of Cherish the Words under the atmosphere of reading, so talking about culture here has far-reaching significance to tradition.

However, the government revised the law in 2016 because Article 95, paragraph 1, of the Electricity Act stipulates that nuclear review will be abolished in 2025. The government is hobbling on a road to a non-nuclear homeland and energy transition. Under the hasty and blind planning, it has been declared that 20% of the country's energy will be renewable. However, under the green energy policy, renewable energy conflict with the saltpans and wetland ecosystems in southern Taiwan. Many projects occupy 70% of saltpans. As of 2018, Taoyuan City occupies 17 larger ponds in Taoyuan to build photovoltaic panels and may be continued to build in the future (Song et al. 2018).

Compound semiconductor thin-film solar cells used in solar photovoltaic panels are composed of four raw materials: copper (Cu), indium (In), gallium (Ga), and selenium (Se). Because the aforementioned elements have good light absorption capacity, high power generation stability, high conversion efficiency, and high overall power generation, solar photovoltaic panels will infiltrate substances such as selenium (Se), gallium (Ga), indium (In), and thallium (Tl). After years of embroidery of the photovoltaic panel, toxic metals will seep into the water of the farm ponds, polluting the water source of the city. Because farmers raise fish in farm ponds, harvesting in June every year, a large number of fish in farm ponds could be entered the fish market for sale, forming carcinogens that are harmful to the human body.

This case tells us that because Taoyuan Farm ponds is a wintering area for birds in winter (Hsu et al. 2019), after a large-scale survey of the birds in the ponds from 2003 to 2019, at least one hundred species of birds were discovered in the birds. In 45 ponds, at least 15,053 birds are found every four months in winter (Figs. 7.38 and 7.39).

In addition, the number of birds has been calculated to be zero in the area where the 9 photovoltaic ponds in Taoyuan are installed. After ecological damage, winter birds in the Taoyuan Farm Ponds, including ducks, have no longer been able to use the photovoltaic ponds (Fig. 7.40).

In addition, the electricity generated by photovoltaic panels are and transmission to

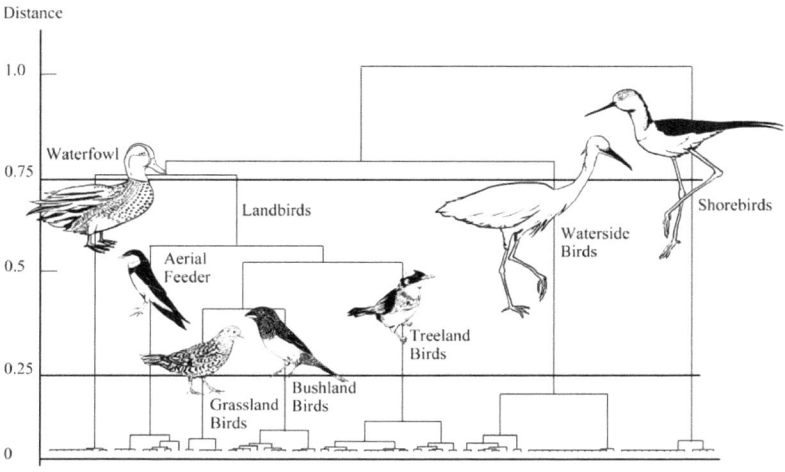

Fig. 7.38 Bird category in Taoyuan Farm Ponds. Please refer to: Hsu et al. (2019). Habitat selection of wintering birds in farm ponds in Taoyuan, Taiwan. Animals 9, 113; doi:10.3390/ani9030113. Illustrated by Yi-Te Chiang

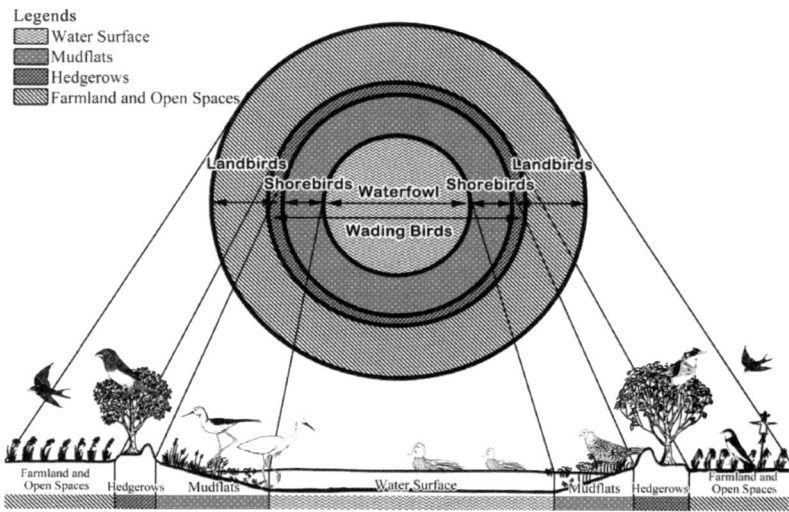

Fig. 7.39 Pond and birds. Please refer to: Hsu et al. (2019) Habitat selection of wintering birds in farm ponds in Taoyuan, Taiwan. Animals 9, 113; doi:10.3390/ani9030113. Illustrated by Yi-Te Chiang

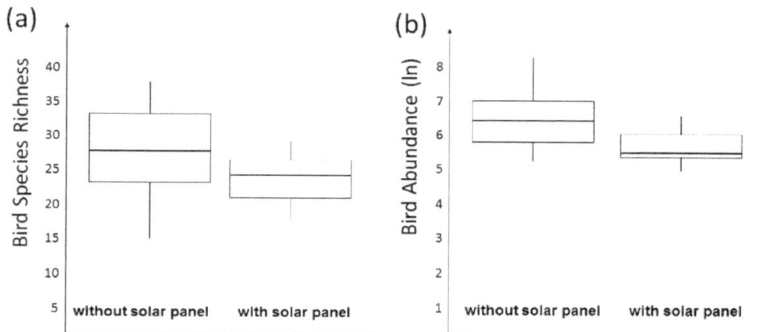

Fig. 7.40 **a** Ponds with solar panels, bird species richness was significantly less (*t*-test: $t = 3.174, p = 0.01$). **b** In the pond with solar panels, ln (the number of all birds with and w/out solar panel) was significantly lower (t-test: $t = 2.846, p = 0.02$). Illustrated by Wei-Ta Fang

the urban areas is far. Moreover, as is the case with transmission systems the loss and waste of electricity are substantial (Wu et al. 2016a, b). In addition, photovoltaic panels are not efficient in the energy conversion process; the efficiencies of compound semiconductors were estimated to be 28% as a solar energy conversion process from Oxford Photovoltaics (PV) reports in UK (Paleocrassas 1975; Ross and Hsiao 1977; Extance 2019), and

during the rainy season in northern Taiwan, power generation is limited (Shih et al. 2016). Because the floating photovoltaic power plant covered part of the pond surface (Su and Lo 2021; Hsiung 2022), the proposed optimization model could only reach up to 15.1% in hydrofloating photovoltaic power output (Zhou et al. 2020). The installation of PV on fish ponds may have a moderate negative impact on fish production, due to a reduction

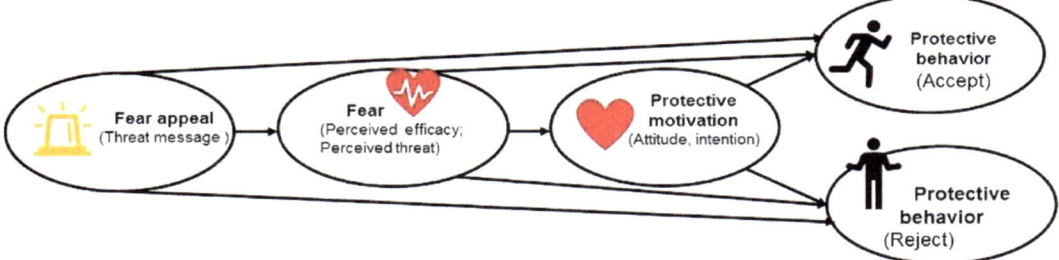

Fig. 7.41 When these information values have an impact on information, the more people find it unexpected, representing the "green electricity" in the original stereotype, the performance of green and environmental protection is incompatible with each other. The greater the magnitude of the pressure caused. This discovery will form a shock to society and great pressure on the incumbent governors. Illustrated by Wei-Ta Fang

in dissolved oxygen levels (Château et al. 2019). In addition, the release of the heavy metals from the panels into the farm ponds (Fabini 2015), create environmental and human health problems, entering humans through the food chain and affecting the safety of the lives and property of the residents in Taoyuan.

We know that environmental education has a long way to go. The green or renewable energy banner may be as meaningless as clean coal. We should be focused on finding renewable energy sources, but it can't be at the expense of the environment. Environmental education and educators now have the difficult task of educating stakeholders on the negative impacts of renewable sources of energy on the environment (see Fig. 7.41), especially at a time when the public has been told renewable energy will solve many of our problems, including climate change.

7.6.1 Learning Interface for Information Transfer

Environmental education is a teaching approaches that transmits environmental information from the source to another, while exchanging/moving information. The dynamic interaction of three elements in the process of environmental education information transmission: learner, domain/expert, and intermediate medium of instructional model is presented in Fig. 7.42. The "reappearance" and interpretation of the significance of environmental education that academia pays attention to needs to be based on the definition of the feedback generator and action assessor for an intrinsic value in self-revelation (Cohen 1983).

Learning environmental education issues through the interface of the learning environment requires integration of relevant courses, including integration of environmental education issues into thematic and multi-disciplinary courses. The basic laws of environmental information transmission need to be defined. The main points are as follows:

7.6.1.1 Planning Arrangement

- **Planned commentary**: Guided commentary according to the needs, time, and location of the ecological learner.
- **Arrange in advance**: Consider the time and space conditions in the explanation, and make proper arrangements in advance.
- **Proficiency in knowledge**: Perseverance in collecting local materials and ecological stories, reading and studying how to apply them.
- **Memorize numbers**: When explaining the age, area, height, and length, you should say the numbers.

Fig. 7.42 Interface of the sender and receiver. Illustrated by Wei-Ta Fang

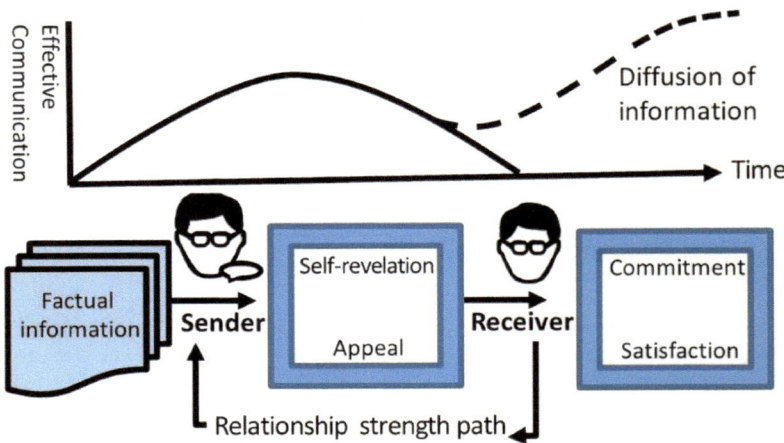

7.6.1.2   On-site Mastery

- **Flexible and flexible**: the explanation varies from person to person, according to time, and according to local conditions. Adjust the content of the commentary promptly in different seasons, climates, occasions, and atmospheres.
- **Projection of sight**: When conducting an ecological explanation, the sight should be directed to each learner who is listening.
- **Modest tone**: Avoid using inappropriate words such as "education" and "you" to avoid discomfort to the listener.
- **Moderate tone**: Do not speak too fast or too slowly when speaking, and whisper in the wild to avoid disturbing the quiet habitat of wild animals.
- **Explain clearly**: The ecological explanation should be detailed and clear, and should not be omitted deliberately, but it should be simple and concise, not overly verbose.
- **Positive interaction**: Let learners express their thoughts or opinions promptly.
- **Collective action**: Pay special attention to the safety of learners, and traffic and accidents are strictly prohibited.

7.6.2   The Transfer Process

In environmental communication, the processing of information is structured to influence and control each other, resulting in complex interaction results. In a complex system, the state of balance and change will occur between different input and output situations. In the system, each part depends on another and is also limited by it. In addition, feedback loops and self-regulating loops are often part of such a system and have a non-linear relationship. The system also strengthens the situation by receiving input, processes the teaching process, produces and outputs results, and interacts with the environment. For example, the results of teaching can create job opportunities, establish growth and satisfaction situations. In the process of information transmission, simpler systems can be embedded into more complex systems.

The impact of environmental communication is also discussed based on phenomenological methods. The basic assumption is that human beings understand the world and the meaning is derived from direct experience of phenomena. Therefore, phenomenology has the following three principles:

7.6.2.1   Knowledge is Created by Direct Experience

It could be learned with the world under conscious experience.

7.6.2.2   The Meaning of Things is Connected with the Underlying Factors of Personal Life

Things have a strong potential influence on life and will give more important meaning. Therefore, meaning is directly related to function.

7.6.2.3 Convey Meaning from Language

Therefore, the experience of channeling knowledge is created by the language channel.

Through the process of interpretation, the world is constructed by individuals. Such an interpretation is an active process of the mind, with the characteristics of fluctuating back and forth between states of experience, and the meaning given to world experience.

This process is also called the hermeneutic cycle. We first experience something, then we interpret and give meaning, then from the next experience to another test, re-interpretation, and so on. The greater the cognitive impact on an individual, the greater the cognitive change and re-interpretation. Make thought corrections.

Box 7.6: Case Analysis: Climate Warming

The importance of phenomenology to environmental communication lies in the great impact on personal experience. For example, half of Americans, often see climate warming as an "abstract phenomenon" that is difficult to connect with personal experience. That is because the continental United States of America (USA) is so vast that it cannot feel the actual pressure of climate change (Karl et al. 1996; Eckstein et al. 2021). The increase of the Global Conflict Risk Index (GCRI) is not large enough to unequivocally. The Long-Term Climate Risk Index (CRI) from the top three countries most affected from 2000 to 2019 should be listed as Puerto Rico, Myanmar, and Haiti, respectively (Eckstein et al. 2021).

For example, Hurricane Maria was devastated the northeastern Caribbean in September 2017, particularly Dominica, Saint Croix, and Puerto Rico, and Haiti. It is regarded as the worst natural disaster in recorded history to affect those islands (Zorrilla 2017). Regarding to the worst natural disasters, however, many teachers from Florida and Puerto Rico Secondary School in science could be held naïve views about climate change (e.g. that ozone layer depletion is a primary cause of climate change) and, poor idea of climate change science before Hurricane Maria hits (e.g. that it must be based on controlled experiments for it to be valid)(Herman et al. 2017). Rafael et al. (2021), however, surveyed citizens with general knowledge of global climate change increased from 43 to 62%, before and after an extreme weather event, specifically Hurricane Maria. Puerto Ricans trust non-profit institutions and the scientific community more than state authorities. After Hurricane Maria hits, citizens increased trust in the scientific community (Rafael et al. 2021). The question is: How can we taught knowledge without direct experience with the world under conscious experience relying as it does on relational values (Reid 2019)? As Karl et al. (1996) noted, the increase of the Global Conflict Risk Index (GCRI) is not large enough to quite significant in a practical sense from American.

According to Yu et al. (2020), the students with lower levels of knowledge were significantly more likely to find uncertainties related to climate change to be a greater obstacle to engaging in pro-environmental behaviors. As lived in one of the island nations, Taiwanese people believe in climate change because the island is small and experience several typhoons in a year. The area of Taiwan cannot be compared to the States of the East Coast of the United States that experience multiple hurricanes every year. Might be we need to go way to put the country size and beliefs into context to compared to all island nations. These impacts in our studies include the negative environmental information transmitted by global climate change, and Taiwan islanders are more likely to have negative connections. In other words, phenomenology also emphasizes the relevance of the individual to information generation from risk perception and response toward climate change (Yu et al. 2020).

When human beings consider environmental issues to be related to themselves, because of the disaster brought about by climate change, they have a negative impact, then human beings are more convinced of the fact of climate change. In addition, typhoon-affected households living in the immediate vicinity will be more empathetic than humans or ecosystems in distant countries that have been affected by hurricanes or tsunamis.

We feel that the empathy of the typhoon victims will be stronger than that of the tsunami victims. This is because of the typhoon for Taiwan residents, the frequency and intensity of the correlation are quite large. Therefore, environmental phenomena are affected by personal experience and personal exposure scenarios. In addition, human beings have a learning connection with environmental disasters that "this is the impact of climate change." These concepts are deeply entrenched and cannot easily be changed by the once and twice information transmission. This is also the main focus of environmental education that we can extract from phenomenological theoretical conventions.

7.6.3 Capacity Building

Capacity building is a concept of behavior change. In the realization that we will understand the obstacles to environmental development goals and measure the results that can achieve sustainable development, environmental education capacity building is the process by which individuals acquire environmental protection knowledge, attitudes, and skills development. In the process of environmental education capacity building, many organizations explain the

connotation of community capacity building in their own way. The United Nations Office for Disaster Risk Reduction defines capacity development in the field of disaster risk reduction (DRR) as: "the process by which humans develop their capabilities through the systematic organization of society." This process is effective to achieve social and economic goals, including improving knowledge, skills, systems, and institutional social and cultural environments. Below we distinguish between capacity building as community capacity building and individual capacity building.

7.6.3.1 Community Capacity Building

The first international organization to talk about community capacity building was the United Nations Development Programme. Since the 1970s, the United Nations Development Program has provided fundraising for the construction of training centers, contact visits to the Third World, construction of offices for local talent cultivation, on-the-job training,, and consulting services. In the process of capacity building, the country's financial resources, manpower, science and technology, organizational system, and resource management are used. The goal is to solve problems related to national policies and development directions while taking into account the constraints and needs of human development in national development. However, the formation strategy of the third world country's development capacity building model requires an independent model. Because once the UN funds are stopped, self-funding must be used to build the country, so as to prevent excessive reliance on long-term international assistance and the systematic intervention of international power.

7.6.3.2 Personal Ability Building

Personal level capacity building requires participants to "strengthen knowledge systems and technical capabilities so that individuals can actively participate in the process of learning and adapting to changes in the environment." According to the development of individual actions, feedback from the learning system is required, and learning can be carried out on an equal and mutually beneficial basis. Therefore, capacity building is not simply a matter of talent training or human resource development, but a change of thinking. In addition, improving personal capabilities is not enough to promote the sustainable development of society. It also needs the cooperation of the system and the organizational environment to make it work.

In management training, Broadwell (1969) described the capacity building model as four-level learning system. These four stages indicate that humans did not know have much knowledge at the beginning of learning and, that they "have no self-awareness or awareness of their ignorance and incompetence." When human beings realize their incompetence, they self-will consciously acquire skills, and then consciously use this skill. In the end, this skill may be used without being consciously thinking about the situation. In other words, the individual has already become familiar with this set of technology, and already has "unconscious competence."

Introduce elements of learning, including helping learners understand "what they don't know" or recognizing their blind spots. When the learner enters the learning state, they pass the four mental states shown in Fig. 7.43 until they reach the "unconscious ability" stage. By understanding the model and training plan for environmental education, the needs of learners can be determined, and learning goals can be set under specific environmental education topics based on the learner's goals.

- **Unconscious Incompetence**: In the state of "unconscious incompetence," the learner is unaware of the existence of skills or knowledge gaps.
- **Conscious Incompetence**: In the state of "conscious incompetence," learners are aware of the gap in learning skills or knowledge and understand the importance of acquiring new skills. It is at this stage that learning can really begin.
- **Conscious Competence**: In conscious competence, learners know how to use skills or perform tasks, but doing so requires frequent practice and conscious thinking and hard work (Fig. 7.44).

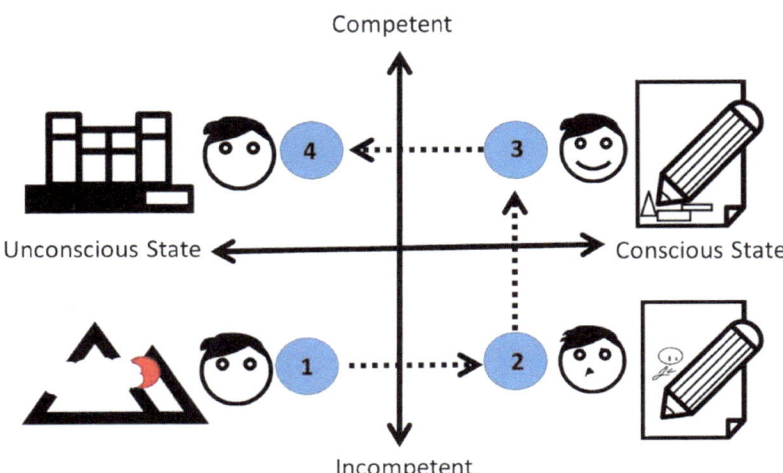

Fig. 7.43 The four stages of capacity building, modified and revised after Broadwell (1969), Curtiss and Warren (1973). This model helps teachers understand the emotional state of learners. For example, unconscious and incompetent learners respond to the curriculum differently than conscious and incompetent learners. If someone does not know that they have a problem, they are unlikely to participate in thinking about environmental solutions. On the other hand, if someone has conscious abilities, the individual may only need additional practice rather than intensive training. Illustrated by Wei-Ta Fang

Fig. 7.44 The recognition of these environmental issues has created strong personal belief, and promoted the establishment of environmental impact assessment systems, such as buying a forest for carbon sinks. Forests provide a "carbon sink" that world's forests could absorb a net 7.6 billion metric tons of CO_2 per year (Harris and Gibbs 2021). At a foundation unit, the areas could be protected from 288,149 square meters of green spaces (Baoshan Hsinchu, Tongluo Miaoli, Yuli Hualien, and Puli Nantou). The Carbon Budget Model of the forest uses the best available information on forests and tree growth sponsored by Hetong Culture and Art Foundation, Tainan, Taiwan (Right person: President Ms. Ileana Lee). Photo courtesy of Hetong Culture and Art Foundation, Tainan, Taiwan

- **Unconscious Competence**: In the unconscious ability, individuals have sufficient experience and can easily perform skills. When they perform, they are skilled in the unconscious state.

In addition, in the four stages of learning ability, the teachers can choose the topic content that helps the learners to enter the next stage by knowing which stage the learners is in, in order to improve their (teachers' and learners') skills based on competency-based education. It is even possible to use an assessment model to prove to the learners their ability gaps, thereby transferring learners from the first stage to the second stage of learning.

7.7   Communication Media

In modern society, communication through the media has become a means of disseminating or obtaining information or a distraction. Therefore, in environmental education and communication, media publications such as formal publishing, communication publishing, and the Internet have become information disclosed path. The mass media has a major task in environmental education. Countries around the world make extensive use of the media to disseminate scientific knowledge about environmental protection and stimulate the general public's awareness of environmental consciousness, such as environmental pollution,

Fig. 7.45 Social media can help environmental education promoters keep in touch with their peers around the world, such as the virtual conference of environmental education held by the Chinese Society of Environmental Education during 2021. Photo by Wei-Ta Fang

soil degradation, resource depletion, species extinction, and the general public about environmental health, public health, and nutrition information. The recognition of these environmental issues has created strong public opinion, stimulated social resistance movements, and promoted the establishment of environmental protection laws and environmental impact assessment systems, all of which are the effects of mass media.

In developing countries, the use of radio and television in the past has a special educational effect for the public, especially the invention of the transistor radio, which made the radio the most widely used mass communication tool in the twentieth century. In addition, after the 1990s, the Internet made the world a realm without borders. Communicate with the "process model" of the blog, can provide feedback on online networking, and help develop ideas and promote teaching strategies and methods for deep learning (Chao et al. 2020).

In addition, more and more environmental education promoters use Facebook groups, Google Meet groups, WeChat groups, and Line groups to share ideas. The heavy use of social media can help environmental education promoters keep in touch with their peers around the world (Fig. 7.45).

7.7.1   Formal Media

7.7.1.1   Research Community

The research community's communication channels include a series of books, monographs, journal articles, seminar papers, posters, and institutional repositories (Fig. 7.46).

Many universities and research institutions have in-house online research archives of open research files. These archives are called repositories and are listed as a publication. If a researcher places work in an open research file, it should then be considered published. However, we all know that this will result in more nonsense being made available to the public due to copyright issues (Kodaneva 2021). We could mean the downfall of peer-reviewed papers. If it is a previously unpublished work, storing it in an

Fig. 7.46 Assembly photos; The research community's communication channels include a series of books. The language could be used in local languages, such as: left side: Traditional Chinese visions (published by Wu-nan Culture Enterprise in Taiwan, ROC) and; right side: Simplified Chinese visions (published by the Social Sciences Academic Press (China), SSAP, in People's Republic of China (PRC)); The SSAP, established in 1985, is one of the academic publishing institutions for the humanities and social sciences directly affiliated to the Chinese Academy of Social Sciences (CASS), PRC. Photos by book authors, Wei-Ta Fang

archive and making it public may cause copyright attribution when these materials are published in the future. If the work has already been published, the original publisher may retain these rights and may not redistribute the work through archival. However, the archive provides an online version for public reading, which is conducive to the outside world (Kodaneva 2021).

7.7.1.2   The General Public
Wikipedia, feature articles/access/talk; open access journals and books release.

7.7.2   Informal Media

7.7.2.1   Research Community
Opening speeches, meeting records, and social media.

7.7.2.2   The General Public
Social media, for example: Twitter, Facebook, blogs (text posting); or Instagram (image posting).

7.7.2.3   Between the Research Community and the General Public
Use research community websites to publish published articles, such as: academia.edu and researchgate.net.

7.7.3   Communication Effect

7.7.3.1   Increased Awareness
Increasing human awareness of environmental knowledge and deepening understanding of environmental knowledge.

7.7.3.2 Informed Choices

Improve the ability to make informed choices among alternatives.

7.7.3.3 Exchange of Information

Increase the degree of exchange of information, themes, or opinions. In the International Environmental Education Project promoted by UNESCO, the following information exchange suggestions have been made for the mass media:

- In regular radio, television, internet, cartoon programs, or live programs, add vivid and interesting environmental issues, such as playing wonderful ecological documentaries or ecological music, and the sounds of animals and birds.
- Inviting people to participate in discussions on environmental issues.
- Radio, Internet and TV programs, and special exhibition should include environmental disasters and environmental degradation (Fig. 7.47).
- Providing radio, internet and television producers to participate in environmental education training, and/or study special exhibition in museums.

Fig. 7.47 Special programs at exhibitions should include environmental disasters and environmental degradation like biological losses. Photo by Wei-Ta Fang

7.8 Summary

J. Robert Cox, an environmental communication scholar who was the President of the National Sierra Club, once mentioned that the field of environmental communication consists of the following seven research and practical fields, including: "Environmental Discourse (Rhetoric and discourse), media and environmental news, public participation in environmental decision-making, social marketing, and advocacy activities, environmental cooperation and conflict resolution, risk communication, and natural expression in pop culture and green marketing" (Cox 2010). From a practical point of view, environmental learning and communication adopt effective communication methods and apply environmental management strategies and technologies to environmental management and protection. We realize that environmentalism begins with environmental learning and communication; however, from the buzz in the 1960s to the twenty-first century, it gradually turned into "environmental skepticism." American society loves economic development, a version of environmental protection. In the 1960s, environmental movements were ignited by the sparks of writers, or more specifically and accurately, by "Rachel Carson's typewriter" (Flor 2004). Therefore, from the perspective of history, there are six basic elements of environmental learning and communication: "ecological knowledge, cultural sensitivity, networkability, ability to use media, environmental ethics practice, conflict resolution ability, and mediation and arbitration ability" (Flor 2004). From the research of scholars, people's indifference is far more complicated than the lack of information. In fact, too much dazzling environmental information today is often counterproductive and confusing. When humans understand the complexity of environmental issues, they will feel overwhelmed and helpless. This will often lead to human apathy for environmental protection or suspicion of environmental science. Therefore, after the rise of artificial intelligence research in the twenty-first century, virtual environments have become the mainstream of consciousness. The conspiracy thesis of environmental skepticism holds that human beings will eventually be combined with technology. Concerning the legitimacy of environmental rhetoric, conspiracy theorists questioned environmental protection and believed that environmental protection was the biggest obstacle to economic development. These myths will pose increasing challenges to environmental protection and sustainable development of society (Jacques 2013). Therefore, how to clear the source and how to dispel chaos anyway depends on the cooperation of environmental research scholars, sustainable development research scholars, and mass communication scholars to convey the correct knowledge and skills of environmental protection.

References

Aguiar OG, Mortimer EF, Scott P (2010) Learning from and responding to students' questions: the authoritative and dialogic tension. J Res Sci Teach Off J Natl Assoc Res Sci Teach 47(2):174–193

Bandura A (1977) Social learning theory. Prentice Hall, Hoboken

Bandura A (1986) Social foundations of thought and action: a social cognitive theory. Prentice Hall, Hoboken

Bisoffi S, Ahrné L, Aschemann-Witzel J, Báldi A, Cuhls K, DeClerck F, Duncan J, Hansen HO, Hudson RL, Kohl J, Ruiz B, Siebielec G, Treyer S, Brunori G (2021) COVID-19 and sustainable food systems: What should we learn before the next emergency. Front Sustain Food Syst 5:650987. https://doi.org/doi.org/10.3389/fsufs.2021.650987

Boyd MP (2012) Planning and realigning a lesson in response to student contributions: intentions and decision making. Elem Sch J 113(1):25–51

Broadwell MM (1969) Teaching for learning (XVI). The Gospel Guardian. wordsfitlyspoken.org

Chao S-H, Jiang J, Hsu C-H, Chiang Y-T, Ng E, Fang W-T (2020) Technology-enhanced learning for graduate students: exploring the correlation of media richness and creativity of computer-mediated communication and face-to-face communication. Appl Sci 10:1602. https://doi.org/10.3390/app10051602

Chao S-H, Jiang J, Wei K-C, Ng E, Hsu C-H, Chiang Y-T, Fang W-T (2021) Understanding pro-environmental behavior of citizen science: an exploratory study of the bird survey in Taoyuan's farm ponds project. Sustainability 13:5126. https://doi.org/10.3390/su13095126

Château PA, Wunderlich RF, Wang TW, Lai HT, Chen CC, Chang FJ (2019) Mathematical modeling suggests high potential for the deployment of floating photovoltaic on fish ponds. Sci Total Environ 687:654–666

Chapman D, Sharma K (2001) Environmental attitudes and behavior of primary and secondary students in Asian cities: an overview strategy for implementing an eco-schools programme. Environmentalist 21(4):265–272

Cheng JC-H, Monroe MC (2012) Connection to nature: children's affective attitude toward nature. Environ Behav 44(1):31–49

Christakis DA (2009) The effects of infant media usage: what do we know and what should we learn? Acta Paediatr 98(1):8–16

Combes BP (2005) The United Nations decade of education for sustainable development (2005–2014): learning to live together sustainably. Appl Environ Educ Commun 4(3):215–219

Cohen JB (1983) Involvement and you: 1000 great ideas. ACR North American Advances, Duluth

Cox JR (2010) Environmental communication and the public sphere, 2nd edn. Sage, Thousand Oaks

Curtiss PR, Warren PW (1973) The dynamics of life skills coaching. Life skills series. Training Research and Development Station, Dept. of Manpower and Immigration

Delors J (2013) The treasure within: learning to know, learning to do, learning to live together and learning to be. What is the value of that treasure 15 years after its publication? Int Rev Educ 59(3):319–330

Eckstein D, Künzel V, Schäfer L (2021) Global climate risk index 2021. Who Suffers Most from Extreme Weather Events, 2000–2019

Egan K (1993) Narrative and learning: a voyage of implications. Linguist Educ 5(2):119–126

Errington E (1991) Role playing and environmental issues. Aust J Environ Educ 7:1–15

Extance A (2019) The reality behind solar power's next star material. Nature 570(7762):429–433

Fabini D (2015) Quantifying the potential for lead pollution from halide perovskite photovoltaics. J Phys Chem Lett 6(18):3546–3548

Fang W-T, Hu H-W, Lee C-S (2016) Atayal's identification of sustainability: traditional ecological knowledge and indigenous science of a hunting culture. Sustain Sci 11(1):33–43

Fang W-T, Ng E, Chang M-C (2017) Physical outdoor activity versus indoor activity: their influence on environmental behaviors. Int J Environ Res Public Health 14(7):797

Fang W-T, Lien C-Y, Huang Y-W, Han G, Shyu G-S, Chou J-Y, Ng E (2018) Environmental literacy on ecotourism: a study on student knowledge, attitude, and behavioral intentions in China and Taiwan. Sustainability 10(6):1886

Fang W-T, Shih S-S, Cheng B-Y, Chou J-Y (2020) Clustered constructed wetland systems in Metropolitan Taipei. Wetland Sci Pract April 2020:96–107

Finn PJ (2010) Literacy with an attitude: educating working-class children in their own self-interest. Suny Press, Albany

Fisk S (2019) Clean out your 'jargon' closet: simplify your science communications for greater impact. CSA News Magazine

Flor AG (2004) Environmental communication: principles, approaches and strategies of communication applied to environmental management. University of the Philippines-Open University, Los Baños

Foulks B, Morrow RD (1989) Academic survival skills for the young child at risk for school failure. J Educ Res 82(3):158–165

Gao F, Noh JJ, Koehler MJ (2009) Comparing role-playing activities in second life and face-to-face environments. J Interact Learn Res 20(4):423–443

Gordon S, Thomas I (2018) 'The learning sticks': reflections on a case study of role-playing for sustainability. Environ Educ Res 24(2):172–190

Gossling S (2006) Ecotourism as experience-tourism. In: Gossling S, Hultman J (eds) Ecotourism in scandinavia: lessons in theory and practice. CABI

Grant J (2002) Learning needs assessment: assessing the need. BMJ 324(7330):156–159

Hansen A (2011) Communication, media and environment: towards reconnecting research on the production, content and social implications of environmental communication. Int Commun Gaz 73(1–2):7–25

Harris N, Gibbs D (2021) Forests absorb twice as much carbon as they emit each year. World Resources Institute. https://www.wri.org/insights/forests-absorb-twice-much-carbon-they-emit-each-year

Herman BC, Feldman A, Vernaza-Hernandez V (2017) Florida and Puerto Rico secondary science teachers' knowledge and teaching of climate change science. Int J Sci Math Educ 15(3):451–471

Hsiung KH (2022) Policy and legal issues of the environmental and social inspection in fishery-solar energy. In: IOP conference series: earth and environmental science, vol 1009, no 1. IOP Publishing, Bristol, p 012009

Hsu C-H, Lin T-E, Fang W-T, Liu C-C (2018) Taiwan roadkill observation network: an example of a community of practice contributing to Taiwanese environmental literacy for sustainability. Sustainability 10(10):3610

Hsu C-H, Chou J-Y, Fang W-T (2019) Habitat selection of wintering birds in farm ponds in Taoyuan, Taiwan. Animals 9:113. https://doi.org/10.3390/ani9030113

Jacques P (2013) Environmental skepticism: ecology, power and public life. Ashgate

Joy S, Kolb DA (2009) Are there cultural differences in learning style? Int J Intercult Relat 33(1):69–85

Jurin RR, Roush D, Danter KJ (2010) Environmental communication: skills and principles for natural resource managers, scientists, and engineers. Springer Science & Business Media, Berlin, Heidelberg

Kahn P, Kellert S (2004) Children and nature: psychological, sociocultural and evolutionary investigations. Environ Values 13(3):409–412

Karl TR, Knight RW, Easterling DR, Quayle RG (1996) Indices of climate change for the United States. Bull Am Meteor Soc 77(2):279–292

Keenan WJ (2020) Learning to survive: Wicked problem education for the Anthropocene age. J Glob Educ Res 4(1):62–79

Kodaneva SI (2021) The transformation of copyright under the influence of the development of digital technologies. Law Digit Econ 4(14):31–38

Kolb DA (1981) Experiential learning theory and the learning style inventory: a reply to freedman and stumpf. Acad Manag Rev 6(2):289–296

Kolb AY (2005) The Kolb learning style inventory-version 3.1 2005 technical specifications. Hay Res Direct 200(72):166–171

Lanahan BK, Yeager EA (2008) Practicing teachers as elementary social studies methods instructors: issues in preparing preservice elementary teachers. Soc Stud Res Pract 3(2):10–28

Liang S-W, Fang W-T, Yeh S-C, Liu S-Y, Tsai H-M, Chou J-Y, Ng E (2018) A nationwide survey evaluating the environmental literacy of undergraduate students in Taiwan. Sustainability 10(6):1730

Liu S-Y, Yeh S-C, Liang S-W, Fang W-T, Tsai H-M (2015) A national investigation of teachers' environmental literacy as a reference for promoting environmental education in Taiwan. J Environ Educ 46(2):114–132

Loh KY (2010) New media in education fiesta (20100906, Day 1). In: Learning journey [Blog spot]. Retrieved from http://lohky.blogspot.ca/2010/09/new-media-in-education-fiesta-20100906_07.html

Martins AA, Mata TM, Costa CA (2006) Education for sustainability: challenges and trends. Clean Technol Environ Policy 8(1):31–37

McGuire NM (2015) Environmental education and behavioral change: an identity-based environmental education model. Int J Environ Sci Educ 10(5):695–715

McGuire SY (2006) The impact of supplemental instruction on teaching students how to learn. New Dir Teach Learn 2006(106):3–10

McGuire WJ (1968) Personality and attitude change: an information processing theory. In: Greenwald A, Ostrom T, Brock T (eds) Psychological foundations of attitude. Academic Press

McLeod SA (2009) Attachment theory. Retrieved from http://www.simplypsychology.org/attachment.html

Norizan E (2010) Environmental knowledge, attitude and practice of student teachers. J Environ Educ 19:39–50

Oh RRY, Fielding KS, Carrasco RL, Fuller RA (2020) No evidence of an extinction of experience or emotional disconnect from nature in urban Singapore. People Nat 2(4):1196–1209

Olivos-Jara P, Segura-Fernández R, Rubio-Pérez C, Felipe-García B (2020) Biophilia and biophobia as emotional attribution to nature in children of 5 years old. Front Psychol 11:511. https://doi.org/10.3389/fpsyg.2020.00511

Otte ML, Fang W-T, Jiang M (2021) A framework for identifying reference wetland conditions in highly altered landscapes. Wetlands 41:40. https://doi.org/10.1007/s13157-021-01439-0

Paleocrassas SN (1975) Photolysis of water as a solar energy conversion process: an assessment. Hydrogen energy. Springer, Boston, pp 243–253

Parker L, Prabawa-Sear K, Kustiningsih W (2018) How young people in Indonesia see themselves as environmentalists: identity, behaviour, perceptions and responsibility. Indonesia Malay World 46(136):263–282

Perkins HE (2010) Measuring love and care for nature. J Environ Psychol 30(4):455–463

Rafael MT, Santos-Corrada M, Sandra M (2021) Perceptions of climate change in Puerto Rico before and after Hurricane Maria. Am J Clim Chang 10(2):153–166

Raymond C (2010) Do role-playing simulations generate measurable and meaningful outcomes? A simulation's effect on exam scores and teaching evaluations. Int Stud Perspect 11(1):51–60

Reid A (2019) Climate change education and research: possibilities and potentials versus problems and perils? Environ Educ Res 25(6):767–790

Rickinson M (2006) Researching and understanding environmental learning: hopes for the next 10 years. Environ Educ Res 12(3–4):445–457

Rickinson M, Lundholm C, Hopwood N (2009) Environmental learning: insights from research into the student experience. Springer, Dordrecht

Ross RT, Hsiao TL (1977) Limits on the yield of photochemical solar energy conversion. J Appl Phys 48(11):4783–4785

Schank RC (1995) What we learn when we learn by doing. Technical Report No. 60. Institute for the Learning Sciences Northwestern University

Scheler M, McAleer G (2017) The nature of sympathy. Routledge

Senecah SL (2007) Impetus, mission, and future of the environmental communication commission/division: are we still on track? Were we ever? Environ Commun 1(1):21–33

Shekarey A (2006) The consequences of the binary opposition/continuation approaches to modernism and postmodernism: a critical educational study. Tamara: J Crit Organ Inq 5(1)

Shih YH, Shi NX, Tseng CH, Pan SY, Chiang PC (2016) Socioeconomic costs of replacing nuclear power with fossil and renewable energy in Taiwan. Energy 114:369–381

Song LY, Yadav R, Liang HC (2018) Research on eco-friendly solar energy generation in Taoyuan pond. In: 2018 IEEE international conference on advanced manufacturing (ICAM). IEEE, pp 264–267

Sorohan EG (1993) We do; therefore, we learn. Training Dev 47(10) 47. Gale Academic One File, link.gale.com/apps/doc/A14536292/AONE?u=googlescholar&sid=bookmark-AONE&xid=279114b1. Accessed 29 Aug. 2022.

Stagg BC, Dillon J, Maddison J (2022) Expanding the field: using digital to diversify learning in outdoor science. Discip Interdiscip Sci Educ Res 4(1):1–17

Sternberg RJ, Williams WM (1996) How to develop student creativity. ASCD

Sternberg RJ, Davidson JE (eds) (2005) Conceptions of giftedness, vol 2. Cambridge University Press, New York

Su PW, Lo SL (2021) Using landsat 8 imagery for remote monitoring of total phosphorus as a water quality parameter of irrigation ponds in Taiwan. Environ Sci Pollut Res 28(47):66687–66694

Tariq S, Sultan S, Choudhary FR (2020) Environmental education and practices in Canada, Turkey & Pakistan at primary level: a content analysis. Res J Soc Sci Econ Rev 1(4):389–400

Thomas I (2009) Critical thinking, transformative learning, sustainable education, and problem-based learning in universities. J Transform Educ 7(3):245–264

Turner J, Paris SG (1995) How literacy tasks influence children's motivation for literacy. Read Teach 48 (8):662–673

Uhlmann K, Lin BB, Ross H (2018) Who cares? The importance of emotional connections with nature to ensure food security and wellbeing in cities. Sustainability 10(6):1844

Wals AE, Brody M, Dillon J, Stevenson RB (2014) Convergence between science and environmental education. Science 344(6184):583–584

Westbury I (1990) Textbooks, textbook publishers, and the quality of schooling. In: Elliott DL, Woodward A (eds) Textbooks and schooling in The United States. NSSE, pp 1–22

Wheeler SM (2006) Role-playing games and simulations for international issues courses. J Polit Sci Educ 2 (3):331–347

Wilson EO (1984) Biophilia. Harvard University Press, Cambridge

Worth NC, Book AS (2014) Personality and behavior in a massively multiplayer online role-playing game. Comput Hum Behav 38:322–330

Wu YK, Ye GT, Shaaban M (2016a) Analysis of impact of integration of large PV generation capacity and optimization of PV capacity: case studies in Taiwan. IEEE Trans Ind Appl 52(6):4535–4548

Wu YK, Ye G-T, Shaaban M (2016b) Impact analysis of large PV integration: case studies in Taiwan. In: 2016b IEEE/IAS 52nd industrial and commercial power systems technical conference (I&CPS). IEEE, pp 1–10

Yu TK, Lavallee JP, Di Giusto B, Chang I, Yu TY (2020) Risk perception and response toward climate change for higher education students in Taiwan. Environ Sci Pollut Res 27(20):24749–24759

Zhou Y, Chang FJ, Chang LC, Lee WD, Huang A, Xu CY, Guo S (2020) An advanced complementary scheme of floating photovoltaic and hydropower generation flourishing water-food-energy nexus synergies. Appl Energy 275:115389. https://doi.org/doi.org/10.1016/j.apenergy.2020.115389

Zipko SJ (1980) Simulation for environmental education and awareness. Sci Activities 17(2):32–38

Zorrilla CD (2017) The view from Puerto Rico—Hurricane Maria and its aftermath. N Engl J Med 377 (19):1801–1803

.

Outdoor Education

8

Process is important for learning. Courses taught as lecture courses tend to induce passivity. Indoor classes create the illusion that learning only occurs inside four walls isolated from what students call without apparent irony the "real world.'

David W. Orr, What is Education for? 1991:52.

Abstract

Environmental Education (EE) promotes the complex interrelationships between human culture and ecosystems. Due to the political nature of environmental decision-making, the field of Environmental Education faces many disputes. For example: What is the correct definition and purpose of environmental education? Should the curriculum include environmental values and ethics, as well as ecological and economic concepts and skills? What is the role of student environmental action in correcting environmental problems? What is the appropriate role for teachers in developing curricula on environmental education? At what age students should understand environmental issues? What types of Environmental Education should urban, suburban, and rural youth receive? What technology can be used to slow ecological damage? Among these problems, Outdoor Education and Environmental Education also face the problems mentioned above. Due to the political factors of human environmental decision-making, Outdoor Education and Environmental Education have been in an undefined state. Educators continually devise better ways to expand the definition of outdoor education to improve the philosophy and practical work of outdoor education. Outdoor education includes earth education, bioregional education, expedition learning and expansion training, ecological education, natural awareness, natural experience, local-based teaching and education, and the use of environmental materials as learning to integrate the local environment.

8.1 Connotation of Outdoor Education

Outdoor education is the organized effort of learning or studying biotic and abiotic aspects of the environment in an outdoor setting that is free of anthropogenic impacts (Fang et al. 2017). For example, how far from a city does a person need to travel to view the stars where there is no light pollution to observe the fauna in their native habitat or the stars? Outdoor education is often referred to as outdoor learning, outdoor schools, forest schools, and wilderness education and these are examples of experiential learning. They combine elements of adventure, environmental, and expedition, and wilderness-based experiences that can sometimes include residential tourist programs. Teachers guide learners to participate in various activities such as hiking, mountain/rock climbing, canoeing, rope lessons, and group games.

8.1.1 History of Outdoor Education

The philosophy of outdoor education has a long history. Such activities have been documented since the seventeenth century and it could reasonably be presumed that the indigenous people of the world that have oral histories and cultures that are thousands of years old are evidence of the efficacy of outdoor education (Gilbertson et al. 2022). From a much more recent and western point of view, the Czech Theologian and educator Johann Comenius (1592–1670) declared in *Didactica Magna* (*The Great Didactic*) that "All people should be allowed to fully learn everything in the world." He proposed that the rule of learning is to study and observe before language intervention and to connect the curriculum with life (Keatinge 2018). Therefore, he was considered the father of modern education at that time (Dent 2021).

In the United States in the late nineteenth century, educators realized that getting students out of the classroom could improve educational skills, attitudes, and values (Wattchow 2006; Öhman and Sandell 2016). These goals were incorporated into the progressive education movement, which was introduced to American schools in the first half of the twentieth century (Dewey 1916/1997, 1938/1997).

In the early twentieth century, the "camping movement" gained traction with American scholars that were focused on the study of nature (Gagen 2004; Cupers 2008). The purpose of camping was to expand the emotional connection between the students and nature as experience the processes of getting food, shelter, entertainment, spiritual inspiration, and other outdoor recreation activities (Nadel and Scher 2019).

This may be partly true. Camping was an inexpensive family activity since people didn't have a lot of money. In Mainland China, the Scout of the Republic of China was founded in 1912 and Boy Scouts of America in 1910. The outdoor training allowed young boys to experience outdoor camping and survival skills.

In the 1940s, the term "outdoor education" appeared, hoping to describe the teaching process of natural experiences through direct experience and to meet the learning goals of students in various disciplines.

In the USA outdoor education promoting contact and experience with nature became prominent in the 1940s. The W. K. Kellogg Foundation created community school camps to grow this type of learning experience. As progressivism began to fade from public schools in the late 1950s, outdoor education became more important (Knapp 1994). Outdoor educators saw the value of immersive teaching (Dillon et al. 2016), so they started programs outside of the traditional inside classroom (Miles 1987; Freeman and Seaman 2020). The outdoor education program builds on community and cultural values, without raises expectations and standards (Bowdridge and Blenkinsop 2011), strengthens connections between students, and develops positive associations around school activities and the outdoors. Outdoor education programs sometimes involve residential or journey wilderness-based experiences in which students participate in outdoor activities. Using the outdoor environment, students can have learning experiences that are far different from that experienced inside a building. These natural connections have also offset the negative effects of urbanization (Pigram 1993; Kellert 2012). The learning process of activities, therefore, associated with camping is closely linked to community activities, and more emphasis is placed on practical knowledge, see Figs. 8.1, 8.2, 8.3, 8.4 and 8.5.

In 1965, the U. S. government promoted the Elementary and Secondary Education Act through additional support from nationally protected areas, education departments, private

Fig. 8.1 Southern Illinois University Campus Aerial View. Photo by Arba'at Hassan

Fig. 8.2 "Welcome" greeting at SIUC. Photo by Arba'at Hassan

education institutions, professional teacher organizations, and other non-governmental organizations (McGuinn and Hess 2005;Gamson et al. 2015; Casalaspi 2017).

Since outdoor activities are often incorporated into professional curricula, outdoor education has influenced education reforms in the early 2000s (e.g., Mikaels et al. 2016). Currently, courses continued to be developed to improve learners' social development and leisure skills on student recruitment, retention, and satisfaction in the curriculum (Andre et al. 2017). Supported by learning environments with vision-based augmented reality (AR) to support studies, many professional societies and universities have developed AR to support outdoor learning in undergraduate ecology and environmental science courses to create virtual objects for

Fig. 8.3 Signboard of the Environmental Center, *Touch of Nature*. Photo by Arba'at Hassan

Fig. 8.5 Water activity. Photo by Arba'at Hassan

meetings annually (Figs. 8.6 and 8.7). In fact, field work and the work that we often perform in the field are extensions of outdoor education programs because more senior/experienced people are passing their collective knowledge to another generation of learners from kindergarten to twelfth grade (K to 12) (Figs. 8.8, 8.9, 8.10, 8.11, 8.12 and 8.13).

8.1.2   Site Planning for Outdoor Education

8.1.2.1   Outdoor Education Bases
An outdoor area that provides teachers and learners with the tools that will allow the learning goals and objectives to be met. The characteristics include the following factors that are summarized below:

(1)   **Ecological Attributes:**
  • **Natural Geographical Environment:** The local landscape features such as abiotic factors, landscape, topography, soil, natural hydrology (wetlands, streams, rivers, lakes, waterfalls, seashores, marine areas) and other landscape structural factors.

Fig. 8.4 Camping ground. Photo by Arba'at Hassan

advanced studies (Chen et al. 2013; Kamarainen et al. 2018).

Until now, outdoor education programs are still considered important at the university, postgraduate, and professional levels. For example, the Society of Wetland Scientists (SWS) has an outdoor education program that it holds during SWS National, International, and Chapter

Fig. 8.6 Exploring environments in SWS field trips (right) and side events (left). Photo by Cheng-Hsiang Liu

Fig. 8.7 Wei-Ta Fang (left) and Ben A. LePage (right) at the joint INTECOL and SWS Annual Meeting in Florida, USA during 2012. Photo by Cheng-Hsiang Liu

- **Man-made Structural Environment:** An environment that is artificially constructed, formed, and managed, such as buildings, agricultural land, paddy fields, dry land, artificial wetlands, etc.
- **Plant Community:** A plant community that is naturally or artificially planted, regenerated or conserved.
- **Animal Group:** It belongs to the animal group that exists naturally or revived (Figs. 8.14, 8.15 and 8.16).
- **Smell:** The floral fragrance of flowering plants and smells we associate with certain ecosystems (e.g., swamps) (Figs.8.17, 8.18 and 8.19).
- **Audio:** The sounds of nature (Fig. 8.20).

The carrying capacity of an ecosystem is the largest population that it can sustain indefinitely with the available resources. It can depend on many abiotic and biotic factors and this needs to be considered when planning outdoor activities. Too many people can simply overwhelm the ability of an ecosystem to absorb the impacts, or our presence could change faunal behaviors.

(2) Management Attributes:

- **Land and Legal rights**: Can the people legally be present on the property and what can or cannot do?
- **Education Supervision and Assessment**: frequency of use of outdoor education bases, fire safety, facility safety, on-site supervision and enforcement by

Fig. 8.8 Outdoor education program that it holds during lakeshore activities (Lake Oeschinen with its turquoise water in Bern, Switzerland, 2009). Photo by Wei-Ta Fang

Fig. 8.9 Senior/experienced people are passing their collective knowledge to another generation of learner for kindergarten kids (Aberdeen Country Park Nature Trail, Hong Kong, 2019). Photo by Denis Woo

government personnel, Environmental education for local volunteers, facility maintenance, and warranty period.

8.1.2.2 Outdoor Education Routes

The outdoor education route has the characteristics of ecological, social, and management attributes. Outdoor education routes and outdoor education designated bases are different. Outdoor education routes have the characteristics of transportation and commuting. They are the air transportation (airspace), waterway (water area), and land transportation (highway) that tourists travel to their destinations and return home (railway, Mass Rapid Transit (MRT) routes, rural roads, lanes, trails, trails). This route is the sum of travelers' journeys from home to their destinations, including rides, short stays, views, and the geographical distance traveled.

Outdoor education however, is complicated and important because it isn't one size fits all. Sometimes data collection is the purpose and other times, just a relaxing walk through the forest is the goal. For example, outdoor education for middle school students, a flower is just a flower. It can be pretty and might have a pleasant smell. For a college student, a flower is the plants with reproductive units and fits with the

Fig. 8.10 Assembly Photos; **a** (top) and **b** (bottom) presented the situations how experienced people are passing knowledge to learners in primary schools' outdoor education (Fulong, New Taipei City, 2022). Photos by Wei-Ta Fang

environment (wind or insect pollinated). The anatomy and morphology of the flower provides important data on the evolutionary history of the plant. These concepts sometimes may be too difficult for a child to understand or figure out on their own, but the concepts can be introduced in a manner that they can understand. It is important to point out that field trips often have a substantial travel element because not all of the

biotic and abiotic characteristics of an ecosystem or community can be seen at one location.

8.1.3   The Content of Outdoor Learning

Outdoor education is learning outside of the classroom (Figs. 8.21, 8.22, 8.23 and 8.24).

Fig. 8.11 Assembly Photos. Outdoor Photos related to wetland activities (Pinglin, New Taipei City, Taiwan, 2010). Photos by Wei-Ta Fang

Fig. 8.12 Rafting on the Upper Colorado River, Colorado, USA. Photo by Wei-Ta Fang

Fig. 8.13 Biking in Bern, Switzerland. Photo by Wei-Ta Fang

Learning results depend on many variables, therefore, humans may or may be not learn 20% of their knowledge by listening, 20% from reading, and 60% from personal experience. Since we really don't remember what we hear or see as much in school contents unless we do hands working on it, therefore, we try to transmit all knowledge to ensure its integrity without skeptical about the information that we receive. Only rare few of people have the learning capability to remember what they glance or hear. Developed in 1946 by Edgar Dale in his Cone of Experience, he provided an intuitive model of the concreteness of various audio-visual media. Dale, however, included no numbers or percentages in his model (Dale 1969, p 107; Masters 2013; 2020).

We tried to investigate the evidence, and evaluate the research regarding to all functions from outdoor education. Most of the Environmental Education takes place outdoors and in the US nature research, conservation education, and school camping were the beginning of environmental education. The techniques used in the

Fig. 8.16 Endangered animal rehabilitation center. Photo by Arba'at Hassan

Fig. 8.14 Bird sanctuary. Photo by Arba'at Hassan

Fig. 8.15 Elephant rehabilitation center. Photo by Arba'at Hassan

Fig. 8.17 Floral fragrance of flowers, Rafflesia. Photo by Arba'at Hassan

study of the environment are combined with academic methods with outdoor exploration (Roth 1978). The learning content is recommended to students to have these criteria or action strategies, such as building on cognition, emotion, psychological satisfaction, and pleasure experiences to organize multifunctional value of education, recreation, training, sports, healing, leisure, and tourism (Fig. 8.25).

Fig. 8.18 Rare flora. Photo by Arba'at Hassan

Fig. 8.19 Smell from wetlands. Photo by Wei-Ta Fang

8.2 Motivation for Outdoor Education

In recent years, society has become more urban, resulting in social trends such as declining birthrates, more urbanization, and digitalization (Egidi et al. 2021). Therefore, the process of standardizing the indoor curriculum in school-

age education has begun to cause problems such as poor learning motivation, which is a symptom of Nature-Deficit Disorder (Louv 2005).

If education is a way to build student expectations for activities, then it can and should improve student motivation for learning. Outdoor education can be said to induce students' learning motivation and enthusiasm, create a more relaxed learning environment, and help students to learn at their own speed and terms, which is believed to be more effective (Gilbertson et al. 2022). Outdoor education is remarkable and a highly effective place for learning. The MOE (Taiwan) managed the 2018 National Outdoor Education Expo to show, and promotion the outdoor education results and they declared that outdoor education can inspire learning and motivation, develop children's curiosity, exploration abilities, and improve learning efficiency. The MOE's outdoor education policy promises "learning out of the classroom and letting children's dreams take off" (National Academy for Education Research 2016). Therefore, the learning environment does not need to be limited to the classroom and should include offsite facilities such as national parks, nature education centers, historical sites, and museums to inspire students' learning motivation and improve learning efficiency. Outdoor education provides students with the opportunity to experience and learn about nature while improving their view on the functional values provided by the environment and to build appropriate outdoor learning knowledge, skills, attitudes, and motivations that further support environmental stewardship and sustainability. The following examples of outdoor education highlight the motivation for learning.

8.2.1 Hometown Learning

Outdoor education is compatible with local education if the outdoor education concepts and guidelines can be integrated with specific learning goals and concepts. This allows students to directly experience nature and the concepts being presented in a natural setting that can be complex

Fig. 8.20 The sound of Taiwan Blue Magpie (*Urocissa caerulea*) is described as a high-pitched cackling chatter, kyak-kyak-kyak-kyak. Photo by Max Horng

Fig. 8.21 Bird watching. Photo by Arba'at Hassan

Fig. 8.22 Discussion about wildlife at site. Photo by Arba'at Hassan

or as simple as one's own backyard. Materials or concepts learned outside of the classroom have higher retention rates among students and in many cases may generate a new set of probing questions (Brown 1998; Bell and Chang 2017). Therefore, outdoor teaching can facilitate learning about environmental awareness, while stimulating their curiosity, emotions, and consciousness.

8.2.2 Research

Students working in an outdoor environment may choose topics or become interested in elements of the environment that they are concerned about, collect the data needed to support/refute their hypotheses, investigate the cause of the problem, and start thinking, planning, designing, interviewing, and conducting research sampling in the field (Figs. 8.26, 8.27, 8.28, 8.29 and 8.30).

Fig. 8.23 Investigating crawling animals. Photo by Arba'at Hassan

Personal experience makes a difference, and our experience is that it's easier to arouse interest in learning. By working in the field (outside of the classroom) students can experience working and learning in outdoor atmosphere and develop experience in outdoor research activities. Returning to the classroom then allows a time to reflect, discuss, and explore new problems.

8.2.3 Experience Learning

Experience is a motivation of human curiosity and children are generally naturally curios and like to experience the world around them as they grow and develop intellectually. Our perception of nature is based on our knowledge and other aspects such as language are built around what we know to be reality at that time. As such, the world or nature is obscured by abstract concepts and if we want to understand nature, it's important to experience the world around us with the fewest anthropogenic factors that would influence such experiences. The most effective way to learn is through participation in outdoor activities, so we should create opportunities for people, especially children to participate in outdoor learning. It's only been in the past hundreds of years that have humans entered classrooms for pen and paper studies.

8.2.4 Emotional Learning

Social emotional learning is an element associated with learning, especially when outdoor activities are included as part of the experience. Frey et al. (2019: 8) consider it to be a form of empowerment. This is a way to interact with others when dealing with physical, social, and

Fig. 8.24 A collared scops owl (*Otus lettia*) from south Asia. Photo by Max Horng

Fig. 8.25 The learning content criteria. Illustrated by Wei-Ta Fang

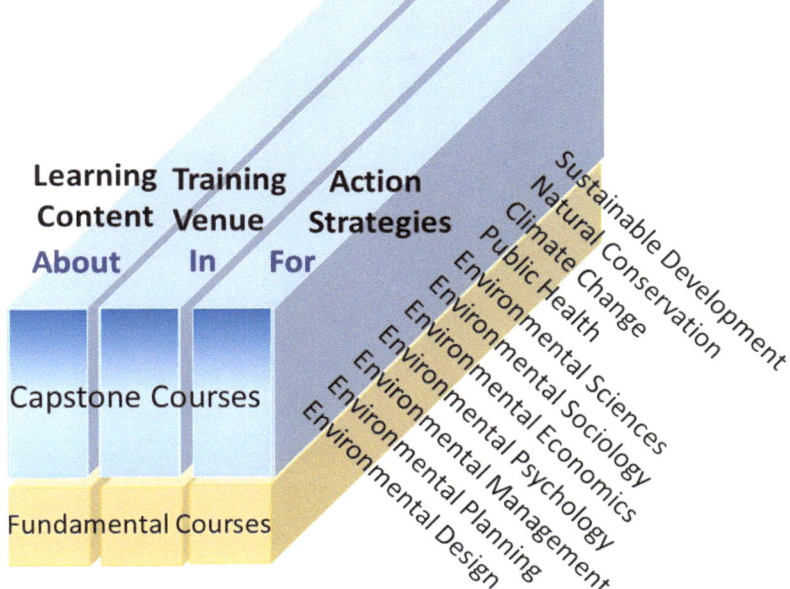

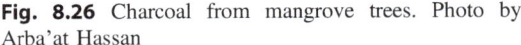

Fig. 8.26 Charcoal from mangrove trees. Photo by Arba'at Hassan

Fig. 8.27 Interviewing the owner. Photo by Arba'at Hassan

emotional issues. Social emotional learning helps cultivate learner identities and confidence in their ability to learn, overcome challenges, and to affect the world around them.

Human emotion can strongly influence learning (Fig. 8.31). This type of social-emotion learning system is based on the "reward-discipline system (tiger mother system, Chua 2011), and uses dopamine, a powerful neurotransmitter to receive rewards evoking increased free-choice actions (left side). The brain responds to the dopamine-producing neurons that could promoting learner performance, such as: (1) assisting learners to identify, describe, and regulate

Fig. 8.28 Talk and discussion to audience. Photo by Arba'at Hassan

emotional responses; (2) promoting cognitive control skills important for environmental decision-making and problem-solving; (3) developing learner social skills, including teamwork, and

sharing, and the ability to build and repair relationships; and (4) letting learners become informed and involved citizens.

Covey (2020) believed that "climbing the peak" is: Listening to the most profound voice of life and be willing to walk this voice. Among outdoor activities, participating in unknown adventure activities is one of the most common emotional challenges in physical education and teamwork in outdoor education. Outdoor education programs sometimes involve residential or journey wilderness-based experiences in which students participate in a variety of adventurous challenges and outdoor activities such as hiking, climbing, canoeing, ropes courses and group games. It doesn't matter if these activities are formal, add-on, or random accidental adventures. After thoughts are endless learning situations and challenges.

Following the completion of an outdoor activity, people often reflect on the activity and develop a sense of satisfaction that forms a calming or healing effect (Schoel et al. 1988). Outdoor activities can be a wonderful way for children and adults to understand and manage their emotions through adventure therapy.

Fig. 8.29 The study in the natural sciences and social sciences are the most interesting for diverse cultures, such as tea plantation, cooking, and culture studies experienced by Marinus Otte (second from the left), the Editor-in-Chief, the journal of *Wetlands* (SCI journal, Springer's publication), and Velsen Otte (left) at Pinglin Township, New Taipei City, Taiwan. Photo by Wei-Ta Fang

Fig. 8.30 Assembly photos; Try hard in thinking, planning, designing, interviewing, and conducting research sampling in the fields. Photos by Wei-Ta Fang

Fig. 8.31 A loop model showing learner emotions. Illustrated by Wei-Ta Fang

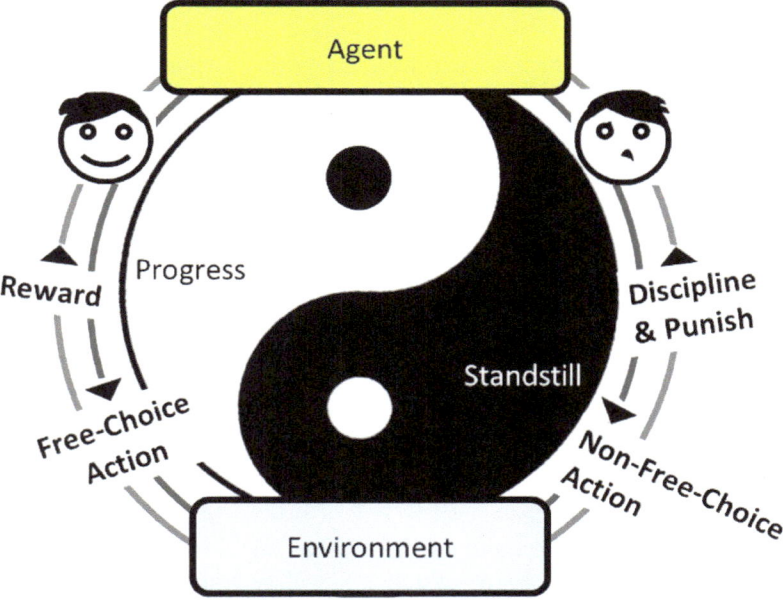

Adventurous education, through teamwork can assist in creating and achieving positive activity goals. Based on each other's feelings in the wilderness, students can express sympathy for others through pain and bitterness and select and maintain a positive and mutually beneficial relationship, thus avoiding making a negative responsibility determination operation.

8.3 Barriers to Outdoor Education

Outdoor education can provide advantages, but the implementation of outdoor education has obstacles such as the lack of teaching resources, resistance to reform, and the rigidity of administrative systems that all contribute to the reluctance of principals and teachers to lead students to implement outdoor adventure education and experiential education activities. In addition, because of the perceived danger of outdoor education, parents today, prefer to keep students at home under their supervision rather than going outside of the home to a classroom or outside experiencing nature.

8.3.1 Barriers to School Education

8.3.1.1 Lack of Teaching Resources

Teachers consider chalk and blackboard (whiteboard pen and whiteboard) learning in the classroom to be the most effective delivery of environmental information for learning. While related, but not always the case, the teachers could have little understanding of the materials that need to be taught or they simply are not good teachers. In addition, limited resources may impact outside activities. There is also the legal risk that a school does not want to carry, so it's easier and possesses less risk to the school to keep the students in a room.

8.3.1.2 Inertia to Resist Reform During Covid-19 Pandemic

Outdoor education requires considerable external support such as tour buses, funding, or salaries/stipends. Teachers usually have on-line teaching experience because of the problem of taking classes in the classroom during Covid-19 pandemic. Therefore, outdoor and environmental education the pandemic has been particularly difficult, with lockdowns meaning that programs have had to be cancelled, and in many places these lockdowns continue from 2020s (Quay et al. 2020). Although they intentionally take students outdoors to "fly the air," they have no way to carry out education outdoors during Covid-19.

8.3.1.3 The Rigidity of the Administrative System

Outdoor education involves student's safety and insurance issues; in addition, there are also problems with the school's budget for transportation matters, and parents'concerns about student safety. School principals have raised questions about whether teachers can safely lead a group of students on a trip, which is their biggest concern. Therefore, under the mentality that one thing is not as good as one thing, all the students in the school should be on a campus with "iron gates and walls which is the principal's "policy to protect the safety of students."

8.3.2 Barriers to Family Education

8.3.2.1 Psychological Obstacles

COVID-19 appears to have had a major impact on physical activity behaviors globally. Staying at home (lockdown) is being recommended to manage the spread of the virus at the expense of outdoor physical activities (Hammami et al. 2022). If young parents are not letting their children outside because there are dangerous situations out in the world, then the parents themselves really don't understand nature or they believe education on environmental matters is the school's responsibility.

However, the three authors of this book were born between the 1950s and 1960s and were expected (or had the freedom) to play outside, unsupervised, and often until it got too dark or dinner was on the table. We imagine that if we were to follow took this approach today, we'd likely be arrested for child neglect or endangerment. Clearly our social values on this aspect have changed substantially in less than 100 years.

Due to the influence of declining birth rates (McDonald 2006), parents take great care of their children, and because of risk aversion they are reluctant to allow their children to engage in outdoor diverse activities (Nichols 2000; Harper 2017). This sense of parental/societal worry affects children's opportunities to participate in many outdoor activities. According to a survey of 5,500 people in the USA, 40% of Americans considered outdoor space to be unsafe when they were 30 years old. This is an obstacle, especially for Americans and the people from other countries to go outdoors. Americans love to participate in outdoor activities are facing these types of psychological barriers and people in other countries are even more afraid of being outdoors (NatureofAmericans.org).

8.3.2.2   Physical Obstacles

Most humans now live in cities and many schools are far from green areas. Because the fields and bases of outdoor education are far from the city, they are located in remote areas and form obstacles to transportation. In particular, some students with a physical disability can't participate in outdoor activities in these places because of personal factors without accessibility facilities. However, some children also in cities have never seen or interacted with nature (Louv 2005).

8.3.2.3   Economic Obstacles

At present, the cost to implement outdoor education programs is becoming increasingly more expensive and can no longer be included as a teaching tool/medium. Administrative, transportation, insurance, and other fees present substantial barriers to include outdoor education in the program (Hanna 1992). In some cases the parents are expected to support these activities (Aarts et al. 2010; Cook 2021). This is especially prevalent in urban communities, which contributes to the environmental justice (EJ) issues these communities face (Fisher et al. 2006; Garzón et al. 2013). However, because some governments encourage outdoor learning, there may be subsidies that can help offset some of the costs. Moreover, local industry can step up and take some of the fiscal responsibilities for low-income communities. The economic status of a community shouldn't determine or dictate whether it's children and residents are provided a good environmental education and opportunities to make a difference. When the government and

community care about the environmental literacy of the next generation of students, outdoor learning is a more effective method of teaching/learning compared to the classroom. Economic obstacles often contribute to the parents' low willingness to pay and even fail to meet the educational needs of individual students studying in the overall environment and future physical and mental development (Jensen 2009).

8.4   Outdoor Education Field

The outdoor environment is rich and diverse in ecosystems and habitats that provides ample opportunities for education that a person can experience. More and more teachers and children like to engage in natural ecological leisure tourism, mainly to get close to the beautiful natural scenery and rich wildlife environment as these locations provide visitors with a natural environment (Sobel 2008). The goal of outdoor education is to show teachers and students the importance of the environment by experiencing the environment so that they are able to develop an understanding and appreciation of the environment. Some habitats may be biologically and/or physically unique (i.e., one of a kind) and provide an opportunity for educators to capitalize using these environments as an excuse to teach environmental awareness, environmental responsibility, and the economic benefits of giving back by supporting the local community. The following categories are important for implementing outdoor education programs.

8.4.1   Field-Dependent Development

Understanding the physical location of the site in the regional geographic setting helps contribute an ecosystem or habitat to a sense of place in the regional environmental mosaic. Sense of place is a cornerstone of environmental education and helps maintains the significance and value of ecosystems for individuals (Fig. 8.32). Field selection for outdoor education is based on the

Fig. 8.32 Field-dependent development theory (Revised and modified after Morgan 2010; Marvin et al. 2002). Illustrated by Wei-Ta Fang

following theories (Morgan 2010; Marvin et al. 2002).

8.4.1.1 Exploration-Assertion Motivation System

The exploration-assertion motivation system is an act of adventure and amusement that occurs during the growth of adolescents because of their curiosity and longing for a place that is special to them. In the process of exploration, there is the arousal or awakening of one's self. The development of an adolescents' natural experience is divided into stages of dependence, enlightenment, exploration, and autonomy.

8.4.1.2 Attachment-Affiliation Motivation System

The attachment-affiliation motivation system is the psychological tendency of adolescents to attach themselves to specific characters in their sense of security for self-growth. For example, frustration in the environment requires a mother's or a caregiver's psychological comfort (Ainsworth et al. 1978). This kind of comfort can provide a sense of connection and calm one's mood (Marvin et al. 2002). When young people grow up, they also develop the phenomenon of local dependence on other people.

8.4.2 Development of Field Selection

The public and private sectors should provide low-cost or free educational facilities for promoting outdoor education.

8.4.2.1 Administrative Management System

Depending on the country, outdoor education fields are classified according to their administrative system, such as national parks, national scenic areas, county and city scenic areas, forest recreation areas, beaches, historical and cultural sites (Figs. 8.33, 8.34, 8.35 and 8.36).

(1) National recreation areas, including national parks, natural parks, or national forests;
(2) Regional recreation areas;
(3) General scenic areas;
(4) Forest recreation areas, nature centers, or forest parks; and
(5) Beaches, fishing grounds, snorkeling areas, or seaside sightseeing and recreation areas.

8.4.2.2 Nature Protection and Scientific Research Areas

The nature protection and scientific research areas are designated as protected areas set aside to protect biodiversity and possibly geological/

Fig. 8.33 Waikiki Hula-Hula Beach Hawaii, USA. Photo by Arba'at Hassan

Fig. 8.34 Grand Canyon National Park, Arizona, USA. Photo by Arba'at Hassan

geomorphological features. These areas are strictly controlled and limited to ensure protection of the conservation values, such as for the studies of scientific research (Figs. 8.37, 8.38, 8.39 and 8.40).

8.4.2.3 Historical Relics and Historic Sites

The historic relics or historic sites are official locations where pieces of political, military, cultural, or social history have been preserved due to their value of heritage (Figs. 8.41, 8.42, 8.43 and 8.44).

8.4.2.4 Industrial Tourism Area

An Industrial Tourism Area is an area mainly for industrial production activities and supplemented by tourism and recreation education such as farms, ranches, orchards, tea gardens, horticultural areas, and urban parks (Figs. 8.45, 8.46, 8.47 and 8.48).

Fig. 8.35 Petrified Forest National Park, Arizona, USA. Photo by Arba'at Hassan

Fig. 8.36 Beaches or seaside sightseeing and recreation areas are attracted tourists to come in rocky beaches (Hong Kong UNESCO Global Geopark, Sharp Island, Sai Kung, Hong Kong, 2020). Photo by Dennis Woo

Fig. 8.37 San Diego Wild Animal Park, California. Photo by Arba'at Hassan

Fig. 8.38 Saguaro Nature Park, Arizona. Photo by Arba'at Hassan

8.5 The Implementation Content of Outdoor Education

The criteria provided in the *Tbilisi Declaration* (1977), suggested that environmental educators should review the main environmental issues that are occurring at local, national, regional, and international levels so that the people understand the state of the environment in the world. In addition, Environmental Educators should use various learning environments and teaching methods, emphasizing practical activities and

Fig. 8.39 Mount Rushmore Protection Park. Photo by Arba'at Hassan

Fig. 8.41   Atomic Bomb Historical Site, Shibaata, Japan. Photo by Arba'at Hassan

Fig. 8.40   The nature protection areas are designated as protected areas set aside to protect biodiversity and possibly geological/geomorphological features from anthropogenic influences (Hong Kong UNESCO Global Geopark, High Island Reservoir, Sai Kung, Hong Kong, 2018). Photo by Dennis Woo

Fig. 8.43 Nakajo Temple, Niigata, Japan. Photo by Arba'at Hassan

Fig. 8.42 Vietnam War Memorial Monument, Washington DC, USA. Photo by Arba'at Hassan

personal experience. Therefore, outdoor education should emphasize the implementation goals and methods.

8.5.1   Goals of Outdoor Education

The goals of learning in an outdoor setting should be focused on using science at the early grade levels. This helps develop personal and social skills, develop a deeper relationship with nature (Fig. 8.49), and understand how nature progresses from one state to another through a process called succession, and the ability to solve problems.

Fig. 8.45 Boh Tea Plantation, Malaysia. Photo by Arba'at Hassan

Fig. 8.44 The historical Mongol buildings of Greater Mongolia are known through Chinese chronicles. After the rise of the Mongols in the thirteenth century, the buildings are now decorated using the styles from central and eastern Inner Mongolia, China. Photo by Wei-Ta Fang

Fig. 8.46 Orchid Garden Danga Bay, Malaysia. Photo by Arba'at Hassan

Fig. 8.47 Toowoomba Cattle Farm, Brisbane, Australia. Photo by Arba'at Hassan

8.5.2 Implementation Methods of Outdoor Education

We do the best that we can and strive to understand the data. It's not only important to understand the background knowledge of outdoor teaching styles/strategies, but the needs of the learners and ecology are crucial.

8.5.2.1 "Flow Learning"

Cornell wrote *Sharing Nature: Nature Awareness Activities for All Ages* and actively promotes outdoor learning (Cornell 2015, 1979). His program for outdoor teaching activities is called Flow Learning and is divided into four stages:

(1) Stage One: Awakening enthusiasm;
(2) Stage Two: Focus Attention (Fig. 8.50);
(3) Stage Three: Offer Direct Experience; and
(4) Stage Four: Sharing Inspiration.

A detailed description of environmental cognition, creates a relaxed and pleasant explanatory

Fig. 8.48 Millennium Park is featuring as a variety of public art in her outdoor space, Chicago, USA. Photo by Wei-Ta Fang

Fig. 8.49 Ourdoor activities develop personal and social skills and develop a deeper relationship with nature (Dan-hai New Town, New Taipei City, Taiwan, 2022). Photo by Wei-Ta Fang

atmosphere (Kulas 2019), allowing learners to focus and gain a deep understanding of outdoor classrooms (Asmara et al. 2016). In addition, learners are encouraged to observe and discover problems carefully to gain knowledge of the issue (Weinstein and Mayer 1983). When a learner conducts an experiment, they can sense (see, hear, smell, feel) the environmental characteristics and learn about the environment in a way that cannot be achieved through a book. Further discussion is a way of sharing the information or interpretation, which could increase a learners' willingness to participate in the learning process.

Fig. 8.50 The stage two of flow learning should be learnt to "Focus Attention". Self-photo by Wei-Ta Fang

8.5.2.2 Eight-Direction Learning

The 8-direction is a method expands on Cornell's (1979, 2015) outdoor teaching approach for outdoor learning activities. Eight-Direction Learning is detailed as the process, such as: pre-planning preparation, the stage of inspiring, cheering, focusing, introspection, collection, reflection, integration, celebration, etc. This outdoor activity curriculum was created to support the academic and emotional needs through outdoor activities. Teachers, therefore, lead learners into the field for social and emotional learning, improving cognitive function while learners have been led to improve their emotional regulation in a wild (Young et al. 2010: 211).

(1) **Inspiration and Excitement:** When explaining the outdoor environment, the teacher is like an actor presenting the information to their audience. Therefore, teachers need to live in the moment (Fig. 8.51).

(2) **Orientation and Motivation:** The learner is full of fear for an unfamiliar base. At this moment, the learner's mood is in the learning motivation situation of "exploration proposition"; but the board game, earth game, or freshman training in outdoor activities taught by teachers It is a directional game that strengthens the motivation of human "attachment and connection" learning.

(3) **Focus:** It is important for students to understand the environment and our relationship with the elements of the environment. In many cases, this information can be dry and boring, so the teachers need to have a sense of humor and incorporate these skills when teaching.

(4) **Relaxation and Internalization:** Let the learners embrace nature and make good use of the learner's sensory perception abilities and curiosity (Fig. 8.52).

(5) **Celebration and Harvest:** This stage allows learners to discover the wonders of nature themselves, make and record observations and collect data.

(6) **Debriefing:** Provide opportunities for the students to discuss as a group what they learned or experienced. Teachers should be involved in this process to understand what was learned and what goals and objectives of the activity were missed.

(7) **Distillation and Integration:** After filtering the data the teacher uses a concise conclusion method to reinforce what was learned.

(8) **Imagination:** Let students celebrate in a relaxed setting. Outdoor activities are an education of love and the most welcoming experience. In a relaxed situation, we strengthen the educational relationship of "attachment

Fig. 8.51 Teachers need to use physical movements. Photo by Wei-Ta Fang

Fig. 8.52 Reflection and relaxation–learners embrace nature. Photo by Max Horng

and connection" of human beings and nature to facilitate the next stage of learning.

8.5.2.3 Development of Outdoor Education and Environmental Education

Outdoor and Environmental Education share common contents and processes but have their own unique characteristics. In the 1960s, Environmental Education scholars once accused Environmental Education administrators of changing the name of outdoor science, natural research, or outdoor education programs to environmental education, but still continuing the same programs as before. Although both are interdisciplinary in nature, the difference is that outdoor education can be applied to any learning goal. Different cultures, history, social values, and special environments make outdoor education activities different experiences in each country.

Environmental Education can take place inside or outside the classroom, or it can be linked to local and global sites. However, the focus of Environmental Education is usually on water, air, and soil quality and the impact of our pollution on these media. Through Environmental Education we hope to understand the disposal methods of solid waste and toxic substances, conduct surveys of urban population expansion, conduct deforestation surveys, understand biodiversity through surveys of endangered animals and plants, or conduct regional studies on drought and floods (Figs. 8.53, 8.54 and 8.55).

Environmental and Outdoor education have similarities and differences between the two. In short, Outdoor education programs are designed to help learners more effectively absorb practical knowledge by seeing, touching, and smelling as opposed to sitting in a classroom (Figs. 8.56 and 8.57). Most Environmental Education programs assist learners investigate or learn about environmental issues through real-world experiences. However, whether students should try to resolve these issues is controversial (Mogensen 1997; Volk and Cheak 2003; Herman et al. 2018).

Fig. 8.54 Landslide along highway, Malaysia. Photo by Arba'at Hassan

Fig. 8.55 Wetland management site, Gongliao. New Taipei City, Taiwan. Photo by Arba'at Hassan

Although both areas advocate the use of a wide range of topics, Environmental Education is usually applied to higher-level students in social science or natural science courses for deeper teaching (Smith 2001). At the elementary school level, outdoor activities often span more learning curricula and combine social and leisure goals to experience teamwork, service, and mutual learning.

Although Outdoor Education and Environmental Education are mainly conducted through schools, nature centers, and outdoor accommodation facilities provide another option.

Fig. 8.53 The result of deforestation, Malaysia. Photo by Arba'at Hassan

Fig. 8.56 Project-oriented environmental learning has been explored in the vertical wetland studies taught by Ben A. LePage at Mount Hemei Trail, Xindian, New Taipei City, Taiwan (left side), and at Zhudong, Hsinchu, Taiwan (right side) during 2021 and 2022. Photo by Wei-Ta Fang

Fig. 8.57 The outdoor education program hopes to help learners more effectively absorb practical knowledge through first-hand materials from nature. Photo by Wei-Ta Fang

Environmental Education and outdoor educators mainly promote learning strategies based on experience, which need to be taught to emphasize the context used in problem-based learning situations and the importance of adopting experiences (Fig. 8.58). Environmental Education scholars hope that when exploring content, students will use a variety of sensory experiences to maximize their learning.

8.6   Summary

In this chapter we emphasized the importance of outdoor education. Gardner (1983/2011, 2006) confirmed that humans possess the natural" intelligence to recognize plants, animals, and other natural environments. Therefore, people that have a predilection for nature and the environment

Fig. 8.58 Most Environmental Education programs assist learners in investigating environmental issues by discovering teaching and real-world experiences, i.e., new technology. This could be a good example to explain corporate social responsibility (CSR) as a company's role of building a "home," not only building a "house" that will last a lifetime for all residents. It has a lifetime bond with the land, as the environment and people live in homelands. Photo courtesy of Ileana Lee, Copyright © H&E Development Company, Tainan, Taiwan. All Rights Reserved)

Fig. 8.59 Assembly photos, Natural intelligence can be summarized as exploratory intelligence (top), including the exploration of society and the exploration of nature, such as knowing facts and information about sea level rise (bottom) (Dan-hai New Town, New Taipei City, Taiwan, 2022). Photo by Wei-Ta Fang

do well in outdoor activities and biological scientific investigations. Natural intelligence can be summarized as exploratory intelligence, including the exploration of society and the exploration of nature (Fig. 8.59).

Naturally intelligent methods of displaying expertise in categorizing plants, animals, and cultural artwork, providing Outdoor Education and Environmental Education into the curriculum and teaching provide important educational justifications. Since the early 1940s, outdoor education has been an important educational reform factor that promotes learning in nature. When Environmental Education emerged in the 1970s, it was focused on local and global knowledge. The predecessors of outdoor education were camping, nature research, conservation education, and adventure education. Outdoor education is an experience-based learning program that lends itself well to meeting the goals and objectives of environmental Education.

References

Aarts MJ, Wendel-Vos W, van Oers HA, Van de Goor IA, Schuit AJ (2010) Environmental determinants of outdoor play in children: a large-scale cross-sectional study. Am J Prev Med 39(3):212–219

Ainsworth MDS, Blehar MC, Waters E, Wall S (1978) Patterns of attachment: a psychological study of the strange situation. Lawrence Erlbaum, Hillsdale

Andre EK, Williams N, Schwartz F, Bullard C (2017) Benefits of campus outdoor recreation programs: a review of the literature. J Outdoor Recreation, Educ Leadersh 9(1):15–25

Asmara CH, Anwar K, Muhammad RN (2016) EFL learners' perception toward an outdoor learning program. Int J Educ Literacy Stud 4(2):74–81

Bell BJ, Chang H (2017) Outdoor orientation programs: a critical review of program impacts on retention and graduation. J Outdoor Recreation Educ Leadersh 9 (1):56–68

Bowdridge M, Blenkinsop S (2011) Michel Foucault goes outside: discipline and control in the practice of outdoor education. J Exp Educ 34(2):149–163

Brown DA (1998) Does an outdoor orientation program really work? Coll Univer 73(4):17–23

Casalaspi D (2017) The making of a "legislative miracle": The Elementary and Secondary Education Act of 1965. Hist Educ Q 57(2):247–277

Chen DR, Chen MY, Huang TC, Hsu WP (2013) Developing a mobile learning system in augmented

reality context. Int J Distrib Sens Netw 9(12). https://doi.org/10.1155/2013/594627

Chua A (2011) Battle hymn of the tiger mother. Penguin Group, London

Cook R (2021) Utilising the community cultural wealth framework to explore Sierra Leonean parents' experiences of outdoor adventure education in the United Kingdom. J Adv Educ Outdoor Learn. https://doi.org/10.1080/14729679.2021.1902825

Cornell J (1979) Sharing nature with children: the classic parents' and the teachers' nature awareness guidebook. Dawn, Nevada

Cornell J (2015) Sharing Nature®: nature awareness activities for all ages. Crystal Clarity, Commerce

Covey SR (2020) Summary and insights of the 8th habit: from effectiveness to greatness. Independently published

Cupers K (2008) Governing through nature: camps and youth movements in interwar Germany and the United States. Cult Geogr 15(2):173–205

Dale E (1969) Audio-visual methods in teaching. Dryden, New York

Dent RA (2021) John amos comenius: inciting the millennium through educational reform. Religions 12 (11):1012. https://doi.org/10.3390/rel12111012

Dewey J (1938/1997) Experience and education. Free Press

Dewey J (1916/1997) Democracy and education. Free Press

Dillon J., Rickinson M., Teamey K (2016) The value of outdoor learning: evidence from research in the UK and elsewhere. In: Towards a convergence between science and environmental education. Routledge, Oxfordshire, pp 193–200

Egidi G, Salvati L, Falcone A, Quaranta G, Salvia R, Vcelakova R, A (2021) Re-framing the latent nexus between land-use change, urbanization and demographic transitions in advanced economies. Sustainability 13:533. https://doi.org/10.3390/su13020533

Fang W-T, Eg E, Chang M-C (2017) Physical outdoor activity versus indoor activity: their influence on environmental behaviors. Int J Environ Res Public Health 14 (7):797. https://doi.org/10.3390/ijerph14070797

Fisher JB, Kelly M, Rommet J (2006) Scales of environmental justice: combining GIS and spatial analysis for air toxics in West Oakland, California. Health Place 12(4):701–714

Freeman M, Seaman J (2020) Outdoor education in historical perspective. Hist Educ Rev 49(1):1–7

Frey N, Fisher D, Smith D (2019) All learning is social and emotional: helping students develop essential skills for the classroom and beyond. ASCD, Alexandria

Gagen EA (2004) Making America flesh: physicality and nationhood in early twentieth-century physical education reform. Cult Geogr 11(4):417–442

Gardner H (2006) Multiple intelligences: new horizons in theory and practice. Basic Books, New York

Gardner H (1983/2011) Frames of mind: the theory of multiple intelligences. Basic Books, New York

Gamson DA, McDermott KA, Reed DS (2015) The Elementary and Secondary Education Act at fifty: aspirations, effects, and limitations. RSF: Russell Sage Found J Soc Sci 1(3):1–29

Garzón C, Beveridge B, Gordon M, Martin C, Matalon E, Moore E (2013) Power, privilege, and the process of community-based participatory research: critical reflections on forging an empowered partnership for environmental justice in West Oakland, California. Environ Justice Apr 2013:71–78

Gilbertson K, Ewert A, Siklander P, Bateset T (2022) Outdoor education: methods and strategies, 2nd edn. Human Kinetics, Champaign, IL

Hammami A, Harrabi B, Mohr M, Krustrup P (2022) Physical activity and coronavirus disease 2019 (COVID-19): specific recommendations for home-based physical training. Manag Sport Leisure 27(1–2):20–25

Hanna G (1992) Jumping deadfall: overcoming barriers to implementing outdoor and environmental education. In: Paper presented at the international conference for the association of experiential education. Alberta, Canada

Harper NJ (2017) Outdoor risky play and healthy child development in the shadow of the "risk society": a forest and nature school perspective. Child Youth Serv 38(4):318–334

Herman BC, Sadler TD, Zeidler DL, Newton MH (2018) A socioscientific issues approach to environmental education. International perspectives on the theory and practice of environmental education: a reader. Springer, Cham, pp 145–161

Jensen E (2009) Teaching with poverty in mind: What being poor does to kids' brains and what schools can do about it. ASCD, Alexandria

Kamarainen A, Reilly J, Metcalf S, Grotzer T, Dede C (2018) Using mobile location-based augmented reality to support outdoor learning in undergraduate ecology and environmental science courses. Bull Ecol Soc Am 99(2):259–276

Keatinge MW (2018) The great didactic of john amos comenius, now for the first time englished—with introductions, biographical and historical. White Press, London

Kellert SR (2012) Building for life: designing and understanding the human-nature connection. Island press, Washington, DC

Knapp CE (1994) Progressivism never died—it just moved outside: What can experiential educators learn from the past? J Exp Educ 17(2):8–12

Kulas KA (2019) Developing An outdoor mindful activity-based curriculum for English language learners. In: School of education and leadership student capstone projects, vol 285. https://digitalcommons.hamline.edu/hse_cp/285

Louv R (2005) Last child in the woods: saving our children from nature deficit disorder. Workman, New York

Marvin RS, Cooper G, Hoffman K, Powell B (2002) The circle of security project: attachment-based intervention with caregiver—preschool child dyads. Attach Hum Dev 4:107–124

Masters K (2013) Edgar dale's pyramid of learning in medical education: a literature review. Med Teach 35 (11):e1584–e1593

Masters K (2020) Edgar dale's pyramid of learning in medical education: further expansion of the myth. Med Educ 54(1):22–32

McDonald P (2006) Low fertility and the state: the efficacy of policy. Popul Dev Rev 32(3):485–510

McGuinn P, Hess F (2005) Freedom from ignorance? The Great Society and the evolution of the Elementary and Secondary Education Act of 1965. In: The great society and the high tide of liberalism, pp 289–329

Mikaels J, Backman E, Lundvall S (2016) In and out of place: exploring the discursive effects of teachers' talk about outdoor education in secondary schools in New Zealand. J Adv Educ Outdoor Learn 16(2):91–104

Miles JC (1987) Wilderness as a learning place. J Environ Educ 18(2):33–40

Mogensen F (1997) Critical thinking: a central element in developing action competence in health and environmental education. Health Educ Res 12(4):429–436

Morgan P (2010) Towards a developmental theory of place attachment. J Environ Psychol 30(2010):11–22

Nadel M, Scher S (2019) Not just play: summer camp and the profession of social work. Oxford University Press, Oxford

National Academy for Education Research (2016) Project of the ministry of education's outdoor education research office. National Academy for Education Research, Taiwan, ROC (in Chinese)

Nichols G (2000) Risk and adventure education. J Risk Res 3(2):121–134

Öhman J, Sandell K (2016) Environmental concerns and outdoor studies. Routledge, Oxfordshire

Orr D (1991) What Is education for? Six myths about the foundations of modern education, and six new principles to replace them. Context 27(53):52–58

Pigram JJ (1993) Human-nature relationships: leisure environments and natural settings. Adv Psychol 96:400–426

Quay J, Gray T, Thomas G, Allen-Craig S, Asfeldt M (2020) What future/s for outdoor and environmental education in a world that has contended with COVID-19? J Outdoor Environ Educ 23:93–117

Roth CE (1978) Off the merry-go-round and on to the escalator. In: Stapp WB (ed) From ought to action in environmental education. SMEAC/IRC, Columbus

Schoel J, Prouty D, Radcliffe P (1988) Islands of healing: a guide to adventure based counseling. Project Adventure

Smith GA (2001) Defusing environmental education: an evaluation of the critique of the environmental education movement. Clear: Environ Educ Res Teach 108 (6):22–28

Sobel D (2008) Childhood and nature: design principles for educators. Stenhouse Publishers, Portsmouth

Volk TL, Cheak MJ (2003) The effects of an environmental education program on students, parents, and community. J Environ Educ 34(4):12–25

Young J, McGown E, Haas E (2010) Coyote's guide to connecting with nature. Owlink Media, Shelton

Wattchow B (2006) Playing with an unstoppable force: paddling, river-places and outdoor education. J Outdoor Environ Educ 11(1):10–20

Weinstein CE, Mayer RE (1983) The teaching of learning strategies. Innov Abstr 5(32):32

Conclusion: How Can We Maintain Our Precious Environment and Its Resources Now and for the Future?

These are not "virtual" generated by artificial intelligence (AI), but all "real."

By Wei-Ta Fang (2022).

Environmental education has been an important component of our personal and professional lives and a topic that we've always wanted to write and share with others. We provide an historical overview of environmental education, its, strengths, weaknesses, and the role of environmental education in the future. Song Dynasty scholar Zhang Zai (張載)(1020–1078 AD) once said: "Set your heart for the heavens and earth, set your life for the living people, sacrifice your studies for the saints, and open up peace for all ages." In realism, seeking a reputation as a saint, and thinking that the foundation of peace and prosperity in a future heyday was the mission of scholars from the past, which is the highest ideal of sustainable development. Confucian scholars mostly hoped that the Holy King was born and created a world of peace and unity.

Buddhist's hope that Metteyya, who is born in the world in the future, will develop a pure land on earth. The so-called "horizontal ground is like a mirror, rain is always at sooner the better," mountains and rivers will be eliminated naturally, and the four seawaters will be reduced by 10,000 L each. We saw the ancient Indians' aversion to the mountains and the sea, their longing for good weather, or the afterlife, attracted to the Sakhavati's favorite as living a pure land in Mahayana Buddhism, associated with the Buddha Amitābha. The Elysium has seven treasure pools, eight merit waters, and the bottom of the pool is covered with golden sand cloth. Four-side steps, gold, silver, glass, and glassy glass synthesis. The lotus in the pond is as big as a wheel. There are also magical motley birds transformed by Amitabha-white cranes, peacocks, parrots, relics, Jialingpinga, and other birds of common life, and out at all days and nights with beautiful music. We saw human regrets about "life is not a prince and noble," but after death, they hope to enjoy the visual shock of the royal family's "gold and silver jewelry" and the desire to enjoy the rare bird with beautiful music.

The Christian Bible is the book of God. the Messiah Jesus Christ and God's chosen humanity will rule the earth for thousand years. At that time, the dead will be resurrected and given the opportunity to have eternal life. Sick people will be cured, and sickness and death will disappear. Isaiah 65:25 states in the Bible: "The wolf and the lamb shall feed together, and the lion shall eat straw like the bullock; and dust shall be the serpent's meat. They shall not hurt nor destroy in all My holy mountain," "*saith the Lord.*" "The wolf and the lamb shall feed together, and the lion shall eat straw like the ox; and dust shall be the serpent's food. They shall not hurt nor destroy in all my holy mountain, saith Jehovah." Christians long for the environment to be harmless, life to last, and the ecosystem food chain relationship to cease to exist. We see God's kindness in the Old Testament.

On the other hand, Islam beliefs the true God is Allah—there is no God, but Allah. Allah is one, not two, not three, not many. He is the God of the Muslims and humans. He is the God of Muhammad (PBUH), one of the messengers (prophets) among others, like Adam, Abraham, David,

W.-T. Fang et al., *The Living Environmental Education*, Sustainable Development Goals Series, https://doi.org/10.1007/978-981-19-4234-1

Joseph, Jesus, Moses. Muhammad was not considered to be changing the world today, but was planning *al-Jannah* gardens (heaven) after death as mentioned in the Qur'an. Paradise has beautiful gardens, valleys filled with trees, huge trees, musky mountains, rivers of water, milk streams, honey streams, and vineyards, as well as spring flavored or ginger. Heaven has delicious fruits in four seasons. There are gold, diamond, silver, pearls, and castles, as well as "white and dazzling" creatures such as white horses and camels, and valleys made of pearls and gems. We see that Muslims are struggling to survive due to the harsh environment in Central Asia. A craving for jewelry, gourmet cuisine, and rich ecological environment and banquet services, the eternal virgin remains a feast for the senses of sight, taste, smell, and touch. This is the imagination of the people living in Central Asia for their ideal living environment and rich source of living.

From the sixth century BC to the sixth century AD, the earth's resources and living enjoyment became increasingly scarce. At the *al-Jannah* (Paradise) described in the Qur'an, the human imagination of Paradise is more and more like the fusion of the rich and rich world—a transformation of nature, not a fantasy world of the "inexplicable." Judging from religious classics, in addition to the need for gold and silver jewelry in heaven until the sixth century BC, real life and even delicious food, easy, and luxurious living will be available.

In addition to the religious worries about moral corruption and war during the twenty-first century, eschatologists (those who believe in judgement day) have also expressed their sadness for society, believing that the end of the world is coming and that mankind must repent. Because humans caused environmental damage, and after the World War II in the twentieth century, when humans were trapped in concentration camps, the Messiah did not come; human beings are transformed into religions for the hope of "heaven or paradise." If the earth is not beautiful, then, we must create a paradise world with human life, eternal species, full of gold and silver ornaments, and charming towers in the clouds.

In the "permanent view of life" of religious theory and the "ecological competition view" of Darwinism, Environmental Education has never deceived the world with the "Gaia Hypothesis," but adopted the cruel reality of the "Media Hypothesis." We hope that through our efforts, the world will transform to the peaceful world we can live in. We adopted a perpetual view of "building the heart for the heavens and the earth, setting the lives for the people, learning from the studies of the saints, and opening the peace for all ages." According to Xun Kuang (Chinese: 荀况; c. 310–c. after 238 BC), "Theory of Heaven": "The heavens (the movement of nature) are constant, and nature is the true law. If you rule, you should be good, if you should be chaotic, you should be fierce." "If you are strong as an increasing agricultural production, and you use it sparingly, then you can't be poor." We have adopted a "different approach to love the people, benevolent people, and love things." At the same time, we understand the "law-awareness" of Gautama Sakyamuni Buddha from the sixth century BC. The "familiarity with the people" came to the background of Muḥammad's "conversion to Allah" in the sixth century. During the 1,200-year-old foundation of the main religion of mankind, it has created the essence of religion, which is to achieve peace in this world and create a real life in the future.

In the twenty-first century, we think of Virtual Reality (VR) and Augmented Reality (AR) visions and auditory hallucinations, "as intoxicated as dream bubbles." We thought of the machine god in Gnosticism of *The Matrix*, letting humans choose to survive in the virtual matrix, through various built-in programs, through the brain and nerve. The connections make the signals of sight, hearing, smell, taste, touch, and psychology to the human brain seem to be the real world of dreams, but also a kind of super-technical world that violates the physical phenomena of reality. We have been imagining technology and religion to lead humanity to a better world. On the surface, technology and religion are in conflict, but their principles are based on the fact that humans are not satisfied with the real space of their physical and eco-logical environment. They hope to use the con-cept of the "Heaven View of Religion" to transform the real world into a Metaverse, including Buddhist pure land, the birth of the Christian Messiah, and *al-Jannah* (Heaven or Paradise).

If modern religions flaunt their dazzling par-adise, pure land, and Eden-style environmental views, it will then affect human attitudes to global "environmental creation." We try to find the answer to "environmental creation" in the film. According to Doraemon's Comics movie of *Nobita's Drift in the Universe* in 1999, due to environmental pollution, the original planet "Mother Star" had no plants. A fleet of surviving astronauts is looking for a planet with green plants that can support life. In the 2009 science fiction movie *Avatar*, the earth was described as a "world without green plants" because of human greed. In the 2018 Marvel Cinematic Universe (MCU) *Avengers: Infinity War*, the villain Tha-nos, in one click, destroyed half of the creatures in the universe, contains half of the Marvel her-oes. We seem to see the abuse of the theory of Deep Ecology. Thanos mistakenly believes that the development of life in the universe exceeds the limits of the growth or carrying capacity of the universe. It is necessary to eliminate half of the life according to Deep Ecology. This is a completely misunderstood universe of Ecologi-cal Theory.

However, in the twenty-first century's envi-ronmental education, we often joke around that environmental education has formed a concept of religion, referred to as "environmental religion." The purpose of environmental education is to "prosper the country and succeed the world," promote "universe love, world peace," and create a living space for sustainable development for "life that the universe inherits." All of these ideas could be based on the ideas of love and peace that are beyond the oligarchs of the past, present, and future. Humans have limited physical space for survival and our imagination of the environ-ment is inherently better than reality. Think about metaverse.

Living Environmental Education discusses the value of ecological conservation, rehabilitation, and education methods to conduct ecological investigation, recording, and reconstruction. We used photography to "capture instantaneously" "that precious moment in space and in time" and a narrative to express the turmoil and confusion of the nation's love for itself and the environ-ment. It can be said that after the aboriginal people came, followed by the people from Fujian, Guangdong and Han, the Dutch, Spanish, Japanese, and Han, we now have a diverse and inclusive island nation that could serve as a model for the rest of the world. While the physical conditions in Taiwan are not ideal because of the typhoons, floods, and earth-quakes, the resiliency, kindness, and generosity of the Taiwanese people is second to none. Despite the current global pandemic, the people accept the inconvenience of keeping the virus at bay, and consider it a civic duty. Despite the colonialism and politics that the people have endured for hundreds of years, Taiwan's most important resource and scenery is people and a rare example of the adaptations of the people when worldwide religious and ethnic conflict were/are globally turbulent.

Living Environmental Education is being published as an open access so that ideas can be shared freely across the geographic, economic, political, and cultural boundaries that we've created as a species. In Mother Nature's eyes, "We (humans) are all just another bug on the plant and can be squashed at will." Our anthro-pocentric points of view and predilection for superiority over the rest of the world must end. We need to start working smartly and collabo-ratively if we want to continue to exist. Change is coming and there isn't much that we can do, except know that we can't change the inevitable and that we need to start to adapt to change. We're all on the same journey, but getting there following different beliefs, languages, and ways of life. Different doesn't mean one way is better/worst than another, it means it's just dif-ferent. Ironically, we go to great lengths to pro-tect an increase the diversity of life on this planet, except for that of our species. How intriguing! "Oh, is that my bus coming? I think I'll take this one."

"The world could never be instantaneous," and the world was stunned and fleeting. We hope that environmental education will continue in the future, and through cultural, religious, and pluralistic analyses of society, we will strive to create environmental aspirations of future generations to settle on Mother Earth, and interpret the mosaic-rich images of world's ecology. By studying the classic "land's love" theory of Eastern and Western scholars, we have created hard-working memories of the ethnic group working in order to construct a collective energy full of knowledge, enthusiasm, and environmental action.

Appendix: Environmental Attitude Scales

Unless a person works with these types of data daily, we can't dedicate the right amount of time and space to better explain these complicated concepts. Understanding the acronyms and meanings for Environmental Attitude Scales themselves is well outside of the scope of this book. The idea of measurement is good, but the readers need to know what we are measuring and why, and what the results mean (e.g., measurement of environmental attitudes such as: Ecological Scale (Maloney and Ward 1973; Maloney et al. 1975); Environmental Care Scale (Weigel and Weigel 1978); and New Environmental Model Scale (Dunlap and Van Liere 1978; Dunlap et al. 1983; 2000). Otherwise, we expect readers will skip over this part because it's technical and the readers may not understand the statements being made.

Environmental Attitude is often considered to be the same as environmental care (Hart 2013). We discussed environmental attitudes because environmental care is seen as a broader psychological dimension. In Chap. 5, we explored a new environmental paradigm scale. Environmental Attitudes are generally measured by one-order constituents, such as caring or not caring. In addition, environmental attitudes adopt a multi-component concept and have been adopted in many studies, i.e., a multi-component intervention at encouraging more sustainable behaviors (Cherdymova et al. 2018; Trewern et al. 2022).

In the 1970s, the New Ecological Paradigm (NEP) concept was critiqued because "the claimed lack of human-environmental focus in the classical sociologists and the sociological priorities". In 1982, Cotgrove proposed a hierarchical structure of environmental attitudes. It consists of two second-order factors that form a bipolar scale (Cotgrove 1982), such as the integration of fact/value, thought/feeling by Cotgrove (1982). It includes the "Dominant Social Paradigm (DSP)" and "New Environmental Paradigm (NEP)." The Dominant Social Paradigm is derived from the "Human Exceptionalism Paradigm" (HEP). It refers to the values, attitudes, beliefs, and knowledge that we have in common with one another in relation to the natural and social environment at a time when material and economic growth, advocating control of nature and emphasizing the control of natural resources to promote economic development were considered important.

However, Human Exceptionalism Paradigm (HEP) has also been criticized. The HEP viewpoint claims that human-environmental relationships were unimportant sociologically because humans are 'exempt' from environmental forces via cultural change. This view was shaped by the leading Western worldview of the time and the desire for sociology to establish itself as an independent discipline against the then popular racist-biological environmental determinism where environment was all. In this HEP view, human dominance was felt to be justified by the uniqueness of culture, argued to be more adaptable than biological traits. Furthermore, culture also has the capacity to accumulate and innovate, making it capable of solving all natural problems. Therefore, as humans were not conceived of as governed by natural conditions, they were felt to have complete control of their own destiny. Any

© The Editor(s) (if applicable) and The Author(s) 2023
W.-T. Fang et al., *The Living Environmental Education*, Sustainable Development Goals Series,
https://doi.org/10.1007/978-981-19-4234-1

potential limitation posed by the natural world was felt to be surpassed using human ingenuity. Research proceeded accordingly without environmental analysis. Sociology scholar Riley Dunlap recognized the limitations of the "Human Exceptionalism Paradigm" (HEP) and proposed a new environmental model. He argued that in the environmental model. Here, Cotgrove (1982) simplified the "Dominant Social Paradigm" and "New Environment Paradigm" to become "Dominant Paradigm" and "Environmental Paradigm," forming a bipolar scale.

References

Cherdymova EI, Ukolova LI, Gribkova OV, Kabkova EP, Tararina LI, Kurbanov RA, Belyalova AM, Kudrinskaya IV (2018) Projective techniques for student environmental attitudes study. Koloji 106:541–546

Cotgrove SF (1982) Catastrophe or cornucopia: the environment, politics, and the future. Wiley, p 166

Dunlap RE, Van Liere KD (1978) The "new environmental paradigm." J Environ Educ 9(4):10–19

Dunlap RE, Grieneeks JK, Rokeach MM (1983) Human values and pro-environmental behavior. In: Conn WD (ed) Energy and material resources: attitudes, values, and public policy. Boulder, Nova Scotia, pp 145–168

Dunlap RE, Van Liere KD, Mertig AG, Jones RE (2000) Measuring endorsement of the new ecological paradigm: a revised NEP Scale. J Soc Issues 56(3):425–442

Hart RA (2013) Children's participation: the theory and practice of involving young citizens in community development and environmental care. London, Routledge. https://doi.org/10.4324/9781315070728

Maloney MP, Ward MP (1973) Ecology: Let's hear from the people: an objective scale for the measurement of ecological attitudes and knowledge. Am Psychol 28 (7):583–586

Maloney MP, Ward MP, Braucht GNŽ (1975) Psychology in action: a revised scale for the measurement of ecological attitudes and knowledge. Am Psychol 30:787–790

Trewern J, Chenoweth J, Christie I (2022) Sparking change: evaluating the effectiveness of a multi-component intervention at encouraging more sustainable food behaviors. Appetite 171(1). https://doi.org/10.1016/j.appet.2022.105933

Weigel RH, Weigel J (1978) Environmental concern: the development of a measure. Environ Behav 10(1):3–15

Index

A

ABC learning method, 85
Abiotic, 8, 26, 28, 29, 49, 57, 98, 154, 229, 232, 233, 235
Academy of natural sciences, 65
Accountability, 30, 31, 107
Acid, 101
Acid rain, 29
Action, 3, 8–13, 17, 19, 22, 25, 30, 41, 42, 46, 57, 59–61, 64, 67, 68, 78–82, 85, 95, 98, 100, 105, 107, 109, 114–119, 122, 123, 130, 132, 136, 140, 146, 155, 163, 164, 166, 168, 169, 179, 198–200, 206, 207, 209, 218, 237
Action art, 122
Action research, 67, 74, 79–81
Activating event, 86, 87
Activism, 164, 165
Activity, 13, 16, 21, 32, 35, 50, 53, 60, 62, 63, 66–68, 72, 77, 83, 94, 98, 107, 108, 113, 116, 120, 130, 136, 138, 141, 143, 158, 165, 177, 184–186, 188, 191–194, 196–198, 200–204, 206, 209, 210, 224, 230, 232–234, 236, 238, 240, 242–245, 247, 248, 251–253, 255, 258
Activity skill, 63, 199
Activity workshop, 17
Adaptive, 140, 141
Adaptive learning, 189, 190
Advancement, 59
Advocacy, 169, 177, 224
Affection, 85, 96, 113, 114, 118, 191, 209, 210
Affective, 75, 85, 97, 101, 112, 118, 119, 130, 143, 199, 208, 209
Affective domain, 85, 113, 209
Agenda 21, 41, 42, 45, 64
Agent, 98
Alienation of self, 145
Alpha reliability coefficient, 72
Altruism, 109, 156, 160
Ambient, 32, 151
America, 26, 217, 230
American, 8, 19, 28, 36, 65, 68, 85, 105, 158, 217, 224, 230, 245
American environmental movement, 28
Animal husbandry, 17, 153
Animal liberation, 156

Animal rights, 153, 154, 156, 157
Animal welfare, 62, 156, 157
Animation, 63
Anthropocene, 178
Anthropocene epoch, 178
Anthropocentric, 152, 153, 155, 264
Anthropocentric value system, 152
Anthropocentrism, 28, 151–153, 156
Anthropogenic, 8, 107, 137, 138, 229, 240, 249
Anthropogenic environmental pressure, 138
Anti-anthropocentrism, 153
Applied ecology, 27
Approach, 11–13, 17, 19, 27, 30, 55–60, 64, 70, 71, 74, 75, 81, 83, 84, 111, 118, 128, 131, 151, 154, 155, 157, 158, 178, 182, 184, 185, 215, 244, 253, 262
Argumentation, 60
Aristotle, 26, 27, 152, 190
Artificial, 72, 121, 180, 188, 224, 233, 261
Artificial Intelligence (AI), 261
Ascription of Responsibility (AR), 157, 163–165, 231, 263
Atmospheric level, viii
Attainment, 115, 182
Attitude, 10, 11, 29, 30, 38, 39, 41, 44, 45, 49, 62, 63, 65, 66, 70, 72, 73, 75, 84–87, 93–102, 105, 107, 110–114, 118, 119, 123, 127, 130, 135, 137, 139, 156, 159, 160, 162, 165, 167, 199, 204, 206, 209, 218, 230, 238, 263, 265
Attitude-Behavior (A-B), 160
Augmented reality, 231, 263
Australia, 43, 65, 140, 251
Australian, 185
Australian Academy of Science, 65
Autopoiesis, 156, 157
Awareness of Consequences (AC), 163, 164

B

Balance, 8, 64, 80, 98, 120, 127, 178, 201, 216
Balance of nature, 7
Bandura, 98–100, 131, 145, 206, 207
Barrier, 85, 140, 244, 245
Behavior, 10, 26, 30, 32, 39, 41–43, 45, 51, 54, 57, 66, 72, 74, 80, 82, 85–88, 93, 94, 96, 98–100, 105,

107, 109, 111, 114, 116–120, 122, 123, 127–129,
   131, 132, 134–138, 140, 145, 153, 158–167, 206,
   207, 209, 210, 218, 233, 244
Behavioral change, 66, 160, 207, 209
Behavioral consequence, 86, 87, 131
Behavioral intention, 100, 130, 135, 137, 160, 162, 165,
   166
Belgrade, 39, 44
Belgrade Charter, 39, 41, 44, 66, 117
Belief, 26, 32, 86, 87, 93, 96, 98, 101, 104, 107, 109, 111,
   113, 119, 127, 130, 142, 151–153, 156, 158, 160,
   162–165, 168, 217, 220, 262, 264, 265
Bilateral, 83, 84
Binary, 179
Binary opposition, 179
Biocentric, 153, 156
Biocentric ethic value system, 153
Biocentrism, 151, 153, 154
Biocidal, 29
Biologist, 28, 29
Biomass, 30
Biophilia, 179, 180
Biophilia hypothesis, 160, 179, 180
Biophobia, 179, 180
Biophobia hypothesis, 179, 180
Biophysical environment, 7, 22, 37, 65
Bio-regional education, 229
Biota, 155
Biotic, 7, 26, 28, 29, 49, 57, 98, 156, 229, 233, 235
Bloom, 8, 35, 49, 85, 86, 96, 113, 114
Bloom-style learning method, 85
Bonn, 46
Bonn Declaration, 46, 64
Bottom-up, 83, 84
Boundary, 52, 84, 146, 178, 264
Boundary trap, 262
Brainstorming, 16, 52, 53
Brazil, 41, 42, 45, 46, 64
Brundtland, 41

C
Camp, 9, 197, 230, 262
Camping, 197, 204, 230, 236, 258
Capacity, 42, 169, 213, 218, 266
Capacity building, 178, 218, 219
Carbon, 168, 188, 199, 200, 220
Carbon dioxide (CO_2), 29, 123, 220
Carbon footprint, 75, 197, 199
Carbon sequestration, viii
Carrying capacity, 30, 31, 98, 138, 233, 263
Carson, 6, 7, 38, 44, 65, 224
Causal, 73, 100, 132, 133, 156, 160, 163
Causal factor, 165
Causality, 165
Center, 9–11, 14, 19, 37, 40, 42, 43, 50, 62, 74, 94, 96,
   100, 117, 153, 191–194, 197, 198, 201, 202, 210,
   232, 237, 238, 246, 255
Chain, 6, 96, 123, 163, 215, 262

Childhood, 55, 103, 142, 160
Childhood education, 55
Children, 6, 29, 32, 35, 36, 67, 75, 78, 79, 84, 103, 104,
   107, 131, 145, 168, 183, 184, 190, 193, 197, 198,
   206, 235, 238, 240, 242, 244, 245
China, 3, 25, 26, 140, 168, 222, 230, 251
Chinese, 17, 26, 27, 53, 152, 153, 222, 251, 262
Chinese Society for Environmental Education, CSEE
   (Taiwan, ROC), 56, 221
Choice, 25, 58, 59, 95, 115, 151, 169, 185, 190, 204, 223,
   241
Christianity, 26
Circle, 21, 67, 102, 181
Circle of re-action, 67
Citizen, 3, 7, 8, 10, 22, 28, 60, 61, 65, 68, 93, 95, 96, 109,
   118, 119, 123, 130, 132, 166, 169, 185, 188, 190,
   191, 199, 217, 242
Citizenry, 7, 37
Civic, 35, 36, 118, 119, 169, 199, 263
Civic society, 160
Civil, 105, 191
Civil right, 28
Civil servant, 99
Classroom, 3, 15, 18, 57, 59, 60, 62, 179, 184, 190, 230,
   235, 238–240, 244, 245, 252, 255
Classroom-based laboratory, 66
Classroom education, 11, 15
Classroom-type laboratory, 71
Co-evolution, 29
Cognition, 49, 85, 87, 96, 109, 111, 113, 114, 127, 128,
   130, 140, 145, 206, 209, 210, 237
Cognitionism, 59
Cognitive, 66, 75, 85–87, 97, 101, 105, 112–114, 119,
   128–132, 162, 163, 183, 206, 207, 209, 217, 242,
   253
Cognitive theory, 127–129, 206, 207
Cognizant, 6
Cold, 104, 133, 143
Cold war, 28
Collaboration, 4, 53
Collective, 29, 75, 93, 103, 111, 123, 146, 166, 199, 212,
   232, 234, 264
Collective action, 11, 67, 129, 216
Collective behavior, 64
Comfortable environment, 145
Communication, 15, 16, 20, 27, 38, 51–53, 55, 56, 58,
   104, 105, 107, 113, 114, 117, 118, 122, 131, 145,
   163, 167, 169, 177, 178, 189, 191, 201, 202,
   220–222, 224
Communication effect, 223
Communication media, 63, 177, 220
Community, 6, 8, 20, 28, 29, 36, 38, 42, 45, 50, 52, 54,
   62, 64, 65, 67, 68, 70, 79, 81, 104, 107, 111, 123,
   142, 154, 158, 169, 177, 179, 188, 191, 197, 201,
   217, 218, 221, 222, 230, 233, 235, 245
Community tour, 62
Competence, 57, 163, 178, 219, 220
Competency-based education, 220
Comprehensive, 16, 17, 19, 42, 85, 167, 185, 206

Concept, 3–6, 8, 10, 19, 20, 22, 25–27, 30, 38, 41, 43, 49–52, 55, 57–59, 61, 65–67, 69, 71, 73, 76, 78, 82, 93–98, 104, 106–109, 111–113, 116, 117, 119, 120, 127, 128, 135, 137, 138, 143, 152, 154–160, 162–164, 184, 185, 188, 189, 191, 204, 218, 229, 235, 238–240, 263, 265
Concern, 3, 11, 21, 35, 41, 44, 102, 106, 107, 111, 137, 151, 153–155, 158, 165, 167, 209, 244
Conference, 7, 11, 36, 38, 41, 42, 44–46, 55, 64, 65, 67, 69, 191, 221
Confucius, 32, 33, 182
Connect, 16, 17, 41, 85, 109, 114, 180, 217, 230
Connotation, 35, 60–62, 93, 114, 120, 178, 218, 229
Consciousness, 26, 93, 100, 102, 106, 107, 109, 116, 123, 156, 164, 224, 239
Conservation, 6, 8, 26, 27, 29, 30, 36, 43, 45, 69, 75, 111, 117, 137, 153, 154, 184, 191–194, 197, 204, 247, 263
Conservation biology, 27, 62, 65, 154, 155
Conservation education, 36, 46, 65, 190, 236, 258
Consolidation, 68
Construction, 57, 59, 65, 72–74, 113, 115, 138, 139, 184, 202, 218
Construction education, 6
Constructivism, 57, 59–61, 121
Constructivist, 59
Construct validity, 72, 73
Consumer, 117, 118, 134
Consumerism, 17, 22, 117, 118
Consumption, 134, 153, 162, 168
Content, 7, 8, 19, 20, 29, 38, 41, 43, 44, 56–58, 61, 65, 70, 72, 80, 83, 113, 129, 130, 135, 138, 152, 162, 177, 178, 184, 186, 188, 189, 198, 199, 201, 204, 206, 209, 216, 220, 235–237, 241, 248, 254, 256
Content analysis, 74, 82
Content knowledge, 18–20
Content validity, 72
Context, 6, 12, 59, 60, 64, 66, 74, 75, 81, 98, 115, 145, 151, 169, 207, 217, 256
Continuum, 9, 80
Control, 29, 52, 70–72, 87, 96, 97, 100, 108, 110, 119, 121, 132, 133, 135, 137, 140, 145, 160, 168, 204, 216, 242, 265, 266
Control group, 66, 71, 72, 136
Coral, 180, 181
Coral reef, 140, 141, 181
Correlation, 71–73, 101, 105, 114, 132, 133, 137, 145, 218
Correlation coefficient, 72, 73
Cost, 32, 70, 114, 123, 166, 245, 246
Cost–benefit evaluation, 166
Council, 39, 192
Council for Environmental Education, 7
Course, 8, 9, 16–18, 32, 36, 51, 53, 57–59, 61, 67, 74, 84, 109, 111, 115, 142, 159, 168, 178–180, 182, 188, 192, 197, 204, 206, 209, 215, 229, 231, 242, 255
Course content, 184, 198
Covid-19, 32, 110, 143, 244
Covid-19 Pandemic, 129, 143, 145, 244

Creator, 26
Critical pedagogy, 169
Critical theory, 60, 66
Critical thinking, 4, 8, 16, 42, 57, 60, 61, 169, 190
Criticism, 54, 60, 61, 66, 67, 115, 121
Cronbach alpha, 72
Cross-disciplinary, 190
Cross-disciplinary education, 190
Cultural landscape, 29
Culture, 6–8, 10, 20, 21, 25, 26, 30, 33, 34, 38, 43, 46, 49, 50, 56, 65, 70, 75, 87, 94, 104, 105, 107, 108, 117, 146, 152, 155, 160–162, 177, 198, 202, 204, 205, 212, 213, 222, 224, 229, 242, 255, 266
Curricula, 3, 11–13, 15, 17, 20, 37, 38, 41, 53, 56, 58, 64, 65, 70, 80, 82, 85, 104, 113, 119, 152, 179, 185, 188, 191, 194–196, 198, 200, 201, 219, 229–231, 238, 253, 255, 258

D
Dadaism, 121
Darwin, 52, 84, 85, 153
Data, 4, 13–16, 26, 33, 36, 49–52, 54, 56, 64, 68–71, 73–76, 78–81, 83, 84, 129, 181, 185, 234, 235, 239, 251, 253, 265
Data analysis, 70, 73, 78
Data comparison, 78
Data compilation, 78
Data query, 68
Debriefing, 253
Decision, 10, 11, 17, 21, 42, 49, 51, 66, 68, 73, 86, 87, 93, 94, 107, 108, 118, 119, 127, 130, 132, 155, 158–160, 162, 163, 165–167, 177, 208, 209, 224, 229, 242
Decision-making, 165
Declarations on the Human Environment, 38
Deconstructivism, 59, 60
Deep ecology, 27, 29, 30, 153, 155–157, 263
Definition, 3–8, 10, 25, 32, 33, 35, 37, 44, 60, 65, 71, 94, 104, 105, 110, 117, 123, 152, 165, 189, 215, 229
Degree, 72, 73, 103, 115, 123, 129, 134, 160, 162, 165, 210, 211, 223
Degree of difficulty, 73
Delayed test, 71
Delphi, 73, 74
Democracy, 152
Demonstration, 17, 20, 66, 188, 189
Derrida, 60
Descriptive norm, 135–137, 162
Design, 6, 56, 58, 63, 71, 72, 78, 115, 130, 179, 184, 201, 204
Development, 6, 11, 17, 18, 21, 29–32, 34, 39, 41–46, 52, 53, 55–57, 64–66, 70, 78, 79, 83, 84, 87, 94, 98, 100, 103, 107–109, 117, 119, 123, 128, 131, 135, 137, 138, 145, 146, 159, 160, 163, 165, 167, 182, 185, 191, 197, 199, 209, 211, 218, 219, 245, 246, 254, 263
Development theory, 78
Dewey, 35, 190, 230

Dialogue, 27, 32, 46, 56, 60, 74, 112, 169
Dichloro-diphenyl-trichloroethane (DDT), 6
Direct observation, 75
Discipline, 6, 8, 10, 17, 19, 56, 58, 62, 83–85, 107, 120,
    138, 139, 146, 158, 230, 241, 266
Discourse, 56, 63, 153, 156, 169, 177, 178, 224
Discrimination, 73, 115
Dissertation, 52, 59, 119
Doctrine, 34, 60, 85, 151, 152
Documentary library, 64
Dominant Social Paradigm (DSP), 158, 167
Drought, 255
Dual, 61
Dualistic, 60
Dualistic structure, 60
Dual opposition, 61

E

Earth, 6, 20, 27–31, 50, 60, 61, 63, 64, 85, 87, 108, 109,
    122, 127, 129, 133, 140, 142, 146, 151–155, 157,
    158, 168, 180, 229, 253, 261, 262, 264
Earth Day, 7, 28
Earth science, 8
Earth Summit, 41, 42, 45, 46, 64
Ecocentric, 151, 154, 155
Ecocentric-anthropocentric continuum, 155
Ecocentrism, 151, 154–156
Ecofriendly, 204
Ecofriendly behavior, 165
Ecological, 3, 6, 7, 11, 26, 27, 29, 30, 41, 43, 57, 59, 60,
    64–67, 95, 96, 98, 108–111, 117, 130, 134, 138,
    146, 151, 152, 154–156, 160, 163, 164, 184,
    191–194, 201, 202, 204, 209, 210, 213, 215, 216,
    223, 224, 229, 232, 234, 245, 262, 263, 265
Ecological economics, 85
Ecological environment, 59, 65, 108, 164, 191, 194, 197,
    262
Ecological ethic, 151, 154, 155
Ecological ethic value system, 153
Ecological guide, 63
Ecological management, 7, 17, 108, 117, 118
Ecological paradigm, 60, 168
Ecological principle, 8
Ecological worldview, 156, 164
Ecologist, 43, 65, 155, 156, 179
Economic development, 17, 30, 51, 112, 224, 265
Economic growth, 21, 32, 137, 146, 167, 168, 265
Economic orientation, 21
Economics, 3, 8, 20–22, 27, 30, 36, 41–43, 49, 54, 56, 60,
    64, 94, 107–109, 111, 117, 118, 129, 138, 152,
    155, 158, 159, 167–169, 218, 229, 245, 263
Ecopsychology, 142
Ecosystem, 6, 8, 20, 25–29, 43, 54, 58, 103, 104, 107,
    117, 118, 139, 142, 151, 153–157, 164, 184, 185,
    191, 197, 213, 218, 229, 233, 235, 245, 262
Ecotourism, 62, 63, 140, 197, 201, 204
Education, 3–15, 17–22, 25, 28–46, 49, 51–62, 64–67,
    69–72, 83–85, 93–98, 100, 103–105, 107,

    111–115, 118–120, 123, 127–130, 132, 140, 146,
    151, 152, 156, 157, 163, 169, 170, 178–186,
    188–202, 205, 206, 208–210, 215, 216, 218–221,
    223, 229–231, 233, 234, 236–238, 243–245, 247,
    253–258, 261–264
Educational interpretation plan, 201
Educational psychology, 85
Education curricula, 65, 103, 185, 197, 199
Education for All (EFA), 42
Education of practice, 66
Effective, 12, 19, 20, 51, 52, 62, 65, 75, 79, 109, 115,
    117, 118, 131, 156, 169, 189, 218, 238, 240, 244,
    245
Effective communication, 4, 224
Effectiveness, 6, 166, 191
Efficacy, 160, 177, 230
Efficiency, 21, 26, 167, 213, 214, 238
Efficiency theory, 27
Eighteenth century, 7, 34, 35, 155
Einstein, 3, 4
Elementary school, 7, 58, 62, 67, 69, 99, 255
Emerson, 27
Emotion, 85–87, 100, 102, 103, 109, 112, 127, 128, 130,
    133, 140, 143, 159, 180, 206, 209, 237, 239,
    241–243
Emotional, 18, 49, 86, 87, 101, 106, 107, 110–113, 115,
    122, 130, 133, 183, 197, 206, 207, 209, 210, 230,
    240–242, 253
Emotional stability, 133–135
Emotional state, 61, 219
Empirical, 55, 56, 60, 66, 71, 72, 81, 112, 169, 198
Empirical research, 66
Empiricism, 26, 27, 112
Enlightened selfishness, 28
Environment, 3–8, 10–18, 21, 22, 25–30, 32, 33, 36–39,
    41, 42, 44, 45, 49, 51–53, 56–62, 64–69, 74, 75,
    79, 80, 82, 83, 85–87, 93–95, 97, 98, 100–112,
    114–123, 127–132, 136, 138–142, 145, 146, 151,
    152, 155–157, 163–166, 168, 177, 180, 184, 185,
    190, 191, 193, 197, 199, 201, 204, 206–208, 210,
    215, 218, 219, 224, 229–233, 235, 237–239, 245,
    246, 248, 252, 253, 255–257, 262, 263, 265, 266
Environmental action, 8, 11, 12, 17, 18, 31, 34, 67, 80,
    82, 95, 96, 100, 104, 116–119, 130, 166, 204, 229,
    264
Environmental attitude, 15, 18, 57, 72, 82, 87, 97, 101,
    107, 110–112, 117, 133, 134, 158–160, 162, 265
Environmental awareness, 8, 10, 12, 26, 28, 39, 57, 100,
    102–107, 112, 120, 123, 135, 140, 184, 197, 199,
    239, 245
Environmental behavior, 8, 9, 18, 52, 87, 93, 96, 97, 99,
    104, 117, 119, 127, 130, 132, 133, 135, 137, 159,
    162–165
Environmental care, 162, 265
Environmental chemistry, 65
Environmental cognition, 127, 128, 130, 145, 146, 251
Environmental cohesion, 93
Environmental communication, 63, 177–179, 217, 224
Environmental consciousness, 93, 220

Environmental culture, 62
Environmental damage, 15, 36, 103, 262
Environmental degradation, 66, 95, 223
Environmental ecology, 63
Environmental economy, 62
Environmental education research, 49, 52, 54, 56, 94, 157
Environmental Education Research (EER) (Taylor & Francis' Journal), 55, 56
Environmental engineering, 62, 65
Environmental ethics, 8, 12, 26, 56, 63, 94, 105, 107, 109, 151, 154, 157, 158, 169, 199, 224
Environmental hazard, 65
Environmental impact, 15, 109, 138, 140, 142, 201, 220, 221
Environmental interpretation, 63
Environmentalism, 28, 29, 165, 224
Environmental issue, 3, 8, 10–18, 29, 38, 50, 51, 57, 60, 63, 67, 69, 95, 97–100, 102, 103, 105, 107, 117–119, 123, 158, 164, 166, 167, 169, 185, 191, 199, 201, 206, 218, 220, 221, 223, 224, 229, 248, 255, 257
Environmentalist, 22, 29, 66, 94, 111
Environmental justice, 61, 245
Environmental knowledge, 7, 8, 30, 31, 49, 50, 57, 62, 63, 73, 96, 97, 101, 102, 105, 106, 127, 128, 130, 159, 160, 184, 185, 204, 222
Environmental learning, 8, 16, 85, 127, 129–131, 177, 178, 206, 224, 256
Environmental learning center, 191, 192, 194–198, 201, 202, 204, 209
Environmental literacy, 20, 22, 28, 30, 61–63, 67, 85, 93–97, 99, 100, 107, 115, 116, 118–120, 123, 127, 129, 191, 210, 245
Environmentally friendly, 117, 132, 134, 162, 165, 167, 202
Environmentally friendly behavior, 3, 7, 85, 97, 107, 135–137, 141, 162
Environmental management, 8, 155, 186, 224
Environmental orientation, 21
Environmental paradigm, 151, 157, 266
Environmental perception, 130, 160
Environmental policy, 61, 111
Environmental politics, 62
Environmental pressure, 137–140, 145
Environmental problem, 3, 6–8, 11, 15, 30, 31, 36, 41, 44, 50, 51, 57, 65, 68, 83, 85, 87, 94, 97, 101, 102, 107, 111, 117–119, 123, 138, 140, 151, 158, 159, 164–166, 168, 185, 190, 229
Environmental protection, 6–8, 10, 18, 19, 21, 22, 25–27, 29, 30, 33, 37, 39, 41, 46, 52, 53, 55, 60, 62, 63, 67–69, 87, 93, 99, 102, 104, 107, 111, 112, 116, 120, 134, 136, 138, 139, 146, 152–154, 160, 162, 165, 167, 185, 186, 190, 191, 196–199, 209, 215, 218, 220, 221, 224
Environmental Protection Administration, 14, 61, 67, 184
Environmental Protection Administration of the Executive Yuan, 14, 61, 67, 184
Environmental Protection Agency, 14, 188

Environmental psychology, 62, 73, 107, 111, 127, 128, 130, 142, 146, 179
Environmental quality, 6, 10, 28, 38, 49, 66, 98, 158, 167, 194
Environmental science, 9, 51, 62, 84, 85, 101, 138, 146, 224, 231
Environmental science education, 62, 66
Environmental sensitivity, 12, 87, 94, 96, 103, 105, 107, 191
Environmental stress, 127, 138, 146
Environmental system, 22
Environmental value, 33, 62, 94, 107, 109, 112, 120, 123, 134, 142, 151, 206, 229
Environment map of amenity, 68
Epistemology, 3, 5, 60, 67, 100, 129, 155
Eternal, 93, 261, 262
Eternal value, 93
Ethnic, 21, 41, 82, 191, 202, 205, 263, 264
Ethnic diversity, 43
Ethnographic, 81
Ethnographic analysis, 82
Ethnographic research, 74, 81
Ethnography, 81, 82
Europe, 26, 35, 122, 129, 140
Executive Yuan, 186, 188
Existence, 5, 59, 102, 108, 111, 113, 130, 142, 155, 219
Existentialist, 59
Expedition, 212, 229, 230
Experience, 8, 9, 11–14, 17, 22, 25, 33, 52, 55, 57–59, 61, 66, 74, 75, 81, 82, 86, 88, 93, 97, 98, 100, 102–104, 108–112, 114, 115, 120, 121, 123, 128–130, 133–135, 139, 142, 145, 146, 159, 169, 178, 179, 182, 188, 192, 198, 199, 201, 204, 208–210, 216–218, 220, 229, 230, 236–238, 240, 242, 244–246, 250, 251, 253, 255–257
Experience-based learning, 258
Experience learning, 240
Experiential learning, 77, 169, 195, 198, 229
Experiment, 13, 49, 50, 59, 70, 71, 117, 131, 135–137, 159, 177, 217, 252
Experimental, 36, 65, 66, 70–72, 158, 190, 196
Experimental design, 67, 71, 72
Experimental group, 66, 71
Experimental object, 71
Experimental research, 70, 71, 73
Exploration, 8, 13, 29, 57, 75, 85, 127, 130, 237, 238, 246, 253, 257, 258
Exploration topic, 21
Exploratory, 130, 257, 258
Exploratory intelligence, 257, 258
Exploratory theory, 130
Ex-post facto, 73
External, 80, 93, 98, 102, 106, 130, 145, 156, 158, 179, 198, 207, 244
External incentive, 98
External stimulus, 5, 85
Extinction, 29, 30, 98, 103, 142, 169, 221

F

Family, 98, 110, 140, 160–162, 190, 196, 197, 200, 230, 261

Family education, 244

Feasibility, 22, 117, 166

Feeling, 11, 15, 41, 59, 97, 98, 101–103, 106, 109, 110, 112, 113, 116, 119, 127, 128, 130, 132, 133, 136, 142, 145, 146, 189, 206, 210, 243, 265

Field, 6, 7, 11, 20, 22, 33, 34, 37, 42, 46, 55–59, 62–65, 67, 71, 74, 75, 78, 81, 82, 84, 85, 87, 96, 107, 113–115, 120, 162, 168, 169, 177–179, 184–190, 197, 209, 218, 224, 229, 232, 233, 235, 239, 240, 243, 245, 246, 253

Field-dependent development, 245, 246

Field selection, 245, 246

Field studies, 212

Focal-point, 179

Focus group, 75

Forest, 13, 25, 26, 28, 101, 108, 117, 118, 127, 129, 130, 142–144, 210, 220, 229, 234, 246, 247

Formal education, 6, 41, 45, 64, 83, 84, 197

Formal media, 221

Fragility, 21, 103, 107

France, 36, 44, 45, 122, 129, 140

Frankfurt, 66, 162

Freedom, 21, 59, 133, 153, 244

G

Gaia, 27–29, 156

Gaia hypothesis, 27–29, 262

Game, 17, 18, 35, 103, 114, 158, 196, 202, 206, 208, 230, 242, 253

Gender, 66, 108

Gender equality, 21, 66

Generation, 7, 17, 20, 26, 33, 35, 38, 41–43, 51, 53, 64, 70, 81, 123, 132, 164, 167, 168, 190, 213, 214, 218, 232, 234, 245, 264

Generational justice, 97

Geologic epoch, 29, 30

Germany, 35, 46, 105, 121, 162

Gestalt psychology, 129

Gland, 45

Global change, 52, 64

Global communities, 128

Global Education Monitoring Report (GEM report), 42

Globe, 54, 68

Goal, 3, 11, 12, 20–22, 36, 38, 41, 42, 44–46, 50, 52, 54, 56–58, 65, 72, 79, 81, 85, 87, 94, 95, 98, 103, 110, 112, 113, 115–119, 123, 127, 128, 130, 142, 167–169, 179, 190, 198, 199, 201, 204, 207, 209, 218, 219, 230, 232, 234, 238, 243, 245, 250, 253, 255, 258

Graduate, 40, 58, 82, 198

Graduate Institute of Environmental Education (GIEE), 50

Graduate school, 62

Greek, 26, 28, 29, 105, 158

Green, 6, 7, 29, 42, 46, 104, 134, 177, 184, 188, 213, 215, 220, 224, 245, 263

Greenhouse, 197

Green personality, 134

Green personality trait, 133

Green school project, 184

Grounded theory, 74, 76, 78, 79

Groundwater, 29

Group, 9, 11, 21, 51, 65, 71–73, 75, 79, 81, 82, 107, 109, 111, 117, 118, 122, 123, 133, 135, 136, 144, 160–163, 190, 191, 198, 208, 221, 230, 233, 242, 244, 253

Group evaluation, 60, 61

Group learning, 16

H

Habermas, 66

Habit, 112, 162, 163, 204, 207

Habitat, 7, 27, 28, 36, 155, 188, 189, 211, 213, 214, 216, 229, 245

Half-reliability method, 72

Harvard, 36

Harvard University, 36, 194

Healing, 8, 127, 142, 143, 145, 146

Healing environment, 127, 142–145

Healing power, 142

Heat, 140, 180, 184

Heat wave, 129, 130, 140

Heaven, 59, 60, 109, 168, 180, 261–263

Hermeneutics, 60, 66, 217

Hetong culture and art foundation, 220

High school, 62

Historical, 25, 32, 34, 56, 64, 66, 168, 169, 238, 246, 247, 249, 251, 261

Historical trajectory, 64

History, 4, 6, 8, 25, 32, 34, 37, 44, 46, 47, 49, 50, 56, 64, 66, 103, 138, 158, 159, 168, 191, 192, 201, 202, 204, 217, 224, 230, 235, 247, 255

History-hermeneutics, 66

Homeostatic, 29

Hometown, 22

Hometown learning, 238

Human, 4–7, 20–22, 25–34, 36–39, 41–45, 50, 52, 54, 59, 62–67, 74, 77, 81–83, 85–87, 93, 94, 96, 98, 104, 107–112, 116, 117, 119–122, 130–132, 135, 137–140, 142, 144–146, 152–156, 158–160, 164, 165, 168, 177, 180, 193, 206–208, 213, 218, 219, 222, 224, 229, 240, 241, 253, 261–263, 265, 266

Human being, 5, 6, 8, 21, 29, 30, 33, 35, 36, 38, 41, 51–53, 57, 59, 65, 66, 75, 85–87, 93, 95, 98, 102, 104, 105, 108–112, 114–116, 128, 129, 140, 142, 143, 145, 146, 152, 153, 155, 156, 163, 168, 179, 180, 216, 218, 219, 224, 254, 262

Human-centered, 29, 151–153

Human dimension, 38, 64

Human health, 6, 30, 101, 107, 110, 123, 143, 215

Humanitarian, 26

Humanity, 8, 22, 28, 30, 39, 84, 85, 98, 104, 105, 127, 146, 152, 154, 178, 182, 183, 202, 222, 261, 263
Humankind, 4, 32, 42, 146, 168
Human knowledge, 34, 66
Human sense, 26, 75, 76
Hungerford, 11, 12, 49, 85–87, 95–97, 117–120, 123, 200
Hungerford-style learning method, 85
Hunting, 26, 111
Hydrogen, 29, 30
Hydrogen sulfide (H$_2$S), 29, 30
Hypothetic-deductive method, 76

I

Icehouse, vii
I Ching, 180
Identification, 30, 33, 66, 128
Impact, 21, 26, 29, 30, 51, 74, 80, 82, 87, 96, 105, 110, 111, 117, 119, 120, 129–132, 135, 137–140, 142, 153, 156, 159, 164, 165, 184, 191, 211, 214–218, 229, 233, 244, 255
Incheon, 42, 46
Incompetence, 219
Indirect action, 159
Induced event, 86
Industrial revolution, 28
Industry, 62, 140, 245
Inference, 64, 66, 71, 103, 105
Informal, 55, 56, 64
Informal media, 222
Information, 4, 8, 13, 15, 33, 37, 38, 57, 58, 63, 68–71, 73, 81, 86, 94, 100, 102, 104, 105, 109, 113, 114, 128, 130–132, 159, 177, 191, 201, 202, 209–211, 215–218, 220, 223, 224, 236, 244, 252, 253, 257
Information exchange, 41, 83, 84, 156, 157, 177, 223
Information processing, 209
Information transfer, 215
Inherent, 72, 93, 123, 158
In-house, 221
Injunctive norm, 135–137
Inquiry, 57, 66, 74, 75, 185
In-service education, 62
In-service Education and Training (INSET), 62
Installation art, 122, 123
Instructivism, 57–61
Instrument, 71
Instrumental research, 51, 66
Integration, 17, 59, 60, 70, 84, 85, 162, 209, 215, 253, 265
Integrationism, 60, 61
Intellectual, 104, 112, 158
Interdependence, 41
Intergovernmental Biosphere Conference, 65
Intermediary, 87, 131, 132
Intermediary variable, 162
Internal, 71, 73, 98, 107, 119, 131, 159, 198, 204
Internal driving force, 98
International, 8, 36, 38, 41–45, 52, 55, 56, 64–66, 94, 178, 191, 192, 218, 223, 232, 248

International Environmental Education Programme (IEEP), 39, 44
International Environmental Workshop, 39, 44
International Union for Conservation of Nature (IUCN), 8, 36, 38, 44, 45, 94
International Women's Year, 66
Interpretation, 5, 25, 51, 61, 71, 73, 74, 102, 103, 105, 152, 154, 201, 202, 204, 210, 215, 217, 252
Interpretive, 37, 74, 169, 178, 209
Intervention, 61, 66, 71, 104, 121, 218, 230, 265
Interview, 13, 14, 68, 74, 75
Intrinsic, 104, 128, 156
Intrinsic motivation, 98
Intrinsic value, 29, 107, 153, 154, 156, 215

J

Japan, 26, 42, 46, 68, 139, 194, 249, 250
Jevons paradox, 168
Johannesburg, 41, 45
Journal, 37, 54–56, 119, 142, 221, 222, 242
Journal of Environmental Education Research (Copyright © Chinese Society for Environmental Education, Taiwan, ROC), 56
Junior, 62
Junior high school, 62

K

Kant, 35, 111, 112, 152
Keele university, 44
Killing theory, 27, 28
Kindergarten, 17, 19, 35, 58, 62, 232, 234
Knowing-how, 169
Knowledge, 7, 8, 11, 18–22, 29, 32–34, 36, 41, 44, 45, 49, 52, 57–63, 66, 71, 73, 75, 79–81, 83–87, 94–98, 100, 104, 105, 111, 113–115, 118, 119, 123, 128, 130, 146, 154, 159, 160, 162, 163, 169, 184, 188, 190, 191, 194, 196, 198, 199, 201, 206, 215–220, 224, 230, 232, 234–236, 238, 240, 251, 252, 255, 256, 258, 264
Kolb, 198
Korea, 4, 31, 42, 43, 46
Kyoto, 139

L

Land, 36, 63, 64, 107, 111, 138, 151, 152, 181, 193, 201, 202, 211, 233, 234, 257, 261, 263, 264
Land ethics, 30, 110, 151, 152, 155
Landscape, 121, 122, 138, 146, 193, 210, 232
Landscape art, 121, 122
Learned, 3, 4, 6, 18, 34, 50, 57, 59, 61, 87, 113, 114, 127, 130, 178, 208, 209, 216, 239, 253
Learned skill, 62
Learning, 4, 8, 11–13, 15–20, 22, 25, 32, 33, 35, 42, 47, 56–59, 61, 62, 66, 68, 79, 83–87, 93, 96–98, 100, 101, 108, 110, 112–115, 123, 127–132, 140, 160, 167, 169, 177–179, 181, 182, 184–186, 189, 191,

193, 196–198, 201, 206–210, 215, 218–221,
229–232, 235–242, 244, 245, 248, 250, 251,
253–256, 258, 262
Learning interface, 215
Learning method, 3, 12, 17, 49, 58, 59, 61, 83, 85, 86,
115, 131, 132, 208
Learning model, 87, 115, 208
Learning plan, 50, 177, 197
Learning platform, 177
Learning process, 15, 32, 36, 57–59, 61, 93, 98, 113, 115,
127, 130–132, 157, 178, 179, 208, 209, 230, 252
Learning theory, 49, 85, 98, 131, 198
Leisure, 14, 56, 145, 212, 231, 237, 245, 255
Leisure education, ix
Leopold, 27, 28, 111, 152, 154, 155
Lesson, 3, 4, 47, 159, 169, 178, 230
Lesson plan, ix
Likert, 70
Limits to growth, 263
Literacy, 74, 93–95, 97
Literacy-based pedagogy, viii
Livestock, 25
Living, 4, 5, 27–30, 45, 62, 64, 96, 110, 116, 153,
155–157, 168, 179, 184, 191, 218, 261–263
Living environment, 25, 36, 59, 69, 84, 102, 104, 155,
169, 170, 184, 262
Living media lab, 175
Living world, 28
Local, 7, 25, 36, 52, 56, 62, 68, 69, 81, 99, 107, 114, 130,
191, 193, 194, 197, 198, 201, 202, 204, 216, 218,
222, 229, 232, 234, 238, 245, 246, 248, 255, 258
Local communities, 28, 128, 245
Lockdown, 244
Locke, 35
Locus, 119
Locus of control, 119
Lovelock, 27, 28

M
Malaysia, 105, 202, 212, 250, 251, 255
Mammal, vii
Marine, 139–142, 180, 181, 232
Medea, 29
Medea hypothesis, 27, 29, 30
Media, 10, 38, 55, 58, 63, 73, 120, 123, 169, 177, 188,
189, 191, 197, 201, 202, 204, 215, 220–224, 236,
245, 262
Mencius, 26, 27, 152
Metacognition, 87
Metacognitive, 87
Metadata, 64
Metaphor, 189
Metaverse, 263
Methane, 29, 30
Mind, 3, 4, 7, 17, 76, 84, 87, 127, 129, 142, 182, 183, 217
Mindset, 18, 101, 104, 132
Ministry of Education (Taiwan, ROC), 69, 70, 184, 201,
238

Mirroring, 60, 61
Mitigation, 21
Model, 18, 21, 22, 30, 33, 41, 54, 60, 67, 76, 83, 84, 87,
95, 96, 101, 104, 105, 119, 120, 128, 137, 151,
156, 158–167, 179, 190, 202, 207, 215, 218–221,
236, 243, 263, 265, 266
Modeled-play, 17
Modelling, 157, 162, 169
Moral, 29, 57, 94, 107, 109, 110, 136, 152, 153, 156, 159,
160, 164, 262
Moral creature, 152
Motivation, 11, 18, 41, 44, 73, 93, 96–100, 102, 112, 116,
123, 130, 162, 163, 165, 178, 196, 197, 206, 238,
240, 246, 253
Motivation-Opportunity-Abilities (MOA), 162, 163
Mt. Tai, 60
Muir, 27, 28
Multi-disciplinary, 8, 215
Multimedia, 63, 121

N
Næss, 27, 29, 30, 155, 156
Nagoya, 42, 46
Nanyang Technological University, 43
Narrative, 18, 59, 204, 263
Nation, 7, 8, 22, 30, 36, 38, 39, 41, 42, 44, 45, 62, 64, 66,
96, 217, 218, 263
National Dong Hwa university, 39, 40, 183
National Environmental Education Act, 7, 41, 94
National Environmental Education Association, 28
National Hualien University of Education, 39
National Little Environmental Planner, 67
National Outdoor Education Expo, 238
National primary school, 71
National Science Exhibition, 68
National Taiwan Normal University (NTNU), 39, 40, 43,
50, 53, 56, 77, 78, 185, 198, 212
Natural, 3, 6, 8, 13, 14, 26, 27, 30, 35–37, 39, 51, 57, 59,
62, 65, 69, 84, 85, 93, 104, 105, 107, 108, 110,
117, 121, 130, 132, 138, 140, 142, 143, 146, 151,
153, 155, 156, 161, 166, 168, 183–185, 192, 197,
198, 204, 207, 217, 224, 229, 230, 232, 238, 242,
245, 246, 254–256, 265, 266
Natural environment, 6, 7, 10, 21, 26, 32, 36, 38, 41, 52,
57, 63, 65, 75, 101, 105, 110, 120, 121, 130,
137–140, 142, 146, 183, 185, 195, 199, 202, 245,
256
Natural intelligence, 257, 258
Naturalist, 27, 36, 37, 212
Natural process, 65
Natural resource, 7, 8, 21, 22, 25–27, 36, 37, 39, 40, 56,
65, 94, 105, 111, 138, 184, 265
Natural scientist, 65
Nature, 3, 5, 7, 9–11, 13, 16, 25–30, 35–42, 44, 45, 49,
56, 57, 59, 60, 62, 85, 88, 93, 102–104, 107–112,
117, 120–123, 127, 134, 135, 139, 142, 143,
145, 151–153, 155–160, 164, 168, 177–180, 183,
184, 188, 190–193, 196–199, 209, 210, 230, 233,

234, 236, 238, 240, 244–246, 248–258, 262, 264, 265
Nature connectedness, 59, 135, 183
Nature conservancy, 38
Nature-deficit disorder, 238
Negative, 21, 73, 85, 87, 102, 104, 109–111, 117, 119, 145, 159, 164, 214, 215, 217, 218, 230
Negative attitude, 102
Neolithic epoch, 104
Neural network, 104
Neuron, 131, 241
Neutral attitude, 102
Nevada, 8, 38, 44
Nevada Declaration, 87
New Ecological Paradigm (NEP), 158, 168, 169, 265
New Environmental Paradigm (NEP), 158, 159, 265
New expressionism, 121
Newton, 156
New world, 27
New York, 45, 123
Nineteenth century, 3, 27, 35, 230
Non-formal, 42, 63, 64, 189, 191
Non-formal education, 41, 83, 197
Non-governmental enterprise, ix
Non-governmental organizations (NGOs), 231
Non-human, 82, 155, 156
Non-objectivism, 59
Non-structured interview, 74
Norm, 16, 17, 50, 51, 94, 99, 101, 118, 127, 136, 137, 156, 159, 162–164, 166
North American Association for Environmental Education (NAAEE), 14, 28, 43
Novel, 3, 12, 35, 52, 66, 123, 156, 206
Nurture, 110
Nurture environment, 110

O
Obstacle, 217, 218, 224, 244, 245
Off-school, 68, 197
Off-school teaching, 68, 197
Ohio State University (OSU), 37–40, 43
One planet, 61
Ontology, 3, 60, 67
Open-ended, 17, 185
Open-ended learning method, 59
Open mind, 3
Open University (Malaysia), 105, 202, 212, 250, 251, 255
Optimization, 27, 214
Orientation, 46, 106, 202, 253
Our Common Future, 41, 45
Outdoor, 3, 9, 10, 13, 14, 18, 32, 57, 61–63, 68, 103, 104, 122, 127, 142, 143, 145, 179, 188, 191, 194, 197, 203, 209, 229–233, 236, 237, 239, 240, 242, 244, 245, 250–256, 258
Outdoor education, 11, 13, 14, 32, 57, 63, 190, 229–236, 238, 242, 244–246, 248, 250, 251, 254–256, 258

Outdoor education route, 234
Outdoor learning, 194, 229, 231, 235, 238, 240, 245, 251, 253

P
Pandemic, 4, 32, 98, 129, 169, 244, 263
Paradigm, 52, 54, 60, 61, 104, 151, 155, 158–160, 167–169, 265, 266
Paradigm shift, 151, 159, 167
Paris, 36, 44, 45, 140, 178
Participatory, 75, 81, 83, 84
Participatory learning, 196
Participatory observation, 75
Pedagogical content knowledge model, 19
Pedagogical knowledge, 19
Pedagogic approach, 183
Pedagogic reason, 98
Pedagogy, 6, 11, 22, 52, 151, 169, 178
Perceived behavioral control, 160, 162
Perceived seriousness, 165
Perception, 5, 13, 34, 75, 95, 100, 102, 104, 111, 115, 122, 128, 129, 136, 145, 160, 163, 165, 218, 240, 253
Personality, 36, 98, 100, 102, 119, 120, 133, 135
Personality trait, 73, 113, 119, 120, 127, 132–135, 145, 146
Persuasion, 17, 117, 118
Philosopher, 5, 26, 27, 34, 60, 112, 152, 158
Philosophy, 25–27, 29, 33, 36, 46, 54, 56, 62, 63, 85, 95, 107, 120, 122, 143, 151, 153, 155, 229, 230
Physical, 13, 25, 29, 52, 62, 64, 66, 84, 86, 94, 102, 110, 113, 114, 116, 118, 123, 127, 133, 138, 142, 143, 206, 210, 240, 242, 244, 245, 254, 263
Physical world, 66
Phytoplankton, viii
Pinchot, 27–29
Planet, 18, 25, 29, 51, 52, 64, 98, 108, 129, 154, 263, 264
Plato, 26, 27, 35, 158
Political, 4, 8, 41, 54, 60, 102, 104, 105, 107, 117, 118, 155, 156, 158, 191, 229, 247, 264
Political nature, 229
Polychlorinated biphenyl (PCBs), 29
Population, 29, 30, 52, 61, 64, 67, 73, 93, 95, 98, 107, 123, 138, 146, 155, 168, 177, 178, 233, 255
Positive, 8, 9, 11, 38, 51, 73, 85, 87, 101, 104, 105, 109, 111, 113–115, 117–119, 130, 133, 135, 137, 160, 169, 184, 190, 206–208, 216, 230, 243
Positive attitude, 101
Positivism, 60, 61, 66, 70
Postnatal environment, 145
Post-positivism, 66
Post-structuralist approach, 64
Post-test, 71
Practical education, 66, 71
Practical learning, 8, 13
Pragmatism, 8, 110
Prerequisite, 163

Prerequisite concept, 20
Presentation, 52, 68, 73, 83, 188
Preservation, 27, 28
Preservation theory, 27, 28
Pre-test, 71
Problem, 7–9, 11, 13, 20, 22, 29, 32, 37, 41, 45, 47,
    50–52, 56, 57, 59, 60, 63–65, 67, 76, 79, 80, 85,
    96, 102, 107, 109, 111, 114, 115, 119, 123, 138,
    151, 165, 168, 169, 177, 178, 185, 191, 197, 199,
    206, 209, 215, 218, 219, 229, 238–240, 242, 244,
    250, 252, 256
Problem-Based Learning (PBL), 185
Problem solving, 4, 29, 79, 100, 208
Problem-solving learning, 59, 85
Problem-solving process, 79
Process, 3–5, 7, 8, 10, 16, 17, 20–22, 25, 27, 29, 33, 34,
    38, 49–54, 56, 57, 59–65, 73–77, 79–81, 83, 86,
    87, 93–98, 100, 105, 107, 111, 112, 114–116, 118,
    119, 122, 123, 127, 128, 130–132, 142, 146, 154,
    155, 158, 165, 177–180, 182, 185, 190, 196, 201,
    204, 206–210, 214–218, 221, 229, 230, 238, 246,
    250, 253, 254
Producer, 62, 223
Production, 6, 66, 167, 168, 204, 214, 247, 262
Pro-environmental behavior, 18, 116, 117, 120, 123, 132,
    134, 135, 140, 145, 159, 160, 162–166, 217
Pro-environmental personal norm, 164
Project, 11, 13–15, 38, 39, 56, 57, 68, 82, 184, 213, 223,
    256
Project planning, 63
Prototype, 54, 59
Psychology, 6, 8, 56, 85, 98, 108, 119, 263
Psychomotor, 112, 116
Psychomotor domain, 85, 114
Public, 3, 6, 10, 28, 29, 50–52, 54–56, 64, 67, 94, 102,
    107, 109, 111, 118, 123, 129, 137, 158, 162–165,
    177, 191, 195, 197, 199, 215, 220–222, 224, 230,
    246, 252
Public economics, 27
Puritan, 27
Purpose, 3, 4, 7, 11, 17, 21, 22, 26, 41, 55, 66, 67, 72, 74,
    79, 87, 105, 116–118, 140, 145, 146, 164, 184,
    187, 193, 197, 199, 201, 229, 230, 234, 263
Purpose-framed play, 17

Q
Qualitative, 13, 56, 75, 76, 83, 111
Qualitative research, 49, 74–76
Quality, 10, 11, 39, 51, 52, 54, 65, 69, 94, 102, 104,
    117, 130, 138, 143, 155, 158, 168, 185, 191, 201,
    255
Quality education, 42, 191
Quantitative, 49, 70, 75, 83, 111
Quasi-experimental design, 71, 72
Questionnaire, 13, 14, 68, 70, 72, 73, 106, 202

R
Radiation, 28, 106
Rational, 39, 51, 87, 104, 111, 112, 121, 209, 210
Rationalism, 27, 156
Rational thinking, 59, 111, 112
Re-action, 67, 80, 82
Recording, 75, 80, 263
Recreation, 30, 56, 198, 237, 246–248
Reflective, 67, 80, 206
Reflective thinking, 68
Reiteration, 71
Relevance, 52, 159, 196, 217
Relevance research, 73
Reliability, 56, 70, 72
Reliable, 71, 84, 133
Religion, 26, 87, 201, 262, 263
Renewable resource, 27
Replica, 72
Republic of China (Taiwan), 68
Republic of Korea (South Korea), 4, 31, 42, 43, 46
Responsibility, 8, 18, 20–22, 26, 43, 45, 50, 57, 59–62,
    65, 85, 93, 94, 97, 105, 107, 109, 119, 129, 130,
    133, 135, 155, 164, 165, 169, 190, 191, 197,
    243–245, 257
Responsible environmental behavior, 57, 61, 87, 96,
    118–120, 132, 133
Responsible environmental steward, 107
Retest, 72
Retrospective research, 73
Rio de Janeiro, 41, 42, 45, 46, 64
Risk, 123, 133, 140, 143, 165, 217, 218, 244
Risk communication, 177, 224
Romanticism, 26
Roth, 38–40, 94, 95, 237
Rousseau, 8, 35, 36, 190

S
Sabah, 202, 205
Sampling, 71, 80, 239, 243
Sampling distribution, 71
Sand county, 27, 28, 111, 152
Sapien, 4, 108
School, 3, 4, 7–11, 16, 17, 33, 35–40, 42, 44, 54, 58, 62,
    65–68, 71, 76, 99, 104, 105, 121, 144, 160, 162,
    178, 184, 186, 188, 190, 197, 201, 208, 217, 229,
    230, 234–236, 238, 244, 245, 255
School curriculum, 8, 11, 17, 44, 65, 66, 189
School education, 38, 65, 69, 105, 193, 244
School environmental education, 11, 184–186, 188, 193,
    194
School of Environment and Natural Resources (SENR),
    37
Science, 4, 8, 10–12, 30, 31, 37, 38, 42, 44, 51, 52, 54,
    56, 62, 63, 65, 66, 70, 84, 85, 102, 104, 105, 107,
    108, 110, 121, 123, 128, 138, 151, 155, 156,

158–160, 167, 168, 184, 188, 197, 200, 217, 218, 242, 250, 254, 255, 263
Science education, 11, 36, 37, 42, 65, 66, 85, 105, 183, 191
Scientific consensus, 81, 129
Scientific jargon, 188
Scientist, 29, 36, 43, 49, 50, 52, 54, 64–66, 104, 109, 111, 112, 129, 138–140, 142, 181, 232
Sea-level rise, 96, 257
Seasonal Affective Disorder (SAD), 143
Self-adjust, 87
Self-awareness, 5, 59, 142, 219
Self-course, 98
Self-directed learning, 58, 61, 177, 192
Self-evaluation, 98
Self-evolution, 27, 29
Self-expression, 98, 122
Self-formation, 66
Self-learning, 58, 59
Self-learning theory, 59
Self-mastery, 87
Self-mindedness, 4
Self-ontology, 4
Self-reflection, 60, 61
Self-reflective circle, 67
Self-regulation, 28, 38
Self-renewing ecosystem, 152
Semi-structured interview, 75
Sentiment, 102, 103
Sex, 129
Sex identification, 66
Short-term memory, 4
Silent spring, 6
Singapore, 43, 46, 161
Site, 26, 202, 216, 233, 238, 239, 245–247, 249, 255
Site planning, 232
Site survey, 68
Social, 6–8, 10, 11, 16, 20–22, 29, 30, 36, 41, 46, 49, 51, 52, 54, 58, 60, 62–64, 66, 67, 69, 74, 78, 79, 82, 98–100, 102, 104, 105, 107–109, 111, 112, 116, 117, 119–121, 123, 127–129, 131, 132, 135, 138, 140, 144–146, 151, 152, 155, 158, 160, 163, 164, 167–169, 177–179, 184, 188–193, 195, 197, 206, 207, 210, 211, 218, 221, 222, 234, 238, 240–242, 247, 250, 252, 253, 255, 265, 266
Social actor, 116
Social cognition, 98, 206
Social consumption behavior, 21
Social context, 66, 67, 74
Social development, 18, 167, 206, 231
Social emotional learning, 18, 240
Social field, 64
Social impact, 98, 108, 160, 209, 210
Social influence, 160, 162
Social issue, 65, 123, 127, 146
Social justice, 8, 21, 43, 67
Social learning, 98, 99, 132
Social Learning Theory, 131, 206
Social marketing, 177, 224

Social meaning, 81
Social network, 155–157, 162, 167, 177
Social norm, 35, 127, 135–137, 145, 146, 159, 160, 162, 166
Social orientation, 21
Social protocol, 98
Social relationship, 85
Social science, 8, 51, 71, 74, 76, 79–81, 84, 85, 104, 158, 210, 222, 242, 255
Social theory, 127, 131
Social value, 32, 169, 244, 255
Society, 8, 10, 12, 19, 21, 22, 29, 33, 39, 41–43, 50, 52, 56, 57, 60, 62, 65, 82, 83, 94, 95, 98, 99, 105, 107, 108, 111, 121–123, 128, 130, 140, 152, 156, 158, 160, 161, 166, 168, 169, 177, 191, 195, 215, 218–220, 224, 231, 232, 238, 257, 258, 262, 264
Society of Wetland Scientists (SWS), 43, 50, 232, 233
Socioeconomic, 98, 118
Sociology, 6, 8, 56, 87, 107, 116, 158, 266
Soil, 6, 22, 43, 56, 81, 123, 168, 185, 221, 232, 255
Soil compaction, 75, 81
Soil hardness, 81
South Africa, 42, 45
Southern Illinois University, 196, 200, 231
Southern Illinois University at Carbondale (SIUC), 196, 200
Soviet Union, 7
Species, 26, 28, 29, 52, 82, 98, 108–110, 112, 129, 142, 153, 155, 157, 158, 191, 192, 204, 212, 214, 221, 262, 264
Staffordshire, 44
Stagnation, 78
Stakeholder, 3, 28, 51, 52, 54, 79, 151, 163, 215
Stanford university, 98
Statistical, 73
Statistical analysis, 70
Statistical measurement, 71
Stockholm, 38, 44
Stockholm Declaration, 44, 64
Storm, 181
Structural, 120, 133, 159, 162, 232
Structural environment, 233
Structuralism, 66
Structuralist, 64
Structure, 5, 52, 59, 70, 80, 93, 108–110, 114, 119, 139, 155–157, 163, 168, 178, 197, 201, 265
Structured interview, 74, 75
Style, 85, 86, 100, 101, 121, 184, 190, 198, 206, 251, 263
Subject, 17, 19, 20, 37, 56, 64, 70–72, 85, 88, 96, 130, 132, 136, 138, 182, 184, 185, 189
Subjective, 59, 74, 93, 109, 111, 112, 120, 121, 131, 133, 135, 137
Subjective norm, 135–137, 160–162, 166
Subjective probability, 160
Subject matter, 19, 88, 188
Subsidiary, 22
Survival, 25, 28, 29, 41, 108, 110, 128, 130, 138, 142, 178, 230, 263

Sustainability, 8, 21, 26, 27, 42, 54, 55, 57, 94, 97, 138,
     140, 151, 154, 164, 178, 179, 197, 238
Sustainability paradigm, 169
Sustainable behavior, viii
Sustainable development, 10, 20–22, 41–43, 45, 46, 51,
     54, 56, 60–64, 93, 94, 96, 97, 135, 137, 146, 156,
     164, 169, 206, 219, 224, 261, 263
Sustainable development education, 12, 20, 21, 42, 46,
     55, 64
Sustainable Development Goals (SDGS), 22, 42, 56
Sustainable future, 14, 42, 43, 45, 70
Sustainable society, 94, 169
Sweden, 38, 44
Switzerland, 9, 35, 45, 193, 234, 236
Symbolic, 109, 110, 139, 196
Symbolic research, 52

T
Taipei, 5, 9, 10, 15, 18, 19, 34, 82, 83, 129, 139, 143, 144,
     157, 184, 190, 199, 203, 208, 211, 212, 235, 236,
     242, 252, 255–257
Taiwan, 5, 8–10, 14, 15, 18, 19, 22, 34, 35, 40, 43, 53, 55,
     56, 61, 67, 68, 82, 83, 99, 100, 118, 121, 129, 139,
     140, 143, 144, 146, 157, 170, 183–185, 190, 192,
     198, 199, 201–203, 208, 211–214, 217, 218, 220,
     222, 230, 236, 239, 242, 255–257, 263
Taiwan EPA, 14, 67–69, 184, 186, 188, 212
Take-action, 68, 69
Tao, 156
Tao of Physics, 156
Tbilisi, 7, 11, 41, 45
Tbilisi declaration, 41, 45, 64, 66, 96, 97, 248
Teacher education, 64, 200
Temperature, 29, 129, 130, 140, 142, 185, 211
Texas, 50, 180
Texas A&M University, 50
Textbook, 60, 71, 72, 178, 188–190
The Journal of Environmental Education (JEE) (Taylor &
     Francis' journal), 37, 55
Theory of Planned Behavior (TPB), 120, 160–162, 166
Thinking mode, 59
Thoreau, 27, 142
Time, 5, 6, 11, 18, 21, 26–30, 32, 33, 35, 44, 50, 52–54,
     57, 60, 62, 65, 67, 72, 74, 78, 81, 82, 87, 95, 96,
     98, 102, 104, 107–109, 111, 112, 114, 115, 117,
     120–123, 128, 130, 132, 135, 138, 142, 145, 146,
     153, 156, 163, 168, 169, 180, 184, 185, 193, 196,
     197, 201, 207, 215, 216, 230, 234, 240, 261–263,
     265, 266
Time series, 72, 108
Tourism, 13, 56, 134, 140, 193, 210, 237, 245, 247
Traditional Ecological Knowledge (TEK), 7, 154
Traditional Ecological Management (TEM), 7
Trait, 87, 134, 266
Transcendence, 26, 27
Transfer process, 216
Translation, 6
Treemap, 16

Triadic Reciprocal Determinism (TRD), 206, 207
Triage, 68
Try-and-error, 208
Twentieth century, 4, 28, 30, 36, 46, 103, 120, 121, 129,
     142, 152, 230
Twenty-first century, 22, 29, 30, 36, 41, 51, 56, 120, 142,
     179, 196, 224, 262, 263
Two-phase Decision-Making Model, 165

U
U.S. Congress, 41
U.S. EPA, 14, 37, 38
U.S. Fish and Wildlife Service, 38
UN Conference on Environment and Development
     (UNCED), 45, 46
UN Decade of Education for Sustainable Development
     (UN DESD 2005–2014), 42
Undergraduate, 62, 99, 231
UN General Assembly, 42, 45
United Kingdom (UK), 7, 39, 42, 44, 56, 214
United Nations (UN), 41, 42, 44–46, 96, 218
United Nations Educational, Scientific and Cultural
     Organization (UNESCO), 7, 8, 10, 11, 20, 22, 38,
     39, 41, 42, 44–46, 49, 65, 97, 105, 223, 248, 249
United Nations Environment Programme (UNEP), 39, 41,
     44, 45, 105, 152
United States of America (USA), 7–9, 36–38, 43, 45, 55,
     81, 94, 194–196, 217, 230, 233, 236, 245, 247,
     250, 252
Universe, 26, 27, 109, 112, 142, 263
University, 7, 9, 10, 21, 22, 36, 43, 46, 50, 55, 62, 105,
     106, 146, 158, 185, 190, 192, 194, 209, 231, 232
University of Arizona, 55
University of California, Berkeley, 55
University of Chicago, 98
University of Colorado, 107
University of Exeter, 42
USDA forest service, 38
Utilization, 27, 51, 52, 66, 111, 167

V
Validity, 56, 70–74, 107, 166
Value, 7, 10–12, 15, 17, 20, 21, 27–30, 32, 33, 35, 38, 41,
     42, 49–52, 56, 57, 61, 62, 65, 72, 73, 81, 85, 87,
     93–98, 100, 107–111, 113, 114, 118–121, 123,
     127, 129, 131, 146, 151–156, 158–160, 162–165,
     179, 206, 211, 215, 217, 230, 237, 238, 245, 247,
     263, 265
Value-Belief-Norm theory (VBN), 163, 165
Value clarification, 8, 15, 16, 29, 208
Variable, 51, 70, 71, 73, 80, 96, 120, 131, 135, 137,
     162–164, 166, 167, 236
Verification, 49, 50, 56, 71, 72, 107
Vietnam, 28, 250
Virtual Reality (VR), 61, 179, 263
Visual, 75, 100, 121, 123, 127, 132, 177, 202, 236, 261
Visual Participatory Methods (VPMs), 75

Vitalism, 27
Vitality, 133
Vitality theory, 28
Vocational, 10, 64, 189
Voluntary, 113
Voluntary environmental learning behavior, 207

W
Walden, 27, 142
Wellness, 143
West, 42, 66, 138, 142
Western, 25, 27, 32, 36, 60, 65, 104, 107, 152, 153, 158, 167, 183, 230, 264, 266
Westerner, 27
Western science, 7
Wetland, 13, 15, 18, 43, 50, 77, 82, 111, 188, 202, 213, 232, 233, 236, 238, 242, 255, 256

Wetlands (Springer's journal), 242
Wilderness, 27, 29, 230, 243
Wilderness ecology, 27
Wilderness education, 229
Wilderness preservation, 27
Wise use, 27, 111
Working knowledge, 3
World Earth Day, 37, 40, 191
World Environment Day, 191
World Wetland Day, 191
World Wildlife Fund (WWF), 45

Y
Yale University, 107
Yugoslavia, 44